INTERCONNECT-CENTRIC DESIGN FOR ADVANCED SOC AND NOC

Interconnect-Centric Design for Advanced SoC and NoC

Edited by

Jari Nurmi
Tampere University of Technology, Finland

Hannu Tenhunen
Royal Institute of Technology, Sweden

Jouni Isoaho
University of Turku, Finland

and

Axel Jantsch
Royal Institute of Technology, Sweden

KLUWER ACADEMIC PUBLISHERS
BOSTON / DORDRECHT / LONDON

A C.I.P. Catalogue record for this book is available from the Library of Congress.

ISBN 1-4020-7835-8 (HB)
ISBN 1-4020-7836-6 (e-book)

Published by Kluwer Academic Publishers,
P.O. Box 17, 3300 AA Dordrecht, The Netherlands.

Sold and distributed in North, Central and South America
by Kluwer Academic Publishers,
101 Philip Drive, Norwell, MA 02061, U.S.A.

In all other countries, sold and distributed
by Kluwer Academic Publishers,
P.O. Box 322, 3300 AH Dordrecht, The Netherlands.

Printed on acid-free paper

Cover: Artwork by Sonics

Printed in the Netherlands.

Contents

Preface

This book was mainly catalyzed by the SoC-Mobinet EU-project (IST 2000-30094), where the editors have acted in 2001 - 2004 to create the European System-on-Chip infrastructure for joint research and education. The topic was seen very important especially after writing the previous book (Networks on Chip) where the higher-level on-chip communication issues were tackled.

In this book, we have tried to create a comprehensive understanding about on-chip interconnect characteristics, design methodologies, layered views on different abstraction levels and finally about applying the interconnect-centric design in system-on-chip design.

We owe very much to the authors of this book, who represent the expertise on different aspects of interconnects, communication and system-on-chip and network-on-chip development worldwide. One major contributing project was also the Complain (Communication platform architectures for gigascale integration) project in the Finnish-Swedish EXSITE research programme, mainly supported by TEKES, Vinnova, Nokia and Ericsson. We were happy to have such a good crew to assist us in creating the book for the interested readers in this field. We hope that you enjoy the book and, even more, wish that it will be of professional benefit to you!

Tampere, Stockholm, Turku
January 2004

JARI NURMI
HANNU TENHUNEN
JOUNI ISOAHO
AXEL JANTSCH

I

PHYSICAL AND ELECTRICAL ISSUES

Chapter 1

SYSTEM-ON-CHIP-CHALLENGES IN THE DEEP-SUB-MICRON ERA

A case for the network-on-a-Chip

Jan M. Rabaey
EECS Dept., University of California at Berkeley
jan@eecs.berkeley.edu

Abstract: Continued scaling of semiconductor technology has created hopes for true system-on-a-chip integration; that is, the integration of a complete system on a single die of silicon. Yet, this prospect is seriously challenged by cost considerations. Solutions that maximize flexibility and re-use may be the only way to address the cost concerns, but necessitate offsetting the energy and performance penalty that comes with software solutions through an aggressive use of concurrency. A truly network-based solution is the only option to resolve the many issues that come with massive parallelism such as reliability, robustness, synchronization, and power management. This chapter presents an insight in some of the approaches and methodologies that are currently under development at the Gigascale Systems Research Center [GSRC].

Keywords: Platform-based Design, Network-on-a-Chip, Deep Sub-micron.

1 INTRODUCTION

It is commonly believed today that from a pure technological perspective there are no reasons for Moore's law to come to a screeching halt in the next decade. The continued scaling of the semiconductor technology creates the potential of true system-on-a-chip (SoC) integration; that is, the integration of a complete electronic system including all its periphery and its interfaces to the outside world on a single die. The reduction in size and cost that would come with this high level of integration opens to the door to truly ubiquitous electronics, and enables the realization of usage scenarios that were deemed to belong to the realm of science fiction not so long ago. An example of such is the "ambient intelligence" application [Boekhorst02, Rabaey03] that relies on wide-spread networks of embedded sensor, control, and actuation nodes, embedded in the daily environment, to enhance a number of daily activities and/or our living environment.

Yet, the enthusiastic embrace of the system-on-a-chip vision has cooled down considerably over the last one or two years, as the approach has run into some major roadblocks and hurdles. While these challenges were already speculated on in the mid 90's, they have come into full bearing today. The combination of deep-submicron effects, manufacturing cost, and complexity have caused a major shift in the economics of IC design, and have marked the beginning of the end of the era of the Application Specific Integrated Circuit (ASIC). The importance of ASICs in shaping the semiconductor success story cannot be overemphasized. Very successful products were brought to market using integrated circuits that were specific to the application, and that were generated using a well understood design flow combining logic synthesis and automatic place and route.

In response to these challenges, a new design methodology called "platform-based design" was proposed, promoting reuse at higher levels of granularity and relying heavily on flexibility and programmability to amortize the NRE costs of complex IC design over a number of designs [Keutzer00]. A summary of the main principles of platform-based design, as developed at the Gigascale Systems Research Center or GSRC, is given in Section III of this paper. For the approach to be ultimately successful in the realization of systems-on-a-chip and to address the challenges raised by further scaling of the semiconductor technology, platform-based design has to come to grips with some major roadblocks which are enumerated in the Section II of the paper. One of the most important ones is the management of concurrency. While solutions that maximize flexibility and programmability may be the only way to address the cost concerns of design, they come with a major penalty in performance and energy efficiency. Offsetting this penalty is only possible through an aggressive use of concurrency. A truly network-based solution is the only option to resolve the many issues that come with massive parallelism such as reliability, robustness, synchronization, and power management. In Section IV of the paper, we explore the advantages of a "network-on-a-chip" (NoC) approach [Sgroi01] and discuss how it helps to address a number of the concerns that were raised in Section II. The paper will be concluded with some future perspectives.

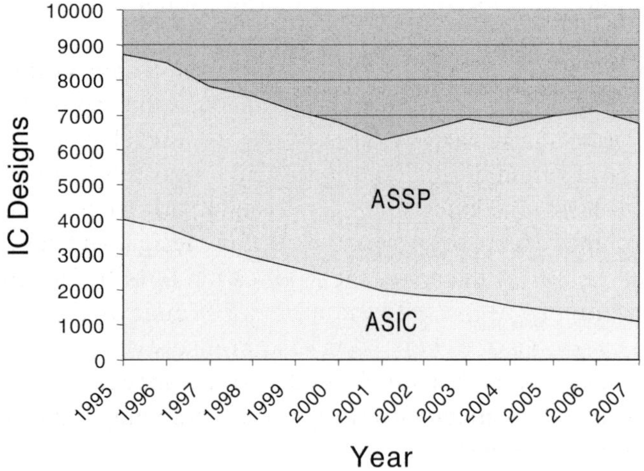

*Figure 2.1.*Number of successful IC design starts per year (Source: Handel Jones, IBS - September 2002). ASIC and ASSP stand for Application-Specific Integrated Circuit and Application-Specific Standard Product, respectively.

2 CHALLENGING THE SOC PROSPECTS

2.1 The Demise of ASIC

While the ASIC design methodology has very successful in bringing cheap electronics into a wide range of applications, economic considerations have dealt it a major blow. This is best illustrated in Figure 2.1., which plots the number of (successful) design starts over the last decade, extrapolating towards the future. The number of ASIC starts dropped by more than half in the period from 1995 to 2002. The reasons for this major decline are manifold:

- **The NRE cost of IC manufacturing.** The cost of mask making alone for sub-micron design is approaching $1 M for the 0.13 μm CMOS technology node, and it has been predicted to range between $2 M and $5 M in 2 to 3 years from now. (See Figure 2.2.). ASIC starts suffer the most from this trend forcing companies to change their business model and product mix. It favors approaches where customization is achieved by software running on embedded microprocessors or configurable hardware (commonly called ASSPs, or application-specific standard products), as is apparent in Figure 2.1.

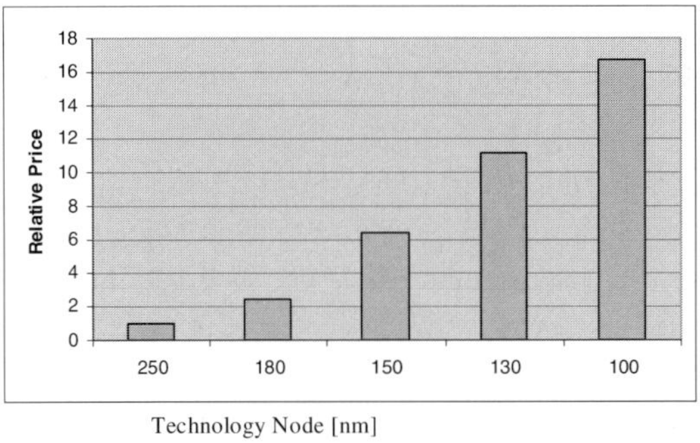

*Figure 2.2.*Cost of mask set versus technology node. Source: Xilinx.

- **Deep-submicron effects**. With deep-submicron technologies readily available through foundries, the ASIC designer became exposed to the myriad of problems that have already plagued the designers the leading-edge microprocessors for a while. Issues such as interconnect delay, crosstalk and supply noise impact the predictability of the final design. All EDA vendors are struggling to solve timing closure problems and the capacity issues due to increased miniaturization of circuit components in the attempt to extend the life of their tools and methodologies. Other effects such as power dissipation, leakage, and variability only compound the problem.Verification has come to dominate the overall design cost.
- **Complexity**. The impact of these physical effects is already bad enough, yet it is further aggravated by the increasing complexity of the IC. Designs with more than 100 million transistors are not exceptional anymore. This complexity increase requires either larger design teams, or raises the time-to-market. Only the aggressive reuse of large modules and a raise in abstraction levels can help to address this challenging issues.

The platform-based design methodology described in Section III goes a long way in addressing the above concerns. Yet, for it to be successful in the System-on-a-Chip arena and to continue to do so in the next decade, some other concerns and roadblocks need to be overcome.

2.2 The Soc Challenges of the Next Decade

In addition to the concerns raised above, we see some important new developments emerging when looking towards the future especially in the area of

SoC:

- **Managing and Exploiting Flexibility (that is, introducing higher levels of abstraction)**

The electronics industry is moving towards programmable solutions to save NRE and design costs and reduce time-to-market. This trend requires thinking of software as an essential component of an embedded system. A radical movement is afoot to move towards a more scientific approach where software design is thought of as a rigorous procedure of refinement from a higher level of abstraction consisting of a mathematical model of the functionality. This approach, which goes under the name of *model-based software design,* is rapidly gaining attention and is now coming to terms with the complexity of designing software that has to satisfy constraints on timing and power consumption, quantities that were almost never an issue in standard software design. It is this dimension that makes embedded software unique. Hence, methods are needed that direct the embedded software design process towards constraint satisfaction and that are aware of both functionality and of the physical platform upon which the software is run. Timing and power estimation for software, software synthesis, timing and power analysis of software programs, formal verification of software are all important components of the design methodology. We believe that the choice of implementation techniques for a set of functions can be approached with the same method at all levels of abstraction, thus linking the various levels in a seamless environment that is based on the tenet of platform-based design. In some sense, software design becomes in this view an implementation choice on the same formal basis as a hardware implementation.

- **Power and Energy**

Power and energy management and minimization have been a concern in CMOS for over more than a decade. Multiple solutions have been developed and have helped to keep the problem within bounds. Yet, considering the projected increases in integration complexity and the unfortunate side-effects of deep submicron transistor scaling such as drain-source and gate leakage, it is fair to state that design for power and energy management has emerged as one of the most dominant roadblocks on the horizon. In fact, we are coming to a situation where power dissipation limits the levels of integration that can be accomplished and hampers the further evolution of the semiconductor industry. With no or little new technological miracle solutions on the horizon, radical new design approaches (from innovative circuit fabrics, to micro-architectures that exploit them, and the system software that controls them), and accompanying design methodologies are a necessity. For instance, it becomes increasingly clear that we might want to scale supply voltages faster than what is predicted in the roadmap. For **very low-voltage design** to be effective, solutions have to be found to problems such as the management of leakage and timing uncertainty. The combination of an increased vulnerability to soft errors (resulting from the scaling of the signal charge) and a reduced signal-to-noise ratio (caused by scaling of the supply voltage) seriously impacts the reliability of future circuits. We believe that power and energy management under these circumstances is best addressed as a system-level problem. However, power optimization should not be performed at the expense of performance. With fewer IC designs serving more applications we believe that the speed demands for these IC designs will actually increase. Application-specific standard parts (ASSPs) are demonstrably replacing application-specific integrated circuits (ASICs). Since ASSPs are catalog parts that compete with other designs for well-defined market niches, speed will remain a key differentiating factor.

- **Reliability and Robustness**

A wide variety of factors are conspiring to dramatically reduce the short- and long-term *reliability of integrated systems*. As already stated, power considerations under increasing integration densities inevitably reduce the signal-to-noise ratio of digital circuits. Under aggressive voltage scaling conditions, soft errors caused by alpha particles and cosmic rays and thermal noise are bound to inject dynamic errors into computations. Combine this with the increased impact of process variations, quantum fluctuations in the operation of very-deep submicron transistors, and the projected proneness to errors of some of the emerging nano-technologies, and it becomes clear that the reliability of computations is bound to be seriously impacted in the years to come. The integration of multiple hybrid and mixed-signal technologies on the same die further reduces the design robustness. While some of these concerns will undoubtedly be addressed by improved technology and manufacturing, design techniques and methodologies that provide *reliable computation in the presence of unreliable circuit fabrics* are highly likely to gain prominence. It is our conviction that a layered top-down design methodology as advocated in platform-based design is the only way of dealing with dynamically occurring computational errors. Error-correction, re-computation or retransmission of faulty results, redundancy, etc. are best addressed at the right level of abstraction, as the communication and networking communities have known for quite some time. The occurrence and detection of dynamic errors further is at the base of an interesting phenomenon: when errors can happen arbitrarily, it is hard to differentiate between design verification and test. We believe that **the verification and test approaches are on a conver-**

gence path and that substantial fractions of these tasks will be turned into **on-line activities**. Enabling and augmenting this convergence are some other interesting directions in verification and test, discussed in the next two paragraphs.

- **Predictability and Timing Strategies**

While approaches such as constructive fabrics [Pileggi02] go a long way toward addressing the *predictability* of the next-generation integrated systems, it is fair to state that timing predictability will continue to decline over the years to come. This can be contributed to increased variations in both device characteristics (declining supply/threshold voltage ratios, quantum fluctuations in current draw, etc.) and interconnect parameters. At the same time, clock frequencies will continue to increase. As a result, the ratio of uncertainty margin over clock period will increase, causing synchronous (that is, worst case) timing strategies to ultimately saturate. The only solution is to step away from the worst-case design by either (i) allowing occasional timing errors to occur—which trades-off reliability for performance and requires reliable computation strategies as stated above—, or (ii) by stepping away from the synchronous paradigm. Meso, plesio, or a-synchronous approaches become attractive when the overhead of the synchronization and handshaking hardware becomes smaller than the uncertainty of the timing prediction. In addition, a pure synchronous approach becomes untenable when supply and threshold voltages of certain parts of the chip are dynamically varied for speed gains or dynamic/leakage power reduction. In short, we predict that **asynchronous (or other non-synchronous) timing strategies** will play an important role in the system-on-a-chip of the next decade, and that appropriate design methodologies and solutions have to be provided.

- **Real-Time Emulation**

The complexity of the integrated systems makes the verification task ever harder. Further abstraction and formal techniques, while partially alleviating the problem, do not suffice to address the rapidly increasing verification costs. **A complete portfolio of complementary verification techniques (including simulation, emulation, on-line checking, as well as formal verification) and an overlaying methodology for their deployment is necessary**. One interesting and attractive opportunity to bring system-level verification of architectural platforms to new levels is offered by the availability of complex heterogeneous field-programmable devices. An engine constructed of these components, combined with a rapid mapping methodology, makes **real-time emulation** of complex systems more affordable and feasible. A fast prototyping path can go a long way in aiding the verification task.

- **Mixed-Everything (Mixed-Signal Mixed-Technology)**

With the ever-decreasing size of embedded or computational systems comes a need for true *integration of multiple hybrid technologies* into a single component or package. In fact, true integration of a complete system requires not only the realization of the digital computational functions, but also the periphery and the interfaces to the external world. These may include sensors (realized as MEMS or other technologies), wired or wireless connections (using RF, Infrared, or high-speed serial links), energy supplies (batteries, energy-scavenging devices, and regulators), actuators, embedded storage, and displays. The tight integration of these components requires a design methodology that considers them in concert. Hence, *mixed-signal, mixed-technology methodologies, re-use strategies and tool-sets* are an essential component of any system-on-a-chip development.

- **Beyond Silicon**

With the roadmap for silicon CMOS moving into what are probably its last stages, new technologies and devices (commonly dubbed *nano-technologies*) are most likely to come into play over the next couple of decades. In fact, some of these technologies (such as organic semiconductors) are already making some in-roads. Other approaches such as molecular, quantum-effect, bio, or nanotube devices might need a longer time to come to fruition (if ever). Many efforts are under way exploring these exciting alternatives. It is, however, certain that these technologies will lead to vastly different circuit fabrics and system architectures. At GSRC, we contend that **the platform-based design methodology, based on a stacked layer of abstractions, is universal and is well-suited to encapsulate late-silicon and nano-technology devices and fabrics**. The layers of abstractions between the implementation layers are precisely what are needed to accommodate dramatically different implementation technologies.

While addressing all of these issues and challenges is beyond the scope of this paper, we will demonstrate that a well-executed platform-based strategy combined with a formal network-on-a-chip approach goes a long way in addressing at least some of the concerns such as programmability, power and energy, reliability, and predictability.

3 THE PLATFORM APPROACH TO SYSTEM-ON-A-CHIP

3.1 Platform Definitions

The platform concept itself is not entirely new, and has been successfully used for years. However, the interpretation of what a platform truly is, has been, to say the least, confusing. In the IC domain, a platform is considered a "flexible" integrated circuit where customization for a particular application is achieved by "programming" one or more of the components of the chip. Programming may imply "metal customization" (Gate Arrays), electrical modification (FPGA personalization) or software to run on a microprocessor or a DSP. These flexible integrated circuits can be defined as members of the *silicon-implementation platform* family. With SOC integration, implementation platforms are becoming more diverse and heterogeneous, combining various implementation strategies with diverging flexibility, granularity, performance, and energy-efficiency properties.

For the case of software, the "platform" has been designed as a fixed micro-architecture to minimize mask making costs but flexible enough to warrant its use for a set of applications so that production volume will be high over an extended chip lifetime. Micro-controllers designed for automotive applications such as the Motorola Black Oak PowerPC are examples of this approach. DSPs for wireless such as the TI C50 are another one. The problem with this approach is the potential lack of optimization that may make performance too low and size too large. A better approach is to develop "a family" of similar chips that differ for one or more components but that are based on the same microprocessor. The various versions of the TI C50 family (such as the 54 and 55) are examples of such. Indeed this family and its "common" programmatic interface is, in our definition, a platform; more specifically *an architecture platform.*

The pure concept of platform-based design, as adopted and promoted by GSRC, is however much broader and has far reaching implications in different design styles and at different levels of design. In essence, it proposes a methodology that advocates and formulates clearly defined and restrictive articulation points. These articulation points between levels of design are critical to achieve a separation of concerns and to break the design process into manageable parts amenable to subsequent automation and optimization. Automation enables us to find high quality acceptable solutions in this restricted design space.

Again, the concept of **a platform as a clearly defined articulation point of restriction and abstraction**, and its role in enabling methodology is not new. The use of standard cells is a classic example of how a restriction in the design space —a limited library of constrained cells— makes it possible to use intellectual and computational capabilities to find high-quality acceptable solutions through synthesis and place and route algorithms. The use of a synchronous timing methodology adds another restriction on the design space, enabling a separation of concerns between the logical and the timing aspects of design. From a user's perspective, the methodology offers a clear and unambiguous Application Programmer Interface (API) —synchronous finite state machines and Boolean algebra—, which makes it possible to capture and implement the intentions of a designer, independent of the library and implementation tools. Synchronous standard cell based design *as a platform,* was the key enabler of the ASIC design methodology.

The following are the salient aspects of the platform based design methodology [Sangiovanni02].

- The platform-based design paradigm is a **meet-in-the-middle approach**. It leverages the power of top-down methods and the efficiency of bottom-up styles. We view the design process as a stepwise refinement of a specification into a lower level abstraction chosen from a restricted library of available components. Components are "computational" blocks and interconnects. This library is a **platform**. In this view, a platform is a family of designs and not a single design. A platform defines the design space that can be explored. Once a particular collection of components of the platform is selected, we obtain a **platform instance**. The choice of the platform instance and the mapping of the components of the specification into the components of the platform instance represent the top-down process. In this process, constraints that accompany the specification are mapped into constraints on the components of the platform instance. Mapping often involves budgeting, since a global constraint may have to be distributed over a set of components.
- The **stepwise refinement** continues by defining the selected platform instance as a specification and using a lower level platform to march towards implementation. Whenever a component is fully instantiated the stepwise refinement stops since we have an implementation for that component.
- When selecting a platform instance and mapping constraints using budgeting, it is important to guide the selection with parameters that

summarize the characteristics of the components of the platform. Delay, power consumption, size and cost are examples of such parameters. When selecting a platform instance it is important to be able to **evaluate quickly and with the appropriate accuracy** what the performance of the design will be. The selection of the parameters to use to guide the platform instance selection is one of the critical parts of platform-based design.

- The component selection process and the **verification of the consistency between the behavior of the specification and the one of the platform instance** can be carried out automatically if a common semantic domain is found where the selection process can be seen as a covering problem. The concepts of platform-based design can be used to describe the ENTIRE design process even when the design style chosen is ASIC. The framework is the same. The platforms are different. The number and the location of the platforms in the design abstractions, the number and the type of components that constitute a platform, the choice of parameters to represent the components are critical aspects of the method.
- **Platforms form a stack**, from design specification to implementation. They demarcate boundaries that are critical in the electronics supply chain, and define demarcation points that warrant special attention. In the paragraphs above, we have already identified the circuit and architecture level platforms. The system level is another potential demarcation point. At the **circuit implementation level** we are redefining the platform from standard cells to fabrics that build on high levels of regularity and predictability to address predictability and manufacturability concerns. At higher levels of design we have defined new platforms at the architecture level and the systems level. The **architecture level** has emerged in our work as a new articulation point in response to the increasing shift from ASICs to Application-Specific Standard Parts (ASSPs) -a result of the variety of pressures facing the ASIC design methodology. We call an architecture platform the articulation point between system architecture and micro-architecture. Micro-architecture can be seen as a platform whose components are architectural elements such as microprocessors, memories, networks, and interfaces. To find the common semantic domain we need to abstract these components via an operating system, device drivers and communication mechanism. In this domain the hardware components are seen as supporting the execution of the behavior of the specification. Similarly, **system level platforms** have emerged as new articulation points in response to the increasing reuse

of components as a way to manage increasing complexity. They provide for describing and designing systems as an integration of diverse (and constrained) functions.

3.2 Some Platform Examples

The platform-concept has been particularly successful in the PC world, where PC makers have been able to develop their products quickly and efficiently around a standard "platform" that emerged over the years. The PC standard platform consists of: the x86 instruction set architecture (ISA) that makes it possible to re-use the operating system and the software application at the binary level; a fully specified set of busses (ISA, USB, PCI); legacy support for the ISA interrupt controller that handles the basic interaction between software and hardware; and a full specification of a set of I/O devices, such as keyboard, mouse, audio and video devices. All PCs should satisfy this set of constraints. If we examine carefully the structure of a PC platform, we note that it is not the detailed hardware micro-architecture that is standardized, but rather an abstraction characterized by a set of constraints on the architecture. The platform is an abstraction of a "family" of (micro)-architectures.

A number of systems and semiconductor companies have fully embraced the platform-concept in the design of integrated embedded systems. An excellent example of such is the Nexperia platform, developed by Philips Semiconductor [Claasen00]. Nexperia serves as the standard implementation strategy for a wide range of video products within Philips. The platform combines a set of processors (MIPS + TriMedia) with a family of peripheral devices, accelerators, and I/O units. Essential is also a set of standardized busses, which form the core of the architecture and ensure that different modules can be connected seamlessly together. Depending upon the needs of a particular product (family), the IC designer can choose to drop/add particular components, as is illustrated in Figure 2.3. The system designers interface however remains unchanged, which allows for maximum reusability and portability. Since all components have been extensively tested and verified, design risk is reduced substantially. Other examples of successful platforms at different articulation points are the Ericsson Mobile Platform [Ericsson Mobile], the TI OMAP architecture platform [OMAP], and the VertexPro Silicon Platform of Xilinx [VertexPro]

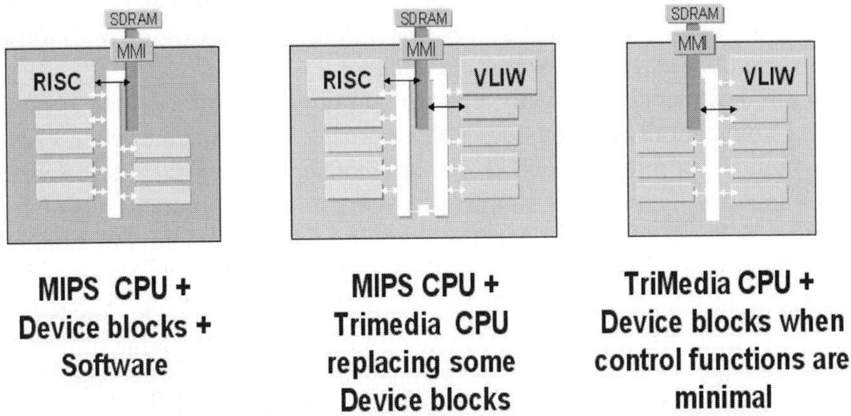

Figure 2.3. Different instances of the Nexperia platform for multimedia applications. Based on the application needs, computational modules can be added or dropped. The core of the platform is formed by the communication network, which consists in this particular case of a set of busses.

4 PLATFORM-BASED DESIGN AND NETWORKS-ON-A-CHIP

Essential to the conception of a platform is not only the design of the computation, *that is* the functional behavior of each core, but also of the communication, *that is* the interaction between the cores. This *orthogonalization of concerns* is essential to the success of a re-use strategy as has been realized in recent years. The platform-based design methodology addresses this concern by placing the computational cores and their interconnect strategy on the same footing. This is, for instance, very clear in the Nexperia architecture, shown in Figure 2.3, where the interconnect architecture acts as the obvious cornerstone and integrating element of the platform.

4.1 Network-on-a-Chip Basics

As was learned by the telecommunications community a while ago, reliable communication between components requires the definition of a protocol that provides a set of rules dictating how the interaction among components takes place, so that the overall system communication and performance requirements are met, while physical resources such as area and energy are minimized. Traditionally, on-chip communication design has been done using rather *ad-hoc* and informal approaches that fail to meet some of the challenges posed by next-generation SOC designs raised earlier, such as *performance and throughput, power and energy, reliability, predictability, synchronization, and management of concurrency*

So far, designers have mainly stuck to traditional techniques such as point-to-point connections, and busses. This approach is acceptable when only a small number of blocks have to be interconnected and the performance/ latency trade-off is relatively simple. For SOCs with their broader sets of constraints, a richer set of interconnect schema should be explored. For example, shared communication resources such as crossbars and meshes can help to provide performance at a reduced energy cost. Solving the latency *vs.* throughput trade-off now requires to take in consideration a large number of design parameters, like the number of pipeline stages, arbitration, synchronization, routing and repeating schemes. In fact, it is our believe that the use of diverse interconnect architectures present a major untapped opportunity.

To address these challenges and exploit these opportunities, it is critical to take a global view of the communication problem, and decompose it along lines that make it more tractable while not restricting the design space at the same time. Communication design has to begin at higher levels of abstraction than the architecture and RTL level. We believe that *a layered approach similar to that defined by the communication networks community* (and standardized as the ISO-OSI Reference Model (RM) [Zimmermann80]) to address the problem of connecting a large number of computers on wide-area networks should also be used for on-chip communication design. The layered approach is well suited to describe protocol functions that operate on data units at different levels of abstraction (in the form of streams, packets, bits or analog waveforms) and that are subject to various time granularity constraints. Each layer may include one or more closely related protocol functions, such as data fragmentation, encoding and synchronization.

Separating the communication protocol functions into layers that interact only via well-defined interfaces allows for a decomposition of the design problem into a set of simpler, tractable problems, and simplifies the synthesis and validation tasks. As amply demonstrated in the communication-network domain, the approach also maximizes re-use. An excellent example in case is the 802.11 wireless local-area network standard, where a single media-access layer supports different physical implementations through a unified interface.

The layered-stack approach to the design of the on-chip inter-core communications has been called the *Network-on-Chip* (NOC) methodology [Sgroi01]. Designing NOCs is not an easy task, and may result in protocol implementations that are incorrect (e.g. due to deadlocks and race conditions), or sub-optimal (e.g. are power hungry or introduce unacceptable latency).

Since its introduction in the early 2000's, NOC design has attracted major attention and multiple approaches have been proposed since then, some of which are enumerated below and are highlighted in other chapters in this book.

- The Virtual Socket Interface (VSI) Alliance [VSIA] has developed a standard interface to connect virtual components (VCs) to on-chip buses. VCs and buses are adapted to this interface wrapping them with appropriate glue logic.
- The Sonics [SiliconBackplane] μ-network approach de-couples the design of the communication among IPs. Each IP core communicates with an agent in the Silicon Backplane using a protocol called OCP (Open Core Protocol) and agents communicate with each other using network protocols. Both protocols can be customized by the SOC designer who can configure parameters such as data width and buffer depth.
- One important aspect of the NOC problem— the reservation of network resources such as buffers and bandwidth—is addressed by B. Dally, who proposes the flit-reservation flow control ([Peh00]). This approach makes use of packets sent over fast control wires to reserve resources in the data connection layer and allows to optimize the use of buffers without penalties in latency.

In parrallel, it is essential to develop a new generation of methodologies and tools to avoid the problems that may result from an ad-hoc application of NOC design. These environments should be centered along the following key tenets:

- By applying a discipline to on-chip communication design transition from ad-hoc SOCs to disciplined IC platforms.
- Are based on formal Models of Computation and support a correct-by-construction synthesis design flow and a set of analysis tools for broad design exploration.
- Maximize re-use with the definition of a set of interfaces between layers.
- Provide an application programmer with a set of APIs abstracting architecture details.

An example of an emerging environment that meets those requirements is the Metropolis Framework, developed at UC Berkeley [Carloni02].

In this remainder of this section, we, first, describe the OSI Reference Model (RM) and discuss its use for NOCs with an example of application, the

Pleiades platform (which was probably one of the first SOCs that fully embraced the NOC approach).

4.2 THE OSI REFERENCE MODEL APPLIED TO NOCs

The OSI Reference Model is a framework that allows us to classify and describe network protocols. Since its standardization, it has been used as a reference for wired and wireless computer-network design. However, the same layering concept and many of the protocol functions can be used also to realize NOCs. Below, we briefly describe the seven OSI layers and for each layer we give examples of on-chip application.

- **Physical**—The physical layer is concerned with the lowest-level details of transmitting data (bits) on a medium. The NOC physical layer protocols define such things as signal voltages, timing, bus widths, and pulse shape. At this level delay and power consumption may be difficult to predict. The floorplan of the chip can have a dramatic effect on both of these metrics, as well as the actual routes chosen for the wires. Also of particular concern at this layer is the synchronization of signals, since IPs may be in different clocking domains or require asynchronous communication.
- **Data Link**—The data-link layer is responsible for reliable transfer of data over the physical link, and may include error detection and correction functions. It must also arbitrate the access to a shared physical medium, like a bus. Examples of Medium Access Control (MAC) protocols are token ring and time division multiplex access (TDMA). Delay predictability, throughput and power consumption may vary significantly depending on which arbitration scheme is adopted.
- **Network**—The network layer provides a topology-independent view of the end-to-end communication to the upper level protocol layers. The connections established in the network could be static, such as offered by the reconfigurable interconnect of FPGAs, or dynamic. Similarly, data routes can be persistent over multiple transactions, or each transaction can be dynamically routed. In the latter case, congestion control may be required to reduce traffic through overburdened links.
- **Transport**—Transport layer protocols establish and maintain end-to-end connections. Among other things, they manage flow control, perform packet segmentation and reassembly, and ensure message ordering. This abstraction hides the topology of the network, and the

implementation of the links that make up the network. Therefore, it is used by the layers above the transport layer to provide components with more formal methods of communication.

- **Session**—Session layer protocols add state to the end-to-end connections provided by the transport layer. A common session protocol is synchronous messaging, which requires that the sending and receiving components rendezvous as the message is passed. The state maintained by the protocol is a semaphore that indicates when both the sender and the receiver have entered the rendezvous. Many embedded system applications utilize this sort of functionality to synchronize system components that are running in parallel. This is especially true when the system components are CPU-like processing elements that execute software programs.
- **Presentation**—The presentation layer is concerned with the representation of data within messages. Protocols at this level convert data into compatible formats. For example, two system components may exchange messages with different byte orderings, so this layer converts them to a common format.
- **Application**—This layer exports to the system components the highest level of abstraction of the underlying communication architecture. For example, in an embedded system that performs video processing the basic communication function may be to transfer a video frame from one system component to another. The application layer would define a function that does exactly this by utilizing the functions defined at lower stack layers. The system components can use these abstract communication functions without concern for the details, thus simplifying the component design.

Communication-based design uses the stack model as a tool to guide the decomposition of the design problem. The separation of computation and communication reveals the communication requirements of the system. The application layer provides the set of communication functions that implement those requirements. It does so by building upon the functionality defined at lower levels in the stack model. In most cases, it is not necessary to implement protocols at all of the OSI stack layers to provide this high-level functionality. One of the benefits of the OSI stack model is that it scales to match the needs of the system components. If the system components do not require connections with state, data format conversion or other features, the corresponding stack layers can be omitted. However, as embedded systems scale in complexity their communication architectures will have to scale in functionality as well.

The layered approach of OSI Model is a useful method for structuring and organizing a protocol at an early stage of the design process. However, on the way to implementation, designers may consider whether the original layering structure should be maintained, or whether performance is optimized by combining adjacent layers.

4.3 A NOC Example: The Pleiades Platform

The Pleiades platform (in its instantiation, the Maia processor) [Zhang00] presents a reconfigurable integrated circuit for DSP applications that demonstrates how the NOC layers of abstraction are applicable to existing designs. The basic Pleiades architecture is a heterogeneous collection of satellites such as arithmetic logic units (ALUs), memories, processors, FPGAs, and multiply-accumulators (Figure 2.4.). This collection of interconnected satellites is analogous to a set of IPs present in a SOC design. From a communication perspective, the computation at each satellite is arbitrary because each is wrapped in an inter-satellite communication interface.

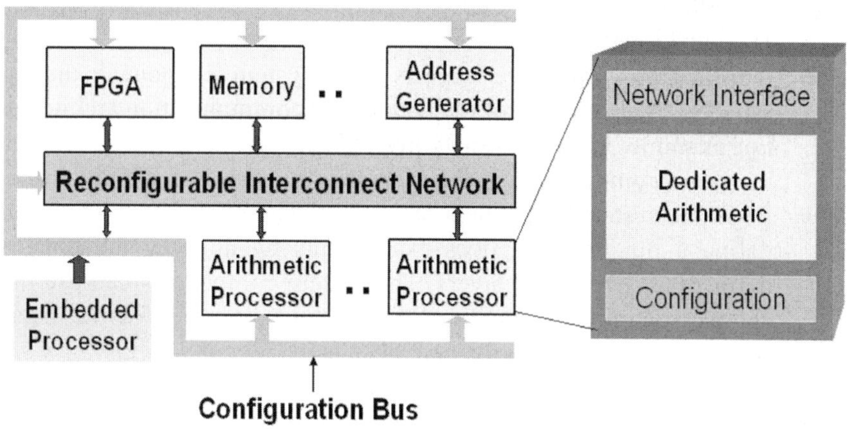

*Figure 2.4.*The Pleiades Platform for Communication Applications.

This interface is actually the **physical layer** in the NOC framework, because it specifies the signal definitions, timing, and synchronization between two satellites. In the Pleiades case, this means that data links are 18-bits wide and have 2 control bits. Additionally, each satellite operates on a local clock, which is not necessarily coincident with the other satellite clocks. For this reason, the interface is self-timed through a two-phase asynchronous handshaking scheme (Figure 2.5.a). This approach makes it possible for each computational core to chose its own frequency, based on computational requirements and workload. This could even be done dynamically, making it possible to save power through dynamic voltage scaling on a per module base. Lastly, the communication reduces power consumption through reduced-swing signaling (Figure 2.5.b). Individual links can be well characterized with predictable delay and energy consumption. Since each link in the Pleiades architecture is dedicated and error-free, no data-link layer is required.

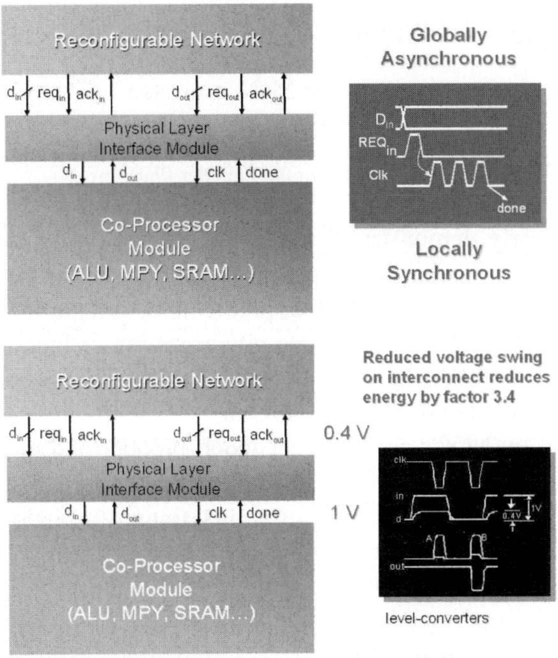

Figure 2.5. The physical layer of the Pleiades Platform protocol stack: globally asynchronous, locally synchronous (a); and reduced swing signaling (b).

It is in the **network layer** that the Pleiades architecture is especially novel. The interconnect consists of a two-tiered hierarchical mesh to provide energy efficiency as well as the required flexibility (Figure 2.6.). At the local level, universal switchboxes provide a method for programmatically connecting wires with a cluster. The global level provides switchboxes connected in a larger-granularity mesh for inter-cluster communication. The switchboxes enable persistent paths while allowing satellites reconnection to implement a different algorithm. A network connection is set up at (re)configuration time, and is rewired every time a new task is over-laid on the configurable fabric Because of this flexibility, the energy consumption and delay are dependent upon the actual path through the switchboxes. From a refinement perspective, higher levels of abstraction cannot know these values a priori so upper bounds or statistical averages can be used to provide early estimations of these metrics. Additionally, constraints can be used in the refinement process to influence the programming of the actual routes.

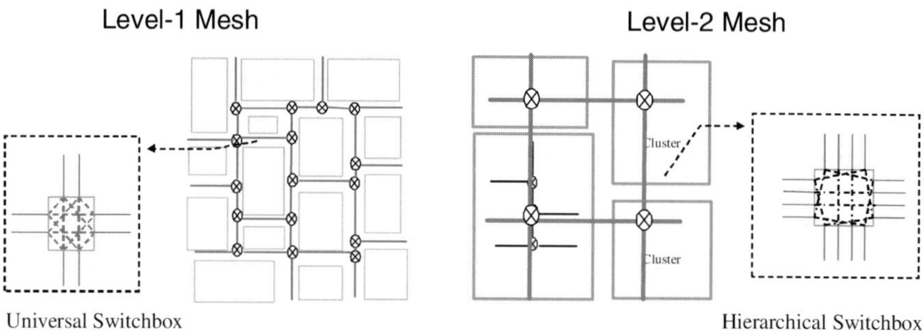

Figure 2.6. Hierarchical Mesh Network as used in the Pleiades Platform.

The Pleiades platform clearly demonstrates how a formal and structured NOC-based approach can help to make the mapping of complex algorithms on concurrent architectures manageable, verifiable, predictable, and automatable. It also illustrates how the approach can enable power-saving techniques such as reduced swings, GALs, and distributed DVS, which otherwise would be hard to implement in a repeatable and reliable fashion.

The opportunities offered by NOCs are however much broader and more far-reaching. For instance, the stack-based approach is one of the premier mechanisms to deal with the DSM reliability concerns. Just to use the analogy with wireless communications again, it is indeed possible to create a wireless link which is in essence error free. This comes however at an unacceptable cost in transmission power. Instead, wireless systems allow errors to happen (to a certain degree), and correct them at different levels of the stack using, for instance, forward-error correction at the data-link layer or retransmission at the transport layer. This not only helps to bring down the power dissipation, but also reduces the interference wrt other neighboring links. The same principles can also be applied to NOCs, as has been demonstrated by a number of researchers [e.g., DeMicheli03].

5 SUMMARY AND PERSPECTIVES

In this chapter, we have argued that the semiconductor industry is facing some unprecedented challenges, which may jeopardize the continuation of its spectacular run. The combination of design complexity, deep-submicron challenges and economics is increasingly undermining traditional design approaches such as ASIC. The platform-based methodology presents an alternative approach, which addresses a number of the concerns through the use of higher levels of abstraction, a higher level of re-use, and a larger role for "soft design". At the core of the platform lies the communication strategy, that is the way how functions interact with each other. In light of the increasing use of concurrency, traditional interconnection strategies have proven to be either inefficient or non-applicable. Hence, the rising importance of a formal network-on-a-chip approach. We have also shown that a structured use of NOC can help to address some crucial DSM problems such as power, reliability, and synchronization. So far however, we have only set the first baby steps, and much progress in research and development is still to be made. However, the main question now is not if the NOC approach will ever become viable, but when and how fast.

References

[Boekhorst02] F. Boekhorst, "Ambient intelligence: The next paradigm for consumer electronics", Keynote presentation, *Proceedings IEEE ISSCC 2002*, San Francisco, February 2002.

[Carloni02] L. Carloni et al., "The Art and Science of Integrated Embedded System Design," Proceedings of the 28th European Solid-State Circuits Conference, September 2002.

[Claasen00] T. Claasen, "First-Time-Right Silicon, but to the Right Specification," Keynote presentation, Proceedings DAC 2000, Los Angeles, June 2000.

[DeMicheli03] G. De Micheli, "Designing Robust Systems with Uncertain Information," Keynote presentation, ASPDAC 2003.

[Ericsson Mobile] Ericsson Mobile Platforms, http://www.ericsson.com/mobileplatforms.

[GSRC] The Gigascale Systems Research Center, *http://www.gigascale.org*.

[Keutzer00] K. Keutzer, S. Malik, R. Newton, J. Rabaey and A. Sangiovanni-Vincentelli, "System Level Design: Orthogonalization of Concerns and Platform-Based Design," IEEE Transactions on Computer-Aided Design of Integrated Circuits & Systems, vol.19, no.12, Dec. 2000, pp.1523-43.

[OMAP] The Omap Platform, http://www.omap.com

[Peh00] L.S. Peh and B. Dally, "Flit-Reservation Flow Control", Proceedings of 6-th International Symposium of High-performance Computer Architecture, Jan. 2000

[Pileggi02] L. Pileggi, H. Schmit, et al, "Via Patterned Gate Array," CMU Center for Silicon System Implementation Technical Report Series, No. CSSI 02-15, April 2002.

[Rabaey03] J. Rabaey, "Ultra-Low Cost and Power Communication and Computation Enables Ambient Intelligence", Keynote presentation, *Proceedings Smart Objects Conference*, Grenoble, May 2003.

[Sangiovanni02] A. Sangiovanni-Vincentelli, "Defining Platform-based Design", *EEDesign*, February 2002.

[Sgroi01] Sgroi M, Sheets M, Mihal A, Keutzer K, Malik S, Rabaey J, Sangiovanni-Vincentelli A. "Addressing the system-on-a-chip interconnect woes through communication-based design." Proceedings of the 38th Design Automation Conference, Las Vegas, June 2001, pp.667-72.

[SiliconBackplane] Sonics Inc., http://www.sonicsinc.com/

[VertexPro] Virtex-II Pro Platform FPGAs, http://www.xilinx.com

[VSIA] VSI Alliance, http://www.vsi.org/

[Zhang00] H. Zhang, et al. "A 1-V heterogeneous reconfigurable DSP IC for wireless baseband digital signal processing," IEEE J. Solid State Circuits, vol. 35, Nov. 2000, pp. 1697-1704.

[Zimmerman80] H. Zimmermann, OSI Reference Model - The ISO Model of Architecture for Open Systems Interconnection, IEEE Transactions on Communications COM-28, No. 4: April 1980.

Acknowledgements

The ideas and concepts introduced in this chapter are the result of a large cooperative effort between many outstanding individuals as a part of the Gigascale Systems Research Center or GSRC (formerly known as the Gigascale Silicon Research Center). Specifically, the author would like to acknowledge the contributions of Alberto Sangiovanni-Vincentelli (who championed most of the core ideas of platform-based design), Richard Newton, Kurt Keutzer and Sharad Malik. The funding of the GSRC by the MARCO consortium, a joined effort of the semiconductor industry and government, is greatly appreciated.

Chapter 2

WIRES AS INTERCONNECTS

Li-Rong Zheng and Hannu Tenhunen
Royal Institute of Technology (KTH), Stockholm, Sweden
lirong@imit.kth.se, hannu@imit.kth.se

Abstract: Deep submicron technology is rapidly leading to exceedingly complex, billion-transistor chips. This has resulted in a new circuit paradigm - system-on-chip (SoC). However, deep submicron physics indicates that wires, not transistors, dominate power and performance. Interconnects have been a key design objective in deep submicron SoC. In this chapter, we review interconnect performance as technologies migrate from 0.25μm to 0.035μm feature sizes. Challenges of deep submicron effects and their impacts on advanced chip design are summarized. Basic concepts of signal integrity and various noise sources in deep submicron SoC are illustrated. Finally, interconnect strategies and interconnect-centric design methodologies are generally described; various design techniques for signal and power integrity in deep submicron SoC are discussed.

Key words: Interconnects, deep submicron CMOS, signal integrity, power integrity

1. EVOLUTION OF MICROELECTRONICS TECHNOLOGY

1.1 Moore's Law and Technology Scaling

In 1958, Jack S. Kilby, an employee of Texas Instruments, created the first integrated circuits [1], which were, of course, very crude as measured by today's standards. Since then, with the rapid development of processing technology, particularly in lithography, the number of components of a chip increased very rapidly. With this pioneer work, Kilby earned the Nobel Prize in 2000.

In the beginning of the 1960's, shortly after the invention of the IC, Gordon Moore, one of the pioneers in Silicon Valley, formulated an empirical law stating that the performance of an IC, including the number of components on it, doubles every 18-24 months with the same chip price. This became known as Moore's law. Remarkably enough, it is still holding up after forty years, and this trend is likely continuing for the next 10 years, as shown in *Figure 2.1*. Today, the mainstream technology for IC fabrication has already been in deep submicron (DSM) regime, and the increased integration capacity has resulted in exceedingly complex chips such as system-on-chip (SoC).

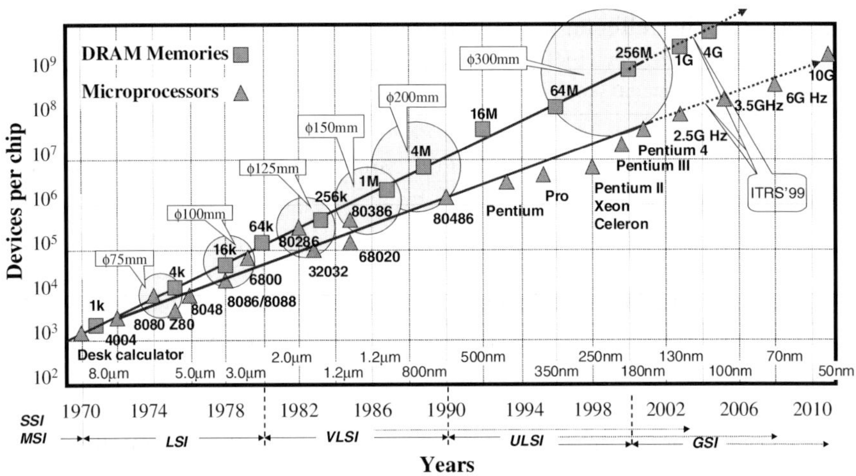

Figure 2.1 * Evolution of microelectronic technology that has so far followed Moore's law very well. For the future, there are different extrapolations, depending upon assumptions on the development of the process technology. Also shown in the figure are the rapid development of process technologies and the size of the silicon wafers.

1.2 Deep Submicron Effects

The system-on-chip design paradigm and deep submicron technology bring two main challenges to the ASIC design community. The first challenge is productivity, i.e., the ability to use millions of gates within the ever shorter time-to-market, which is currently tackled with the design methodology based on Intellectual Property (IP) right blocks. The second challenge is coping with

* The data of this figure are from several sources. Historical data are based on surveys, including that from SCS (www.semiconsulting.com). Future trends are based largely on projections by the NTRS'97 and ITRS'99. The division of LSI, VLSI, ULSI, and GSI is based on some publications.

the dramatically changed technology parameters, where interconnects invite many signal integrity related problems. During the past three to five years, the interconnect issues have been much explored and better understood. This includes efficient parasitic extraction tools, circuit level modeling and analysis, and new design techniques in coping with signal integrity challenges and timing closure problems. However, as the complexity of chips continuously rises, these newly developed tools show limited capacity in dealing with complex SoCs. Therefore, the interconnect problem is still luring and looming and not explored sufficiently yet, especially in relation to IP-based design.

Typical signal integrity effects include **interconnect delay**, **crosstalk**, **transmission line effects**, **substrate coupling**, **power supply integrity**, and **noise-on-delay effects**. In the early days of VLSI design, these effects were negligible because of relative slow chip speed and low integration density. However, with the introduction of technology generations at about the 0.25μm scale and below, there are many significant changes in wiring and electrical characteristics. Interconnect started to be a dominating factor for chip performance and robustness. First, as chip speed continually increases, the inductance effect of interconnects tends to increase, which will cause ringings at signal rise/fall edge and changes of crosstalk characteristics. Second, finer line width and higher aspect ratio (wire cross section becomes narrower and taller) wires are utilized. Because of their enhanced fringing field effects, both capacitive and inductive voltage coupling coefficients increase, which raises the level of coupled noise through wiring hierarchy. Third, the length of global and semi-global interconnect lines when measured in units of wire pitch increases dramatically despite the fact that the maximum on-chip interconnect length grows commensurately with the chip size. Consequently, the circuit delay of these interconnections increases quadratically on RC lines or linearly on LC lines. Transmission line effects, which cause signal reflections at impedance discontinuous junctions, start to matter when the signal rise/fall edge is comparable with or shorter than two and a half times the time-of-flight of the line. Inductive ringing, crosstalk, and transmission line effects combine to produce more noise sources that are coupled to other circuit nodes globally on the chip via the substrate, common return ground, and electromagnetic interference. In addition, more and more aggressive usage of high-speed circuit families (e.g. domino), scaling of power supply and threshold voltages, and mixed-signal integration combine to make the chips more noise-sensitive. Finally, higher device densities and faster switching frequencies cause larger switching-currents to flow in the power and ground networks. As a result, power supply is plagued with excessive IR voltage drops as well as inductive voltage drops (simultaneous switching noise) over the on-chip power network and package pins. Power supply noise not only degrades driver capability of gates (hence introduces additional signal delay), but may also cause false switching of logic gates.

Noise has two deleterious effects on circuit performance. When noise acts against a normally static signal, it can destroy the local information carried by the static node in the circuit and ultimately result in incorrect machine-state stored in a latch. When noise acts simultaneously with a switching node (the signal rise/fall edge), this is manifest as a change in the timing (jitter and skew) of the transient. The effect of noise-on-delay is anticipated to be as large as 30~50% variations of the total signal delay in DSM chips. In today's deep submicron circuits, there is little likelihood of achieving a sound and robust quality design unless power and signal distributions are well managed to minimize these bad effects. Signal and power integrity analysis has been of the same importance as timing, area, and power analysis [2].

1.3 Impact of DSM Wires on VLSI Design Methodology

In deep submicron technologies, degraded interconnect performance and high-speed operation reduce system noise immunity and timing budget which in turn result in faults in system operation. Due to the complexity of the system, chip design poses a difficult challenge to ensure that the resulting system is reliable. The increased difficulty in designing, verifying, and testing the chips has become a larger barrier to provide reliable chips within ever shorter time-to-market than that of providing the technology for manufacturing them.

A major impact of interconnects on chip design is that they blur the traditional distinctions between logic design, circuit design and physical design. In old design methods, unawareness of accurate interconnect information in logic design causes the timing closure problem in DSM regime. *Figure 2.2* illustrates that design tools, which earlier could concentrate on discrete parts of the design problem, now must be aware of a broader range of factors. For example, logic synthesis tools now must take interconnect and placement into account and increasingly accommodate the transistor-level properties of dynamic circuit families and the noise effects at smaller feature sizes and higher frequencies. The design flow must assure functional correctness and timing, power, reliability, manufacturability, signal integrity, and testability requirements.

State-of-the-art VLSI design tools have included signal integrity and timing verifications at the post-layout stage. Parasitics of three-dimensional interconnects are extracted based on pre-characterized interconnect library and/or scalable interconnect parasitic models (built from electromagnetic field simulations). However, the large number of circuit nodes makes it computationally expensive or infeasible. Even if the circuit size is manageable, design problems revealed at this stage are usually very difficult or expensive to fix. More advanced design techniques emphasize to improve *a priori* prediction of interconnects [54]. Instead of finding and fixing the interconnect problems in post-layout, *a priori* estimation improves wire manageability in early design phases and hence reduces the number of design iterations. However, due to the scaling, the total number of wires grows rapidly. Future tools and design methodologies must provide a commensurately increased capability for handling a large number of wires. The exponential scaling of chip complexity will finally destroy the design productivity. Emerging interconnect-centric system architectures such as networks-on-a-chip or NoC, in part, provide a good solution for dealing with interconnect difficulties. Many regular NoC structures specifically develop a global floorplan to accommodate the interconnect constraints. The regularity of these NoC structures eases interconnect design with predictable performance by reducing the design space to a set of uniform wire segments and standard interface circuits.

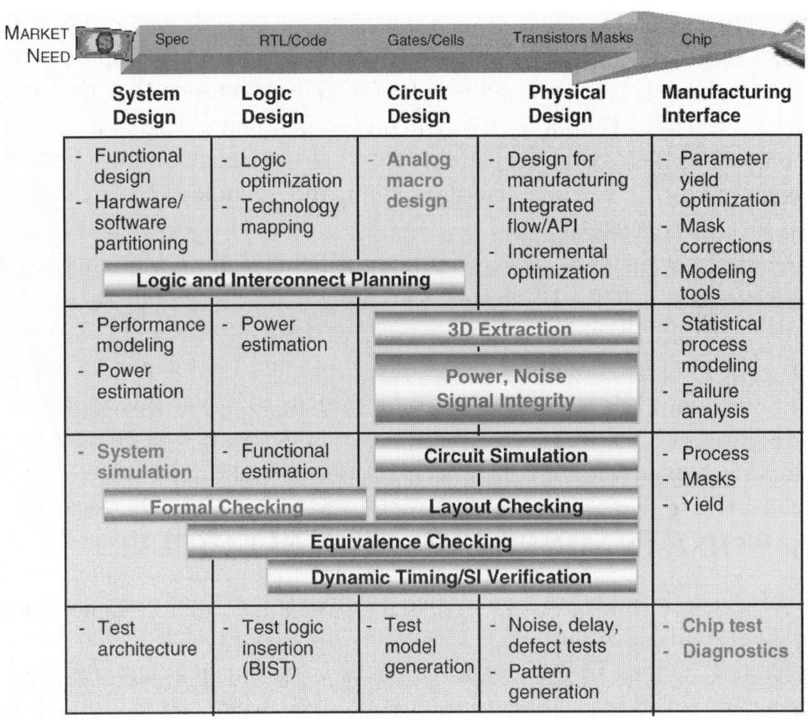

Figure 2.2 Issues in chip design in deep submicron technology in which the interconnect issues blur the traditional distinctions between different design levels [3].

2. SIGNAL INTEGRITY ILLUSTRATION

The signal integrity problem is sketched in *Figure 2.3*. Very basically, a signal takes a certain finite amount of time (t_{tof}) to travel from the driver to the receiver. This time is referred to as the "time-of-flight" (here the LC delay instead of RC delay is assumed). In an ideal case, the received signal has the same quality as its original one. However, in a real case, the received signal quality is degraded because of many reasons such as crosstalk, signal dispersion, and glitches in power supply. As a result, the signal is not considered to be received, or latched in, until it has been stable for a finite amount of time, which is called the "setting time", t_{set}. The setting time varies with various signal qualities, which consequently contributes to timing skew and jitter. If the signal quality is too poor to be judged by the receiver, malfunction may occur.

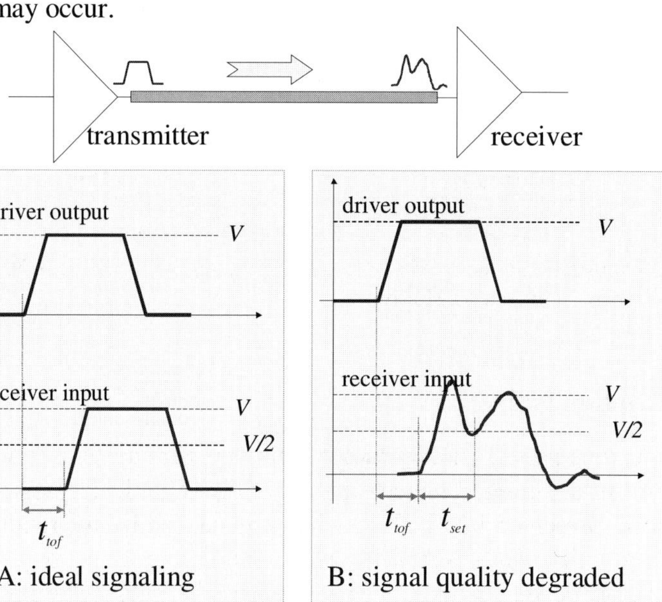

Figure 2.3 Illustration of signal integrity problem in VLSI circuits and systems

3. NOISE IN MIXED-SIGNAL VLSI CIRCUITS

Noise is a deviation of signal from its intended or ideal value. Most noise in mixed-signal VLSI circuits is created by the system itself. Electromagnetic interference could be an external noise source or coupling between subsystems. In deep submicron circuits, the noise created on parasitic components by digital switching exhibits the strongest effect on system performance and robustness. *Figure 2.4* illustrates these noise sources. The following texts summarize these noise sources. Some of these noise definitions are taken from [7].

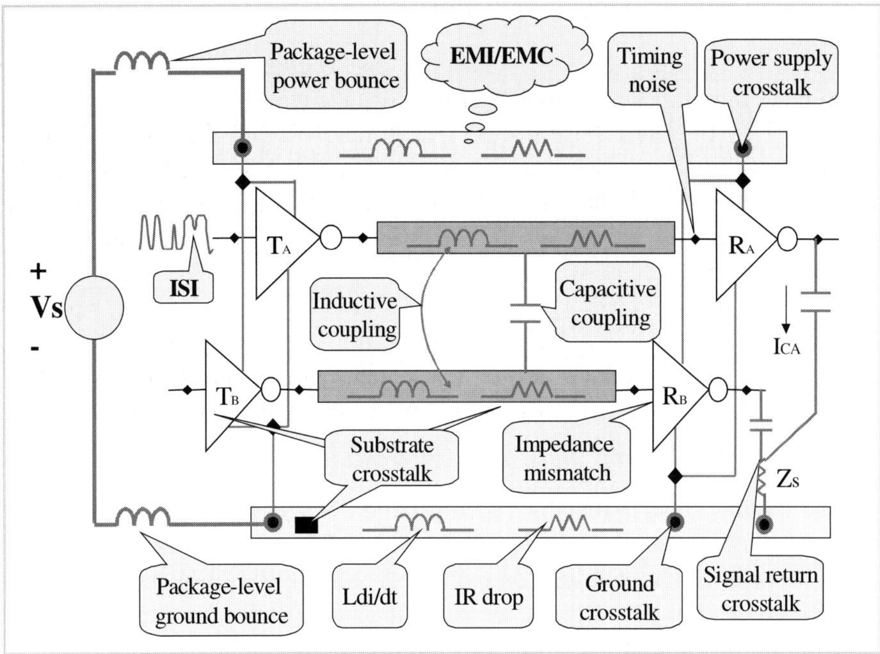

Figure 2.4 Illustration of noise created on parasitic components due to switching current and voltage in deep submicron VLSI Circuits

3.1 Device Noise

The device noise is not a crucial issue in digital systems, but may be very important in analog and mixed-signal circuits. The device noise is the lowest level noise presented in VLSI circuits. It is caused by the random movement of charges through resistance, across transistor junctions, and random fluctuations in the charge recombinations in surface states and in the semiconductor bulk. The level of noise generated and coupled by thermal noise, avalanche noise, shot noise or Schottky noise, and 1/f noise represents a minimal level in coupled noise and all other noise mechanisms treated are usually orders of magnitude worse than these without special design.

3.2 Power Supply Noise

A. Resistive Voltage Drop and Simultaneous Switching Noise

When a current (I) flows through a power line or a ground line, the following voltage drops are induced (see *Figure 2.4*):

$$\Delta V_R = IR \tag{2.1}$$

$$\Delta V_L = L\frac{di}{dt} \tag{2.2}$$

where R is the resistance of the power or ground line, L is the inductance of the power or ground line. ΔV_R is the dc voltage drop or resistive voltage. This is a dominant noise source for on-chip power networks. ΔV_L is known as simultaneous switching noise or inductive noise. It is a dominant power supply noise at package level. However, in deep submicron circuits, because of the increasing switching speed, the inductive voltage drop over on-chip power lines is increasing, which will soon be comparable with *IR* drops [4]. Further reading of power supply noise analysis and power distribution design can be found in many recent literatures such as [4-5] and [56-62].

B. Common-Mode Supply Noise and Differential-Mode Supply Noise

The ground noise is usually called ground bounce, and the noise glitch on the power line is usually called power bounce. When the ground bounce and power bounce are in phase (common-mode noise), they will not affect the local logic cells but will degrade the signaling between distant transmitters and receivers in the circuits. However, when power bounce and ground bounce are out of phase (differential mode noise), they can adversely affect the operation of local circuits, resulting in transmitter and receiver offsets and jitter in timing circuits [5].

3.3 Crosstalk

Noise caused by one signal, *A*, being coupled into a different signal, *B*, is called crosstalk. Crosstalk can occur over many paths, as shown in *Figure 2.5*.

A. Inductive Crosstalk and Capacitive Crosstalk

When interconnects are routed close to each other, signals on the lines can crosstalk to each other via near field electromagnetic coupling, as shown in *Figure 2.5* (a). In circuit theory, the electric field coupling is described as capacitive crosstalk and the magnetic field coupling is described as inductive crosstalk. The influence of capacitive and inductive crosstalk is given by

$$\Delta I_B = -C_m \frac{d(V_B - V_A)}{dt} \tag{2.3}$$

$$\Delta V_B = -L_m \frac{dI_A}{dt} \tag{2.4}$$

with C_m and L_m the mutual capacitance and the mutual inductance between line *A* and line *B*. ΔI_B and ΔV_B will propagate along the interconnect lines and eventually they give rise to noise waveforms on the line terminations. A large number of recent literatures on crosstalk and DSM interconnect design exist. Some of them have been well summarized as books such as [63-66].

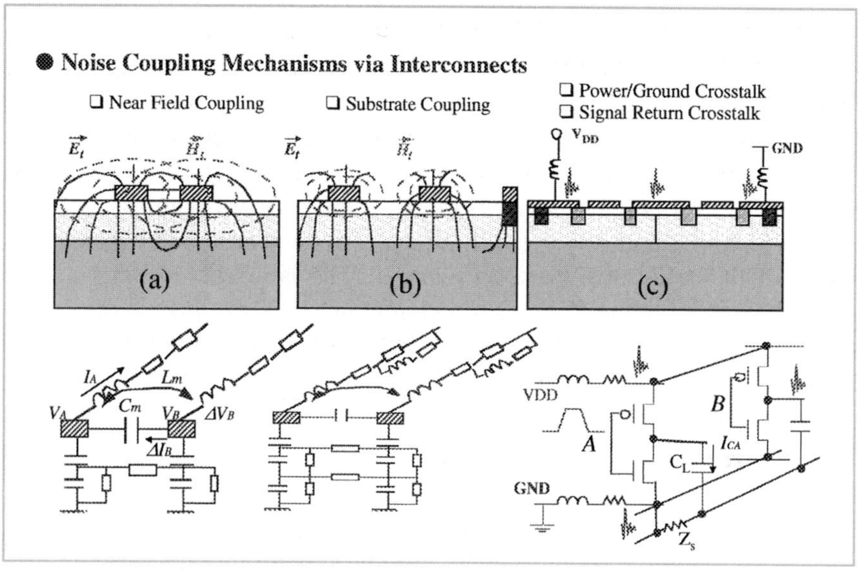

Figure 2.5 Illustration of noise coupling mechanisms via interconnections. (a) Near field coupling via electromagnetic field when interconnect lines are routed close to each other. (b) When wires (or active devices) are located far away from each other, the common substrate can serve as a passive network through which crosstalk from aggressor components to victim components occurs. (c) Due to the circuits sharing common power/ground and/or signal return paths, crosstalk within the system occurs via these shared interconnect networks.

B. Substrate Crosstalk

When interconnects (or other active/passive components) are placed far apart, the common substrate will serve as a channel for signal coupling, as shown in *Figure 2.5* (b). This noise source is known as substrate coupling or substrate crosstalk. The substrate coupling is particularly harmful in mixed-signal ICs where the low resistive silicon substrate can be modeled as a resistive and capacitive network and the noise can spread globally through this network. Substrate crosstalk is closely related to the chip package and the power/ground distribution network. A good power and ground distribution design is most important for reducing substrate crosstalk.

C. Power/Ground Crosstalk

Besides, signals can affect one another via a shared power supply or ground, as illustrated in *Figure 2.5* (c). The figure shows that two output drivers share the same power and ground lines. When driver *A* switches high, the current it draws through the power line creates a glitch that appears on the output of driver *B*, if drive *B* is at quiet high state. This signaling-induced power supply noise is called power supply crosstalk. Similarly, the signaling can also cause ground bounce and the corresponding noise coupling is called ground crosstalk.

D. Return Signal Crosstalk

Whenever a pair of signals, *A* and *B*, share a return path that has finite impedance, a transition on signal *A* induces a voltage across the shared return impedance that appears as noise on signal *B*. This is the return signal crosstalk and is illustrated both in *Figure 2.4* and in *Figure 2.5* (c) ($I_{CA}Z_S$ in the figures). In a typical electronic system, shared wire bonds, package pins, board planes and traces, connector pins, and cables all contribute to the impedance (largely inductance) of the signal return. Unless differential signals are used, or each signal is supplied with its own return, signals share returns, and crosstalk over these returns is a major source of noise.

3.4 Transmission Line Effects

Usually, when a wire is longer than 1/10 wavelength of the signal frequency component that is transmitted, neglecting the wave nature of the signal propagated on it will result in obvious error in analysis. The wave nature of the signal presents a spatial variation of signal amplitude across the interconnect due to a phase difference. Such a line is electrically long and needs to be modeled as a transmission line.

When a wire is a transmission line, the signals that propagate on it behave as traveling waves. As the traveling waves see discontinuities on the line, reflection waves will occur. The reflection coefficient is given by

$$\Gamma = \frac{V_r}{V_i} = \frac{Z_d - Z_o}{Z_d + Z_o} \tag{2.5}$$

where Z_o is the characteristic impedance of the line, and Z_d is the impedance of the discontinuity.

The discontinuity of a transmission line is defined as any change in impedance of the line. A few of the basic discontinuities are series inductance, shunt capacitance, capacitive loads, impedance steps, and unmatched terminations. They are introduced by physical changes of the signal paths such as vias, wire bends, stubs, wire crossovers, bonding wires, package pins, connectors, and non-ideal receivers.

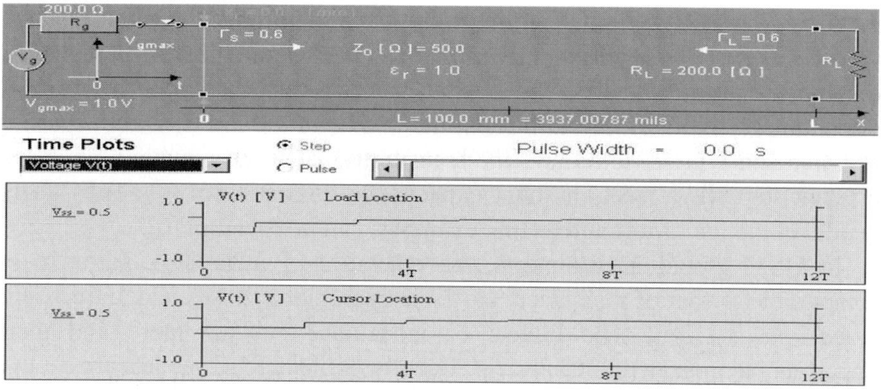

(a) Unmatched terminations at driver and load cause reflection waves.

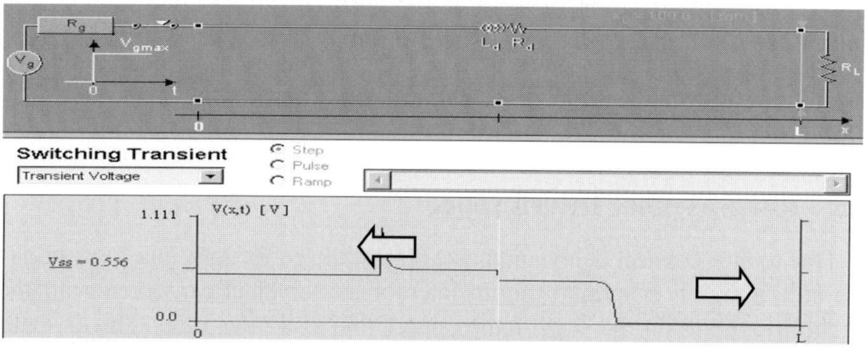

(b) Forward and backward waves after meeting an inductive discontinuity on the line.

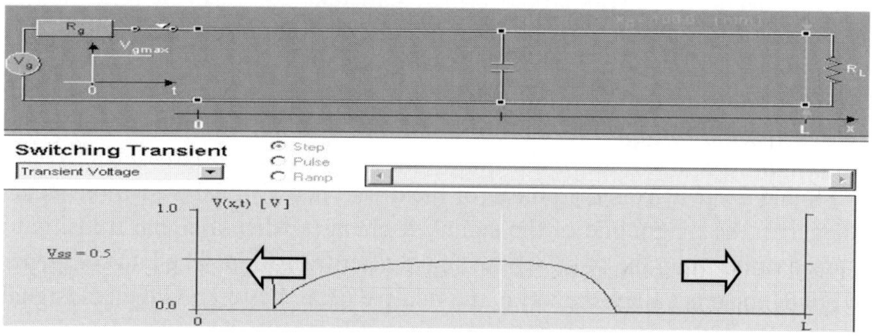

(c) Forward and backward waves after meeting a capacitive discontinuity on the line.

Figure 2.6 Illustration of transmission line effects. T: time-of- flight of the line.

Figure 2.6 schematically illustrates some transmission effects of a 50 Ω lossless transmission line. The incident wave is a step signal. Case (a) shows multiple reflections (wave steps) when the driver and the load are mismatched (200 Ω). Case (b) shows a positive spike is reflected and the transmitted pulse is smoothed when the incident step wave meets an inductive/resistive discontinuity (L_d=1nH, R_d=200 Ω) at the middle of the line. Case (c) shows a negative spike is reflected and the transmitted pulse is smoothed when the incident step wave meets a shunt capacitance discontinuity (C_d=1pF) at the middle of the line. More interesting examples can be found in [6].

In DSM global interconnects, the reactance of inductance starts to be comparable to that of resistance, particularly in global buses, clock trees, and global power lines. The effects of inductance become more and more important, as indicated in Chapter 4. Usually, lumped RLC models are enough to describe the circuit response, few papers use transmission line modes for on-chip wires. However, there are some global interconnects for fast signals and short signal rise edge, distributed circuit models or transmission line models can better describe their circuit behavior. Reference [16] describes the criteria when transmission line effects are important for on-chip wires. Further, most of today's off-chip interconnects behave as transmission lines.

3.5 Inter-Symbol Interference

Due to interconnect delay and noise (crosstalk, reflections in a transmission line etc), a signal or symbol on an interconnect channel can corrupt another symbol traveling on the same interconnect line at a later time. This is called inter-symbol interference (ISI) that occurs as a result of energy from one symbol being stored in the interconnect channel so that it sums with a later unrelated signal [7]. This stored energy can take the form of reflections at discontinuities in a transmission line, LC circuit ring, or charge storage in an RC circuit.

3.6 Timing Noise

Timing noise affects the phase (or the time when a signal transition occurs) rather than the magnitude of the signal. It changes when a signal transition is detected rather than the value of the signal after transition. The DC component of timing noise is called skew (i.e. the relative delay between two clock/signals at two different nodes) and is usually caused by mismatched line lengths and variations in the devices parameters. Jitter can be regarded the AC component of the timing noise, i.e. the cycle-to-cycle variation when a signal transition occurs [7]. Jitter is usually due to power supply noise modulating the delay of active components or additive noise such as crosstalk or inter-symbol interference moving the point when the transition is detected.

3.7 EMI/EMC

Large external electric and magnetic fields can couple into circuit nodes and cause EMI (electromagnetic interference) noise. Excessive emission of electric and magnetic fields from an electronic system can adversely cause other systems or subsystems to fail and violate EMC (electromagnetic compatibility) requirements [8] [9].

In recent years, there is an ever-increasing demand for highly integrated portable wireless communication units in which analog, RF, and digital coexist. Managing EMI/EMC requires isolating not only the external emission sources but also the electromagnetic coupling within the systems. In mixed-signal systems, RF components, high-speed clock nets, global signal and power buses, and any other long on-chip and package-level interconnects can be the sources or receptors of EMI noise. EMI suppression in such systems requires careful engineering to use shielding, decoupling, grounding, and signal/power distributions [8] [10].

4. INTERCONNECT DELAY

As interconnect size is scaled down in DSM VLSI, interconnect delay increases significantly. Currently, there are basically two delay domains – RC delay domain and transmission line delay domain, depending upon circuit parameters. The actual delay is related to the time constant of the circuit response and the definition. For RC delay, usually t_d (0~50%)=0.69τ and t_r (10%~90%)=2.2τ, with τ the RC time constant.

4.1 Interconnect Delay Models

The interconnect delay model in VLSI has been improved as technology is scaling. In early VLSI design, interconnects were modeled as a lumped capacitive load for gates. It ignored resistance of wires. As feature sizes decrease, wire resistance increases, invalidating this approximation. The resistance of interconnects can be taken into account by approximating the distributed RC structure using a lumped RC tree. One of the most popular RC delay models is based on the Elmore time constant [11]. An overview of different interconnect-delay models including RC and RLC lines can be found in Chapters 3 and 4. Reference [67] systematically describes the on-chip inductance effects in deep submicron VLSI circuits.

However, some global interconnects in deep submicron technology below 0.18μm generations and package level interconnects are sophisticated RLC transmission lines. The RC tree model is inadequate to accurately model delay, because RC models cannot model higher-order under-damped systems. A second or higher order RLC model of the interconnect can provide significant improvements over the accuracy of a first-order RC delay model, since RLC trees and higher order transfer functions may have oscillatory output voltages, and are thus able to predict increased delays due to settling times of interconnect waveforms.

A detailed second order RLC delay model for a nonlinear driver and a single transmission line is presented in [12]. Closed-form solutions for the 50% delay, rise time, overshoots, and setting time of signals in a second-order RLC tree are presented in [13] and [67]. Chapter 4 also extensively studied the design methodology of high-order RLC interconnections. However, deriving analytical equations for higher order RLC trees, particularly when parasitic parameters are frequency dependent, is difficult due to the complexity of the systems. Accurate estimates of actual delay for these RLC trees are thus largely dependent upon simulations. Currently, efficient and widely used RLC delay models are derived based on reduced-order-approximation techniques such as reciprocal expansion (REX) and asymptotic waveform evaluation (AWE). Good reviews of these techniques and derivation of high-order RLC delay models can be found in [14] [15].

4.2 RC Delay versus LC Delay

For better accuracy, any interconnects can be modeled as RLC transmission lines in which RC delay and LC delay coexist. When we say one interconnect is an RC line, this implies that the contribution of the inductance to the signal response in this line is negligible. Similarly, an LC line means the resistance in this line is negligible. It was found that if the wire length and wire impedance are chosen such that $R_w l < Z_o$ with Z_o the wire characteristic impedance, the LC delay will dominate the total interconnect delay; otherwise the line delay is dominated by a slow RC response [16] [17]. This is illustrated in *Figure 2.7*.

In practice, overshoots and oscillatory output voltages are usually damped by sizing the driver impedance and utilization of a diode clamp [18][19]. In this case, if an interconnect is an LC dominated line, the signal propagation delay on this line is (if neglecting drive delay)

$$t_{wire} = t_{tof} = \frac{l}{v} = l\sqrt{L_w C_w} \tag{2.6}$$

where t_{tof} is the time-of-flight delay representing the time required for a signal traveling from one place to another at the wave velocity, l is the wire length, v is the signal velocity, L_w and C_w represent respectively the per unit length inductance and the per unit length capacitance of the interconnect. The total circuit delay is a sum of wire delay and driver delay. On the other hand, if a line is a very resistive transmission line, the following empirical formula for adding time-of-flight delay (t_{tof}) and conventional RC delay (t_{RC}) was found well predicting the total wire delay [20]:

$$t_{wire} = \left(t_{tof}^{\ 1.6} + t_{RC}^{\ 1.6}\right)^{1/1.6} \tag{2.7}$$

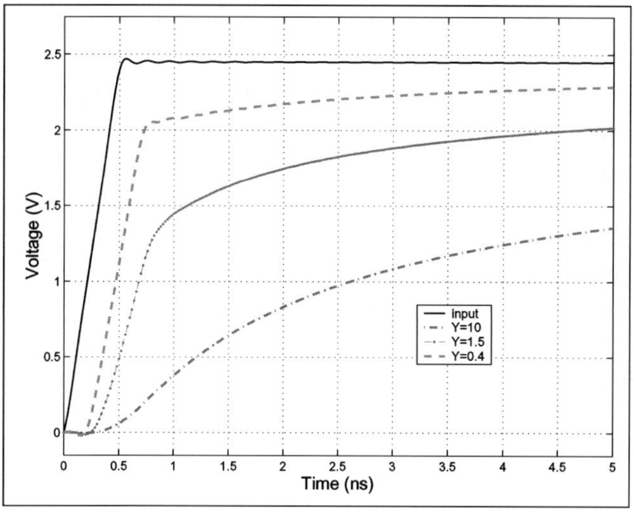

Figure 2.7 Output waveforms of match-terminated transmission lines with various ratio of Y=$R_w l/Z_o$. It is seen that when Y=10, slower RC response dominates the line delay, and when Y=0.4, the waveform first exhibits a very sharp rise edge (LC response) followed by slower RC charging. All these curves will finally reach a saturated voltage level which is determined by the DC voltage divider of this line and the terminations.

4.3 Interconnect Strategies in DSM VLSI Circuits

Figure 2.8 is a popular picture taken from [21], illustrating the interconnect problem in DSM VLSI circuits. The figure shows calculated gate and interconnect delays versus technology generation which illustrates the dominance of interconnect delay over gate delay for aluminum metallization and silicon dioxide dielectrics as feature sizes approach 100 nm. The figure also shows the decrease in interconnect delay and improved overall performance expected for copper and low *k* dielectric constant insulators, attesting the urgent need for such new materials in silicon technology. Utilization of low *k* materials and copper conductors considerably reduces parasitic capacitance and resistance directly. It also reduces the wire aspect ratio compared with the aluminum counterpart. The 1999 and later editions of the roadmap [3] highlight a continued change to the new materials, being introduced at an unprecedented pace. It should be noticed that the "gate" delays shown in this figure are for unloaded single transistors, not for real logic devices. In [22], a scaling analysis was performed and it pointed out that this old curve could be somewhat misleading. After detailed calculations of wire delays and gate delays versus FO4 (fanout-of-four delay), they mentioned that depending on wiring categories, local wires scale in performance and hence are not as bad as this figure shows, while global wires and fixed wires do not scale in performance and hence present more serious problems to designers.

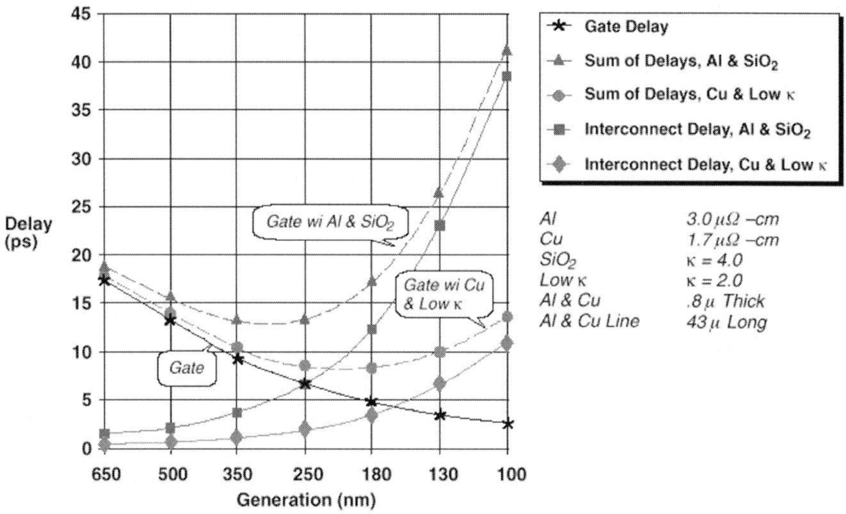

Figure 2.8 Calculated gate and interconnect delay (fixed line length) versus technology generation in submicron and deep submicron technologies [21].

As the complexity of ICs continually increases, the length distribution of on-chip wires has shown separated humps, initially one for local and another for global interconnects but now being in 3~4 tiers [23] (local, intermediate, and global). Local wires, consisting of very thin conductors, connect gates and transistors within an execution unit or a functional block on the chip. These wires usually span a few gates and occupy first, sometimes second, metal layers in a multi-layered interconnect system. Their lengths tend to scale down with the technology. Intermediate wires provide clock and signal distribution within a functional block or inter-module communications between adjacent blocks with typical lengths up to 3~4 mm. Global wires provide clock and signal distribution between functional blocks, and deliver power/ground to all functions on a chip. Global wires, which occupy the top one or two layers, are longer than 4 mm and can be as long as half of the chip perimeter.

Implementation of copper and low k materials allows scaling of the intermediate wiring levels and minimizes its impact on wiring delay, while local wiring levels are relatively unaffected by traditional technology scaling.

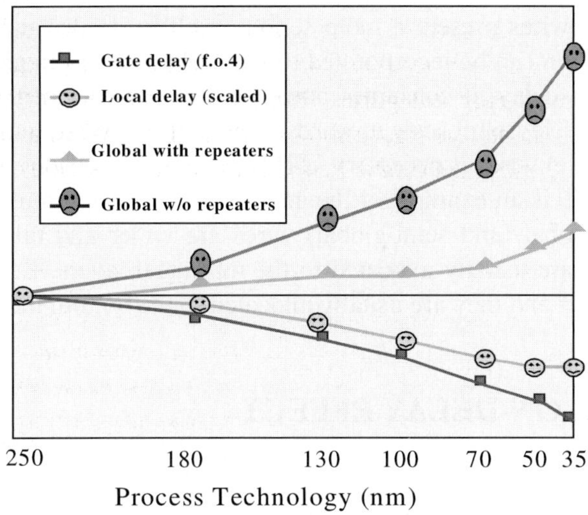

Figure 2.9 Delay for local and global wiring versus feature size. It indicates that local wiring is relatively unaffected by traditional scaling due to its commensurate length scaling. However, delay on global wires becomes worse and dominates chip performance [3].

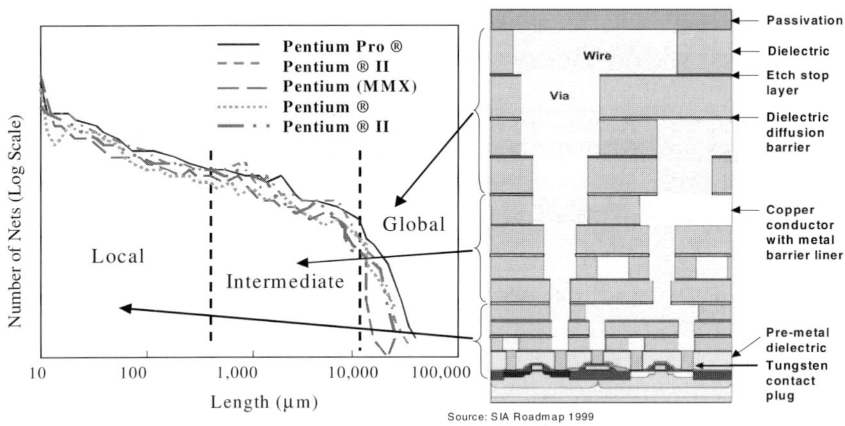

Figure 2.10 Typical interconnect length distribution (left) and chip cross section (right) showing a hierarchical wiring approach with steadily increasing pitch and thickness at each conductor level to alleviate the impact of interconnect delay on performance [3].

However, the benefit of material changes alone is insufficient to meet overall performance requirements as delay is dominated by global interconnection. *Figure 2.9* shows the delay of local and global wiring in DSM technologies. It shows that as technology migrates from 0.25μm to 0.035μm feature sizes, local wires that shorten in length as technologies scale have delays that either track gate delays or grow slowly relative to gate delay; whereas global wires present a more serious problem to designers. Although repeaters insertion can be incorporated to mitigate the delay in global wires to keep a constant delay, it consumes power and chip area whereas the delays relative to gate delays still scale upwards. In addition, reverse scaling of global (and intermediate) wires is necessary [24] [25].

Figure 2.10 is an example of the interconnect strategy in microprocessor design. Here global (and semi-global) wires are wider and taller than local wires and they are usually allocated to the top metal layers. Local wires are narrow and short and they are usually allocated to the bottom metal layers.

5. NOISE-ON-DELAY EFFECT

As mentioned before, noise has two deleterious effects on circuit performance. When noise adds to a normally static signal, it can destroy the local information carried by the static node in the circuit and ultimately result in incorrect operation or bad signal-to-noise ratio. When noise adds simultaneously to a switching signal, this manifests a change in time (jitter and skew) of the transient signal. This noise-on-delay effect is often referred to as pushout.

5.1 Crosstalk-on-Delay

Figure 2.11 illustrates the effect of crosstalk on signal delay. Assuming that the edges are approximately linear, the slew rate is $V_{sw}/t_{r,f}$. The skew is then by first order approximation the noise voltage times the inverse of the slew rate:

$$\Delta t = V_{noise} \frac{t_{r,f}}{V_{sw}} \tag{2.8}$$

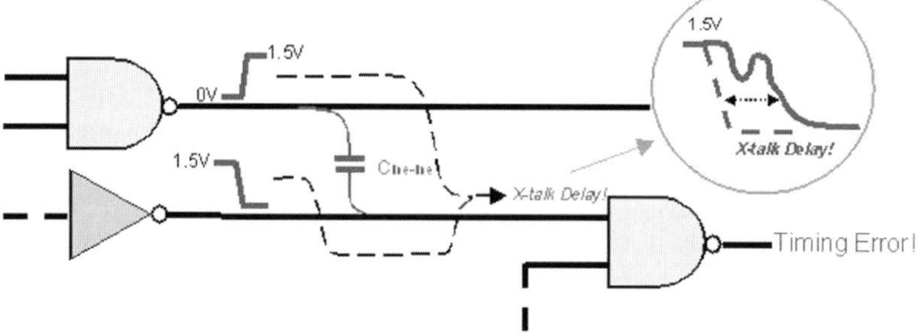

Figure 2.11 A sketch of the effect of crosstalk on signal delay (© Magma [32]).

5.2 Power-Supply-Noise-on-Delay

The delay due to power supply noise is caused by power supply variation modulating the output current of drivers. When a short-channel MOS transistor drives a load capacitor C_L, the first order analysis gives the 50% delay in terms of supply voltage (V_{DS}) as [26] [7]

$$t_d = \frac{C_L V_{sw}}{I_{DS}} = \frac{2C_L V_{sw}}{\beta (V_{GS} - V_T)^2 [1 + \lambda V_{DS}]} \approx \frac{2C_L}{\beta V_{DS} (1 + \lambda V_{DS})} \tag{2.9}$$

where λ is the channel-length modulation constant of the MOS transistors. It is obviously seen that a drop of supply voltage will significantly increase the propagation delay.

6. DESIGN FOR SIGNAL AND POWER INTEGRITY

6.1 General Analysis Approach

Accurate prediction of signal and power integrity of high-performance mixed-signal VLSI requires good modeling of the components including the non-ideal drivers, receivers, signaling interconnects, and power distribution systems. In order to capture the proper physical effects, each model must be a standalone implementation of the correct physics such as bandwidth, frequency dependent effects, and coupling. Awareness of these physical effects and their significance by a first order rapid estimation are necessary for designers. In addition to this, all of the models must work together properly. Since all waveforms are analog in nature, analog simulations of digital signals are required and SPICE or SPICE-like simulators are very commonly used. In order to facilitate analysis, worst-case analysis is sometimes adopted, which however requires construction of a correct worst-case model. After running a number of simulations, the qualities of the signals and the power supply are examined for violations of the specifications.

Nonetheless, accurate physical information of a system is usually available only based on layout database. If the resulting system does not meet the specifications whilst the problems cannot be fixed in the physical layer, information of interconnects and non-ideal effects of components must be back-annotated to higher-level simulators to improve the accuracy of higher-level models. Iterations between higher-level design and physical level design thus occur.

6.2 Interconnect-Centric Design Methodology and Networks-on-a-Chip

As interconnects have been a key limiting factor in VLSI design in DSM regime, chip design must first consider the constraints of interconnects in each design stage. An interconnect-centric design methodology aims to bring the interconnect issues upfront not only in the whole design flow but also within each design phase. System architecture, global communications, clock and power distributions are pre-defined or planned to accommodate the interconnect constraints. We tend to believe that once the physical hierarchy and global interconnects are defined, existing synthesis and placement algorithms can work efficiently at module level up to about 50K gates, which contains mainly local interconnects, as argued in [27] [28]. The interconnect information obtained in the early design phases, will be processively used in logic design and circuit design in each module, and later in physical design, as directives. Finally, a layout can be generated, that confirms the early timing values by carefully tuning the gains of the gates and spreading the delays over the paths. During the design process, a number of point tools should be used for various tasks such as support of partitioning [29], efficient analysis of used communication schemes [30], estimation of wireability and wire capacity under noise and performance constraints [25]. In addition, some emerging tools such as timing driven synthesis, timing driven and noise aware routing, and gain-based synthesis and routing tools [31] [32], will help to improve the design accuracy in the later phases.

One challenge of an interconnect-centric design methodology is to improve the accuracy of *a priori* interconnection estimations [54]. An emerging technique is to ease the global interconnect design by exploring innovative system architectures such as networks-on-a-chip. This brings global interconnect design one abstract-level higher and thus reduces design time and improves interconnect predictability.

6.3 Techniques for Signal and Power Integrity Enhancement

In addition to the efficient models and the new design methodologies, several techniques exist to enhance signal and power integrity. Excellent books on these issues for board and package design include early work in [33], [7] and [34]. This text summarizes some commonly used techniques in VLSI design.

6.3.1 System Architectures and Methods

A. System Architectures and Performance Estimates

An appropriate system architecture is always of paramount importance to signal integrity design. A good example is networks-on-a-chip, as mentioned previously, that facilitates the global interconnect design with predictable performance and integrity by reducing the design space to a set of regular wire segments and standard interfaces. Networks-on-a-chip architectures and their design methods are the main focus in Part 2 and Part 3 of this book.

Due to the increasing importance of interconnects and other physical effects, system level interconnect prediction and good physical models for system performance estimations are crucial to the first success and system optimization. Currently, there have been very successful techniques for system-level performance modeling of microprocessors, early work includes SUSPENS [35] and Sai-Halasz performance estimator [20], and late work includes RIPE [36], BACPAC [37] which was further implemented in GTX [55]. [38] and [39] present interconnect-centric design method and performance estimation for SoC.

B. Signal Integrity and System Partitioning

Based on system performance estimates, optimized system partitioning can be performed with respect to such characteristics as speed, cost, robustness, noise isolation, manufacturability, component availability, and testability. System-level decisions can thus be made with improved accuracy.

Systems are best partitioned so that higher performance can be achieved using high-speed interconnects only where necessary. One example is a microprocessor system where a memory hierarchy can be organized in many ways: moving the L2 cache off the slow system bus and interconnecting the L2 cache and the microprocessor with point-to-point connections can considerably improve performance.

C. Bus Width and Speed

A direct way to increase the information carrying capacity of an interconnect is to simply make the interconnect wider or faster. However, wider buses require more interconnect resources and packaging pins, and the extra drivers use more silicon real estate, dissipate more power and create more noise; faster speed leads to higher power dissipation, noise and radiation can be worse. All these issues must be considered in early decisions of bus width and speeds.

In some cases, special communication architectures and protocols are needed for the global information traffic in SoC. This could result in a system with significant sub-optimal performance and with improved signal integrity [40] [41] [42]. A good example is NoC. Bandwidth maximization in parallel wire structures for NoC will be discussed in Chapter 3. Reference [68] presents an efficient way of reducing effective delay in buses by coding.

D. Topology and Loading

Topology plays a critical role in determining the maximum speed at which an interconnect can be clocked. The fastest interconnect is a unidirectional point-to-point connection between two devices or impedance controlled transmission lines with a matching scheme to absorb any reflection that may be generated by the source or load. Any deviation from this setup will lower the maximum attainable clock speed. In particular, when three or more devices must be connected (such as daisy chain and start cluster) impedance control is compromised due to multiple reflections generated at the impedance discontinuities where extra components must connect. After topology, loading is the next most critical aspect regulating clock speeds on interconnects. Heavy loading results in slow speeds, high power consumption, and bad signal integrity (such as PCI-Bus). Good signal integrity can be achieved by limiting the stub lengths and balancing the capacitive loads.

E. Noise Budgets and Signal Swing

A system designer manages noise by budgeting signal swing against noise sources. In well-designed systems, the noise budget includes all relevant noise sources often with worst-case analysis and leaves a comfortable margin for safety. In cases where an unbounded noise, like over-shot noise, is large enough to be important, we perform statistical noise analysis. It should be noticed that the absolute value of the margin is not relevant; instead, the ratio of the margin to the anticipated noise is the relevant measure of noise immunity.

It is therefore of extreme importance for system designers to be aware of the performance of different circuit families including their typical propagation delay, noise margin, and signal swing. Circuits like static CMOS have larger noise margin, but they also generate larger noise due to large signal swing. The speed of dynamic CMOS is usually faster than static CMOS, but it is more susceptible to crosstalk.

In addition to these, awareness of noise margin constraints to interconnectivity is also important in estimating system wireability and system performance. Reference [25] analyzed noise margin and interconnect delay constraints on interconnectivity in DSM circuits. [43] presented an accurate wire distribution model for VLSI circuits. These two analyses combined together give an accurate early estimation of system wireability for noise immunity and a similar analysis has been presented in [39].

6.3.2 Circuit Levels and Signaling

A. Differential Signaling Techniques

Differential signaling offers the highest signal integrity in which the common-mode noise introduced by the return paths is rejected by the differential receiver. It forms the foundation of many high-performance signaling standards, including LVDS and ECL. The price paid for this is more complex drivers and receivers plus twice the amount of wires. Besides, mixed standards are often used. For example, in many systems the clock is transmitted fully differential to minimize clock skew, but the signals are not differential (termed single-end).

B. Impedance and Termination Control

Control of line impedance and termination can dramatically enhance signal quality for higher speeds by dissipating unwanted reflections. A variety of terminations, series, parallel, Thevenin, RC network, and diode, can be used to minimize the signal distortion due to reflections. In addition, when common-mode current exists, energy will be exchanged between the common-mode wave and differential-mode wave if the termination is not balanced. To avoid this mode-coupling problem, care should be taken to terminate both the differential modes and common modes properly at the receiver. In general, a transmission line with significant return impedance should not be used to carry single-ended signals. They should only be used with differential signals and balanced terminations. The common-mode noise and differential-mode noise on a power distribution network has been well addressed in [5].

C. Slew Rate Control

As slew rate increases both signal coupling and simultaneous-switching-noise increases. This can be seen from Eqs.(2.2)-Eqs.(2.4). On the other hand, a slow slew rate raises the timing skew, as indicated by Eq.(2.8). Repeater insertion is a generally used technique to increase the slew rate by refreshing the signal, as it will be described in Chapters 3 and 4. In addition to extra cost for power and chip area, another drawback of repeater insertion is unidirectional signal propagation. Booster circuits speed up slew rate for bi-directional wires [44]. Recently, there are some load-aware transmitters which adjust circuit slew rate by changing the driver capabilities for different load conditions. Such kind of transmitters hence have much better signal integrity characteristics.

D. Error Correcting Coding

Error correcting coding (ECC) is commonly used in communication systems. In deep submicron VLSI circuits, when the cost of ECCs is comparable with or lower than that of using buffer insertion and wire sizing, ECCs become promising to combat digital noise in VLSI circuits. [45] [46] examined the issues of high speed signing in DSM VLSI circuits and proposed utilization of particular BCH codes to reduce bit error rate in face of crosstalk and power supply noise.

E. Low Noise Power Distribution

In the early days of VLSI, power distribution design was relatively simple – making the wires wide enough to limit the maximum dc voltage drop. Currently, with increased circuit density and speeds, power distribution becomes complex and a number of topologies were employed such as power rings [47], rooted and non-rooted trees, solid plans, meshed grids [48] [49], and a mix of them. In principle, increasing the number of parallel power/ground pins will effectively reduce the package inductance and hence the switching noise. Further improvements can be achieved by appropriately arranging power/ground/signal distributions to minimize the loop inductance and to reduce the *di/dt*. It should be noticed that power impedance is frequency dependent and hence needs to be managed in the frequency domain. Accurate simulation and synthesis of the power distribution network over the whole frequency range are essential to ensure reliable operation of the systems.

6.3.3 Physical and Package Level Techniques

A. Electrical Rule Definition

In a large system, detailed simulation of every net is impractical due to the extremely high cost. Pre-characterization of several topologies can provide a library of acceptable layouts, called wiring rules, which are guaranteed to produce acceptable signal quality and delay. A system designed using such a library is then correct by construction.

In practice however, complex VLSI systems are currently saddled with a "construction by correction" in DSM technologies, instead of having a "correct by construction" flow. This problem as mentioned previously will need to be fixed by increasing the complexity and the flexibility of the electrical rules (wiring rule, termination rule, noise aware, timing aware, and dynamic modeling etc) or using pre-define wire structures such as NoC.

B. Critical Interconnect Length

In addition to the larger signal delay, long interconnects are more susceptible to noise because of the long coupling lengths. Limiting the maximum interconnect length not only reduces total delay but also reduces coupled noise. Repeater insertion is a useful technique to reduce crosstalk both in RC [50] [35] and LC/RLC lines [51] [12] by shortening critical interconnect length. NoC, on the other hand, also shortens the critical length of global wires.

C. Signal Isolation and Grounding

Isolation can dramatically suppress crosstalk. Placing a shielding wire between coupled signal lines can suppress crosstalk by more than 15dB in

DSM circuits [25]. For inter-module communication, using AC-coupled (capacitor or transformer coupling) transmission lines can isolate the DC ground shift between two modules. Optical isolation is commonly used in board-level (or higher) systems. In mixed-signal ICs, guarding rings and special isolation structures are usually made to prevent substrate coupling [52]. Besides, placing ground-plane and/or power-plane closely to signal lines can reduce crosstalk. To suppress EMI, the system and each subsystem must be well grounded, and the impedance of the ground must be low [10].

D. Power Supply Isolation

Power supply isolation includes power-power isolation and power-signal isolation. In power-power isolation, sensitive circuitry (such as analog) uses a clean power supply, isolated from dirty power supplies for digital circuitry. In many cases, the power supply of the digital part is further partitioned into a clean one for core logic cells and a dirty one (sometimes with higher voltage) for I/O and buffer circuits.

In power-signal isolation, supplies are isolated from the signals and the signal returns. To minimize direct coupling of supply noise into signals, signals ideally should be run over a dedicated signal return plane that carries no supply current and is isolated from the noisy supply by a shield plane that is grounded at a single point. Such supply isolation precautions are expensive and are common in analog and mixed-signal systems.

E. Decoupling Capacitors Allocation

Decoupling capacitor allocation is a common technique to suppress switching noise on the power distribution networks. In electronic systems, a power distribution network uses a hierarchy of capacitors, from board and package level to on-chip level, and each stage of the hierarchy filters a frequency band of current transients. Therefore, the final power impedance must be well analyzed in the frequency domain that covers the whole spectrum of all signals. In addition to using *pn* junction capacitors of *n*- or *p*-wells in the silicon substrate, on-chip decoupling capacitors are often implemented by using gate-oxide capacitors and hence consume expensive chip real estate. Reference [53] proposed to use a self-decoupling power distribution technique. This technique uses the mutual capacitance of the wiring hierarchy of the power distribution network, and hence saves chip area. In modern VLSI chips, excessive usage of on-chip decoupling capacitors is not recommended. Implementing 10-15% of chip area for on-chip decoupling capacitors will be regarded as too expensive and hence should be the maximum. To compensate this, decoupling capacitors can also be allocated "near-the-chip" by for example bonding discrete capacitor elements on the top of the chip, or be allocated "off-the-chip" by placing discrete capacitor elements in the chip package module.

F. Low Inductance Package

A direct way to reduce power supply noise is to use low inductance packages. As an example, the parasitic inductance of a package pin in a PLCC or a PGA package is in several nano-Henry, while the inductance of a solder bump in a flip-chip package is less than 100pH. Using such a low inductance package could therefore results in orders of magnitude reduction of simultaneously switching noise.

Besides, flip-chip packages allow area array distribution of the power/ground pins. This solved the large-chip problem in a peripheral counterpart where the center of the chip is susceptible to the excessive voltage drop. The flip-chip connection also provides more I/O pins in large chips. This is a very welcome result because the required pins for signal I/Os and power/ground are rapidly growing as circuit complexity increases. The area array connection combined with self-decoupling provides a very promising power distribution structure in future VLSI chips, as shown in [4].

7. SUMMARY

Concern about the performance of wires in scaled technologies and complexity in interconnect design have led to research exploring new interconnect-centric design methodologies. Despite the fact that local interconnects scale in performance, future design tools will still need more sophisticated capability in dealing with these wires for signal and power integrity as well as timing closure. Global interconnects and some intermediate interconnects that do not scale in length present serious problems to designers. The key focus of interconnect-centric design will be these global wires.

Interconnect-centric design methodology for SoC brings the interconnect issues upfront not only in the whole chip design flow, but also in each design stage. Once the physical hierarchy and global interconnects are defined to accommodate the interconnect constraints, the design tasks start to be focused on module or resource levels and common interface circuit issues. Networks-on-a-chip architectures further facilitate the global interconnect design with controlled predictability by reducing the design space to a set of regular wire and standard interface circuits. Although this may not be an optimal solution in interconnect-centric designs, it significantly reduces design time and wire costs. A number of point tools and models for interconnect analysis are needed at different design stages, for global communication and interconnect design, for logic design and circuit design, as well as for physical design.

Therefore, a better understand of the magnitude of the wire problem and interconnect constraints, better understanding of physical mechanisms of signal integrity and various design techniques at different design levels, are essential to the success of interconnect design in the DSM regime.

REFERENCES

[1] J. S. Kilby, "Invention of the integrated circuits," *IEEE Trans on Electron Devices*, vol.23, July 1976, p.648.

[2] K. L. Shepard, "The challenge of high-performance, deep-submicron design in a turnkey ASIC environment," in *Proc. of IEEE ASIC conference*, pp.183-186, 1998 .

[3] Semiconductor Industrial Association, International Technology Roadmap for Semiconductors, 1999 edition, San Jose, CA, 1999.

[4] L. -R. Zheng and H. Tenhunen, "Design and analysis of power integrity in deep submicron system-on-chip circuits," *Analog Integrated Circuits and Signal Processing*, vol.30, no.1, 2002, pp.15-30. Kluwer Academic, ISSN: 0925-1030

[5] L. -R. Zheng and H. Tenhunen, "Fast modeling of core switching noise on distributed RLC power grid in ULSI circuits," *IEEE Transactions on Advanced Packaging*, , vol. 24, no.3, 2001, pp.245-254.

[6] See e.g. on-line Java programs at http://www.amanogawa.com/. Theory of transmission lines can be found in many books, e.g. S. Rosenstark, "Transmission Lines in Computer Engineering", McGraw-Hill, New York, 1994.

[7] W. J. Dally, and J. W. Poulton, Digital System Engineering, Cambridge University Press, 1998, ISBN 0-521-59292-5.

[8] M. I. Montrose, Printed Circuit Board Design Techniques for EMC Compliance, IEEE Press, 1996, ISBN 0-7803-1131-0.

[9] C. R. Paul, Introduction to Electromagnetic Compatibility, John Wiley & Sons, 1992. ISBN 0-471-54927-4.

[10] R. Morrison, Grounding and Shielding Techniques, 4th edition, John Wiley & Sons, 1998, ISBN 0-471-24518-6.

[11] W. C. Elmore, "The transient response of damped linear networks with particular regard to wideband amplifiers," *Journal of Applied Physics*, vol.19, pp.55-63, Jan.1948.

[12] D. Zhou, F.P. Preparata, and S. M. Kang, "Interconnect delay in very high-speed VLSI," *IEEE Trans. Circuits and Systems*, Vol.38, No.7, pp.779-790, July 1991.

[13] Y. I. Ismail, and E. G. Friedman, J. L. Neves, "Equivalent Elmore delay for RLC trees," *IEEE Trans. Computer-Aid Design of Integrated Circuits and Systems*, Vol.19, N0.1, January 2000.

[14] M. Sriram, and S. M. Kang, Physical Design for Multichip Modules, Kluwer Academic Publishers, 1994.

[15] L. T. Pillage, R. A. Rohrer, "Asymptotic wave evaluation for timing analysis," *IEEE Trans. on CAD*, Vol.9, No.4, pp.352-366, 1990.

[16] A. Deutsch and G. V. Kopcsay *et al*, "When are transmission-line effects important for on-chip interconnections?" *IEEE Trans. Microwave Theory Tech.*, vol.45, no.10, pp.1836-1844, Oct.1997.

[17] L.-R. Zheng, H. Tenhunen, "Transient and crosstalk analysis of interconnection lines for single level integrated packaging modules," in Proc. *IEEE 7th Topical Meeting on Electrical Performance of Electronic Packaging*, pp.120-123, 1998

[18] J. R. Brews, "Overshoot-controlled RLC interconnections," *IEEE Trans. Electron Devices*, Vol.38, No.1, pp.76-87, January 1991.

[19] G. A. Katopis, W. D. Becker, T. R. Mazzawy, H. H. Smith, C. K. Vakirtzia, S. A. Kuppinger, B. Singh, P. C. Lin, J. Bartells Jr., G. V. Kihlmire, P. N. Venkatachalam, H. I. Stoller, and J. F. Frankel, "MCM technology and design for the S/390 G5 system," *IBM J. Res. Develop.*, Vol.43, No.5/6, pp. 621-650, Sept/Nov, 1999.

[20] G. A. Sai-Halasz, " Performance trends in high-end processors,"*IEEE Proceedings* , vol.83, no.1, pp20-36, 1995

[21] Semiconductor Industry Association, The National Technology Roadmap for Semiconductors, San Jose, CA, 1997.

[22] R. Ho, K. W. Mai, and M. K. Horowitz, "The future of Wires", *Proceedings of the IEEE*, vol.89, no.4, April 2001, pp.490-504.

[23] B. Curran, P. Camporese et al, "A 1.1GHz first 64b generation Z900 microprocessor," *International Solid-State Circuits Conference*, Technique Digest, p.238-240, Technical Digest, San Francisco, Feb. 2001.

[24] M. T. Bohr, "Interconnect scaling – the real limiter to high performance ULSI," *IEEE Int. Electron Device Meeting*, Tech. Digest, pp.241-244, 1995

[25] L.-R. Zheng, and H. Tenhunen, "Noise margin constraints on interconnectivity for low power and mixed-signal VLSI circuits," in *Advanced Research in VLSI: Proc. of 20th Anniversary Conference*, pp.123-136, 1999.

[26] J. M. Rabaey, Digital Integrated Circuits: A Design Perspective, Prentice Hall, 1996. ISBN0-13-394271-6.

[27] D. Sylvester and K. Keutzer, "Getting to the bottom of deep submicron," in *Proc. International Conference on Computer Aided Design*, pp.203-211, 1998.

[28] D. Sylvester, and K. Keutzer, "A global wiring paradigm for deep submicron design," *IEEE Trans. Computer Design*, vol.19, pp.242-252, Feb. 2000.

[29] P. Eles, Z. Peng, K. Kuchcinski, A. Doboli, "System level hardware/software partitioning based on simulated annealing and Tabu search," *J. Design Automation for Embedded Systems*, vol.2, pp.5-23, 1997.

[30] B. Svantesson, S. Kumar, A. Hemani, "A methodology and algorithms for efficient interprocess communication synthesis from system description in SDL," in *Proc VLSI Design,* pp.78-84, Chennai, India, Jan 1998

[31] P. Groeneverld, I. Diehl, and R. Smith, "Early timing closure for deep sub-micron design," *Electronic Engineering*, vol.71, no.874, pp.47-49, Nov. 1999.

[32] Available on-line at http://www.magma-da.com, "Gain-based design overview."

[33] J. E. Buchanan, Signal and Power Integrity in Digital Systems, McGraw-Hill, 1995. ISBN 0-07-008734-2.

[34] B. Young, Digital Signal Integrity, Prentice Hall, 2000. ISBN 0-13-028904-3.

[35] H. Bakoglu, Circuits, Interconnections and Packaging for VLSI, Addison-Wesley, 1990. ISBN 0-201-06008-6.

[36] B. Geuskens, K. Rose, Modeling Microprocessor Performance, Kluwer Academic Publishers, 1998. ISBN 0-7923-8241-5

[37] D. Sylvester, and K. Keutzer: "System-Level Performance Modeling with BACPAC - Berkeley Advanced Chip Performance Calculator", SLIP'99: *Workshop on System-Level Interconnect Prediction*, Paper Session pp. 109-114, Monterey, California, USA, April 10-11, 1999.

[38] A. Postula, A Hemani, and H. Tenhunen, "Interconnect centered design methodology for system-on-a-chip integration," in *Proc. 17th IEEE NORCHIP Conference*, pp.197-204, Nov. 1999, Oslo, Norway.

[39] T. Nurmi, L. –R. Zheng , J. Isoaho, and H. Tenhunen, " Early estimation of interconnect effects on the operation of system-on-chip platforms," In *Proc. European Conference on Circuit Theory and Design*, Espoo, Finland, August, 2001

[40] K. Lahir, A. Raghunthan, and S. Dey, "Performance analysis of systems with multi-channel ommunication architectures," in *Proc. of Int. Conf. VLSI Design*, pp.530-537, Jan. 2000.
[41] K. Lahir, A. Raghunthan, and S. Dey, "Evaluation of the traffic-performance characteristics of system-on-chip communication architectures," in *IEEE Proc. Cat. 0-7695-0831-6*, pp.29-35, 2000.
[42] M. Lajolo, M. S. Reorda, and M. Violante, "Early evaluation of bus interconnects dependability for system-on-chip designs," in *IEEE Proc. Cat. 0-7695-0831-6,* pp.371-376, 2000.
[43] J. A. Davis, V. K. De, and J. D. Meindl, "A stochastic wire-length distribution for gigascale integration (GSI) - Part 1: Derivation and validation," *IEEE Trans. Electronic Devices*, vol.45, no.3, 1998.
[44] A. Nalmapu and W. Burlesom, "A practical approach to DSM repeater insertion: satisfying delay constraints while minimizing area and power, " in *Proc. IEEE SoC Conference*, 2001.
[45] D. Pamunuwa, L.-R. Zheng, and H. Tenhunen, "Error-control coding to combat digital noise in interconnects for ULSI circuits," *In Proc. of the 17th IEEE NORCHIP Conference*, pages 275-282, Oslo, Norway, Nov. 1999.
[46] D. Pamunuwa, L.-R. Zheng, and H. Tenhunen, "Combating digital noise in high-speed ULSI circuits using binary BCH encoding," In *Proc. of IEEE International Symposium on Circuits and Systems*, pages IV. 13-16, Geneva, Switzerland, May 2000.
[47] L. P. Cao, Integrated Package, IC and System Design, Ph.D. dissertation of Cornell University, 1998.
[48] Y. L. Low, L. W. Schaper, and S. S. Ang, "Modeling and experimental verification of the interconnected mesh power system (IMPS) MCM topology," *IEEE Trans. Comp., Packag., Manufact. Technol. B*, vol. 20, pp.42-49, Feb.1997.
[49] T. Yamada, Y. Nakata, J. Hasegawa, N. Amano, A. Shiayama, M. Sasago, N. Matsuo, T. Yabu, S. Matsumoto, S. Okada, and M. Inoue, "A 64-Mb DRAM with meshed power line," *IEEE Journal of Solid-State Circuits*, Vol.26, No.11, pp.1506-1510, Nov. 1991.
[50] D. Pamunuwa and H. Tenhunen, "Buffer sizing to minimize delay in coupled interconnects," In *Proc. of 14th Int. Conference on VLSI Design*, January 2001.
[51] Y. I. Ismail, and E. G. Friedman, "Effects of inductance on the propagation delay and repeater insertion in VLSI circuits," *Proc. Design Automation Conference*, pp.721-724, New Orleans, Louisiana, USA, 1999
[52] N. Verghese, T. J. Schmerbeck, and D. J. Allstot, Simulation Techniques and Solutions for Mixed-Signal Coupling in Integrated Circuits, Kluwer Academic Publisher, 1995. ISBN 0-7923-9544-1
[53] L.-R. Zheng, and H. Tenhunen, "Effective power and ground distribution scheme for deep submicron high speed VLSI circuits," in *Proc. Int. Symp. Circuit and Systems*, pp.I. 537-540, 1999.
[54] D. Stroobandt, *A Priori* Wire Length Estimates for Digital Designs, Kluwer Academic Publishers, April, 2001.
[55] A Caldwell, Y Cao, A.B Kahng, F. Koushanfar, H Fu, I.L. Markov, M. Oliver, D. Stroobandt, D. Sylvester, "GTX: The MARCO GSRC Technology Extrapolation System," in *Proc. 37th IEEE/ACM Design Automation Conference*, June 2000, Los Angeles, pp.693-698.
[56] T. Nurmi, L.-R. Zheng, T. Valtonen, and H. Tenhunen. Power management in autonomous error-tolerant cells. In *15th Annual IEEE International ASIC/SoC Conference*, Rochester, New York, USA, September 25-28 2002
[57] A. Dharchoudhury, R. Panda, D. Blaauw, and R. Vaidynathan, "Design and analysis of power distribution networks in PowerPCTM microprocessors." in *Proc. 35th Design Automation Conf.*, pp.738-743, 1998

[58] H. Chen, and D. V. Ling, "Power supply noise analysis methodology for deep-submicron VLSI chip design." in *Proc 34th Design Automation Conf.,* pp.638-643. 1997.

[59] S. R. Nassif and J. Kozhaya, "Multi-grid methods for power grid simulation." in *Proc. Int. Symp. Circuits & Systems,* pp.V. 457-460, 2000.

[60] M. Zhao,; R.V. Panda; Sapatnekar, S.S.; Blaauw, D., "Hierarchical analysis of power distribution networks," *IEEE Transactions on Computer-Aided Design of Integrated Circuits and Systems*, Volume: 21 Issue: 2 , Feb. 2002 pp.159 –168

[61] Y. Shin.; Sakurai, T., "Power distribution analysis of VLSI interconnects using model order reduction," *IEEE Transactions on Computer-Aided Design of Integrated Circuits and Systems*, Volume: 21 Issue: 6 , June 2002, Page(s): 739 –745

[62] S. Lin; N. Chang, "Challenges in power-ground integrity", *IEEE/ACM International Conference on Computer Aided Design*, ICCAD 2001., 2001, Page(s): 651 –654.

[63] C.-K. Cheng, J. Lillis, S. Lin, N. Chang, Interconnect Analysis and Synthesis, Wiley-Interscience, October 1999.

[64] Q. K. Zhu, Interconnect RC and Layout Extraction for VLSI, Trafford, December 2002.

[65] M. Nakhla, R. Achar, Introduction to High-Speed Circuit and Interconnect Analysis, Omniz Global Knowledge Corporation, May, 2002.

[66] S. P. Khatri, R. K. Brayton, A. L. Sangiovanni-Vincentelli, Crosstalk Noise Immune VLSI Design Using Regular Layout Fabrics, Kluwer Academic Publishers, 2001.

[67] Y.I. Ismail, E. G. Friedman, On-Chip Inductance in High Speed Integrated Circuits, Kluwer Academic Publishers, 2001.

[68] P. Sotiriadis , A. Chandrakasan, "Reducing Bus Delay in Sub-Micron Technology Using Coding" *IEEE Asia and South Pacific Design Automation Conf.*, 2001, pp. 109-114.

Chapter 3

GLOBAL INTERCONNECT ANALYSIS

Tero Nurmi*, Jian Liu**, Dinesh Pamunuwa**, Tapani Ahonen***, Li-Rong Zheng**, Jouni Isoaho*, Hannu Tenhunen**

** University of Turku, ** Royal Institute of Technology, Stockholm, *** Tampere University of Technology*

1. Introduction

The rapid development in deep submicron (DSM) technology makes possible to design complex billion-transistor chips. To take full advantage of increased integration density and cope with the difficulties in designing such complex systems, the emphasis of design methodology has changed from gate-level design to the exploitation of intellectual property (IP) blocks. This IP-based design is rapidly becoming the dominating design paradigm in System-on-Chip (SoC) era. IP blocks themselves are usually verified by the supplier for some technology node but the problem is how to ensure the correct performance when the IP block is integrated in the SoC or even in Network-on-Chip (NoC) environment. The problems occur in adapting the block interface into the used communication frame. The main objective is to make computation (IP blocks) and communication independent on each other.

Due to increasing integration density and diminishing wire dimensions, communication using traditional SoC interconnect schemes (such as buses) does not scale up properly compared with system complexity. This leads to the communication scheme where traditional buses and their arbitration are replaced with network switches connecting various IP blocks in different network nodes to each other. Thus, a shift from SoC to NoC is predicted when system complexity scales up on chip level. Network nodes bring inherent pipelining and buffering onto system level which is important when dealing with global wires that have more resistive and inductive nature in current and future DSM technologies. Additionally, undesired transmission errors can be reduced with error-checking, e.g. in each network node. In this case, latency may increase as a result of increased reliability.

In this chapter, we first discuss parasitic modeling in the presence of crosstalk and delay modeling of global wires. Inductance issues are discussed in more detail in chapter 5 and thus we omit them here. Some possible interconnect schemes in SoC and NoC are shortly discussed. In section 3.3 we evaluate cost functions (e.g. power consumption and area) that IP blocks set for the global communication network. We present a method how to evaluate those costs in the early phase of design. By evaluating costs of those resources we can better optimize global interconnects to meet both signal and power distribution challenges. We present one case study example on the cost evaluation. Finally, in section 3.4 we apply methods and theories presented in earlier sections and optimize global interconnects to meet different constraints. The delay in global wires is optimized using repeaters that are sized properly and placed in proper distances so that the overall delay is optimized. Then we present optimal signaling having maximum throughput as a constraint. Last, we present a case study in which both power and signal distribution are simultaneously optimized. This is done by using a method called *interconnect partitioning* and the design constraint in this case is the maximum allowed variation of power supply levels in the power distribution network. The variation depends on the grain size of the power distribution grid and power consumption taking place in IP (or functional) blocks due to simultaneous switching of large amount of logic gates in a very short time interval.

2. Interconnect Modeling

Although there are different high-level abstract interconnect schemes (such as point-to-point interconnect, bus architecture, switched network, etc), the underlying physical global wire structure will be the same, namely the parallel wire structure. For point-to-point connections, each link is realized by a certain number of parallel wires; for the bus architecture, the bus itself is a number of wires in parallel and the same applies to a link connecting two switches in a Network-on-Chip. Therefore, it is of great importance to study the interconnect in general, the parallel wire structure in particular, and model them accurately. The following wire models and the derived delay models [1] serve this purpose.

2.1 Characteristics of DSM Wires

Interconnects in deep submicron (DSM) technologies are typically modeled as RC wires. Recently, the effect of inductance has been added into the models of global, *long* wires. High operating frequencies and fast signal transitions made possible by modern DSM technologies increase the impact of inductance on the electrical behaviour of the long wires. However, we omit inductance in this chapter; chapter 5 deals with inductance issues in more detail.

In order to keep the resistance to a minimum, the aspect ratio (the ratio of height and width) of wires is increased, which gives rise to increased inter-wire capacitance. This inter-wire capacitance results in crosstalk which has an effect on the delay, depending on how the aggressor lines switch. Crosstalk is of special significance in uniformly coupled parallel wires, causing unpredictable delays. A crucial point here is that when the geometry of the wire arrangement changes, the parasitics of the wires change in a highly nonlinear fashion. In particular, the exact manner in which the total capacitance is distributed into a ground component and a capacitance component to the adjacent wires is important, as this dictates the charging/discharging time.

An accurate analysis of interconnects requires solving Maxwell's equations in three dimensions (3-D), which is prohibitively expensive in terms of computation time. However, it is possible to use simplified models in most cases to capture the important effects in the regime of interest [2]. In the following text, parasitic modeling and delay modeling will be described.

2.1.1 Parasitic Modeling. The skin depth at the highest frequency of interest is usually large enough so that the DC resistance (equation 3.1) is quite accurate [3]

$$R = \rho \frac{l}{hw} \tag{3.1}$$

where l, h and w are the length, height and width of a wire, respectively. If second order effects are ignored and the capacitance of a wire is modelled solely by its parallel plate capacitance, changing the width does not affect the RC delay, as a decrease (increase) in resistance by a certain factor is accompanied by an increase (decrease) in capacitance by the reciprocal of the same factor leaving the RC product unchanged. However it is well known that for interconnects in submicron technologies the higher aspect ratio results in the fringing component of the capacitance being of similar or often greater than the parallel plate component [4]. Hence the RC delay does change with the wire width and does so in a highly nonlinear fashion. Further, most of the fringing capacitance is to an adjacent wire, which results in capacitive cross-talk. Hence the accurate distribution of the total capacitance into self and mutual terms is very important. The parasitic capacitance is a very strong function of the geometry and 3D field solvers are required to obtain accurate val-

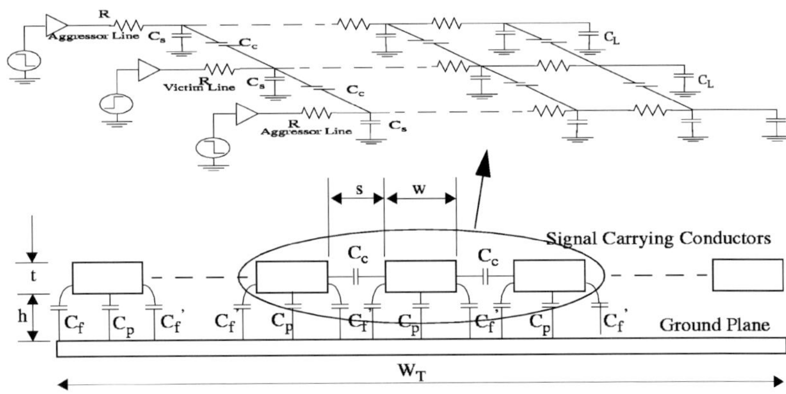

Figure 3.1. Configuration for investigating effect of crosstalk [1]. Geometrical arrangement of the parallel multinet structure (upper figure) and electrical model for delay modeling comprising victim net capacitance coupled to two aggressors in a uniform and distributed manner (lower figure). ©2003 IEEE

ues. However over the years, empirical equations have been developed which have reasonable accuracy and are very important in gaining an intuitive understanding at a system level. The models can be broadly classified into those that consider an isolated rectangular conductor and those that consider a multi-wire structure. The geometrical parameters mentioned below can be identified by referring to Figure 3.1.

The models in the first category describe the self capacitance of the wire, and an overview can be found in [5]. One of the early approaches detailed in [6] gives an empirical formula which decomposes the capacitance of a single rectangular wire over a ground plane into a parallel plate component and a component proportional to a circular wire over a ground plane, and hence has a straightforward physical motivation. The accuracy of this equation however drops rapidly when the ratio w/h falls below values of about 2-3. The trend in modern technologies is to have increasing numbers of metal layers, thus increasing h, and shrinking wire sizes, decreasing w, making the regime below this ratio the most interesting, and hence rendering it unsuitable for on-chip wires. To calculate the capacitance terms shown in Figure 3.1 we use the models proposed in [7]. They use a technology dependent constant β which is calculated from a database of values generated by a field solver, and are defined in equations (3.2) through (3.7). Typical values of β range from 1.50 to 1.75, and 1.65 may be used for most DSM technologies.

$$C_f = \epsilon_k \left[0.075 \left(\frac{w}{h} \right) + 1.4 \left(\frac{t}{h} \right)^{0.222} \right] l \tag{3.2}$$

$$C'_f = C_f \left[1 + \left(\frac{h}{s} \right)^{\beta} \right]^{-1} \tag{3.3}$$

$$C_p = \epsilon k \frac{wl}{h} \tag{3.4}$$

$$C_{s,mid} = C_p + 2C'_f \tag{3.5}$$

$$C_{s,corn} = C_p + C_f + C'_f \tag{3.6}$$

$$C_c = C_f - C'_f + \epsilon_k \left[0.03(\frac{w}{h}) + 0.83\frac{t}{h} - 0.07(\frac{t}{h})^{0.222} \right] (\frac{h}{s})^{1.34} l \tag{3.7}$$

2.1.2 Delay Modeling. From now on, whenever delay is mentioned without further qualification, we are talking about the 50 % point of the step response, which is the delay to the switching threshold of an inverter. The most ubiquitous circuit model in MOS circuits is a lumped capacitance (representing the load) driven through a series resistance (representing the driver impedance), which has a single pole response and a delay as shown in equation (3.8)

$$t_{lump} = 0.7RC \tag{3.8}$$

One of the most prevalent methods of estimating the delay of more complex networks is to model the output by a single pole response, where the pole is the reciprocal of the first moment of the impulse response. This is often referred to as the Elmore delay after the person who first proposed it as an upper bound to the delay in an analysis of timing in valve circuits [8]. Now thin on-chip wires have a high resistance, and are most often modeled by distributed RC lines. Signal propagation along such lines is governed by the diffusion equation which does not lend itself readily to closed form solutions for the delay at a given threshold. However it turns out that a first order approximation results in very good predictions [9], [10]. One way of explaining this is to recognize that a distributed line (which comprises cascaded RC sections in the limit where the number of sections tends to infinity) is a degenerate version of an RC tree, with the step response in consequence having a dominant time constant. This time constant can be well approximated by the Elmore delay, or 0.5RC, which leads to equation (3.9) as the model for the delay of a distributed RC line [9].

$$t_{distr} = 0.4RC \tag{3.9}$$

This is a very good approximation and is reported to be accurate to within 4 % for a very wide range of R and C. Sakurai in [11] reports heuristic delay formulae based on a single pole response which predicts values which are very close to the Elmore delay.

In closely coupled lines the phenomenon of cross-talk can be observed. Cross-talk may be both inductive and capacitive. In coupled micro-strip lines for example, the mutual capacitance couples the time derivative of voltage while the mutual inductance couples the spatial derivative of voltage, so that a signal transition on one line may induce travelling waves on another line [12], [13]. For DSM circuits capacitively coupled lossy lines are the most relevant when the phenomenon of cross-talk causes signal integrity and delay problems. Cross-talk couples a noise pulse onto the victim net which can have two effects: it can result in a functional failure by causing the voltage at a node to switch above or below a threshold, and it affects the propagation velocity of signal pulses on the victim line.

The effect of cross-talk on the delay depends on the switching of the aggressor lines, and can truly be captured only by dynamic simulators which take into account arrival times of different signals and carries out a full transient analysis. It is possible however to limit the aggressor alignment to a few specific cases and develop timing models for static analyses. One such work is [14] where moment matching techniques are used to obtain single pole responses for coupled lines. Most often static timing models which take cross-talk into account are based on a *switch factor*. The capacitance for a line is modelled as the sum of two components, one of which represents the capacitance to ground, while the other represents the capacitance to adjacent nets. This second component is multiplied by a factor which takes the value of 0 and 2 for the best and worst cases respectively. Kahng et. al. in [15] show that 2 does not necessarily constitute an upper limit on the delay in general, where the inputs are finite ramps, and have different slew rates, and that 3 is a better factor for worst-case estimations in such situations.

As we are only concerned with the influence on delay and other performance metrics related to it, such as bandwidth, in the following text, we ignore the inductive effects and use analytic delay models that are very simple, yet model with good accuracy the most important phenomenon in closely coupled wires: that of capacitive cross-talk. The lines are modeled as coupled uniformly distributed RC lines, and a slightly modified switch factor based analysis of delay in long uniformly coupled nets is presented, where the capacitance is distributed over two components and two empirical constants are used to appropriately modify the dominant time constant. This is shown to be more accurate than using a factor of 2 to model the worst case, although the complexity is the same.

2.2 Interconnection Schemes in SoC and NoC

As the number of computational and functional units on a single chip increases it also increases the need for communication between those units. As a consequence, it is important to choose an appropriate interconnect scheme. This design methodology focusing on the interconnect is called *interconnect-centric* design. It demands the network to be scalable, flexible and efficient. An interconnect scheme is scalable if it is able to accommodate growth, i.e., if it is possible to attach additional IP blocks to an existing network without severe degradation of performance. Flexibility means that it should be easy (in a plug-and-play fashion) to attach or remove a component from the interconnect network without modifying the whole network. Other components attached to the network should not be affected by the attachment or removal of one or more IP components. At last, an interconnect scheme needs to be efficient in terms of silicon area, provided bandwidth and power consumption. It should be noticed that one interconnect scheme does not necessarily exclude others. They can coexist to optimize the overall performance. For example, a point-to-point connection may exist between two components that require high bandwidth at the same time as one or both of them are connected to a bus to communicate with other IP components.

In SoC, typical interconnect schemes are point-to-point interconnect, single-bus architecture and multi-bus architecture. Point-to-point interconnects are usually reserved for links demanding high bandwidths. It is also easier to optimize the point-to-point link than the whole bus. However, the point-to-point interconnect is not a very good solution in future SoCs because it is neither scalable nor flexible. The most severe problem with a single-bus architecture is scalability; if more IP blocks are added the bandwidth per IP is decreased and there is also a problem of distributing a synchronous clock signal over the whole chip. Finally, the advantages of a multi-bus architecture are segmentation of one bus into many buses (by using *bridges*) and differentiated services for different traffic types (latency and bandwidth as parameters). Still, the multi-bus architecture suffers from the same limitations as other shared-bus implementations, it is not scalable as the number of IP blocks is increased to tens or hundreds of blocks in the future.

In NoC, an interconnect scheme is a *switch* and IP blocks communicate with each other via these switches. The switches are connected via point-to-point interconnect links (e.g. a 32-bit link) to each other. Information (data, control and address) is embedded in *packets*. The switched NoC fills all the requirements: it is scalable, flexible and efficient. The last property follows from packet-type communication. A tiny disadvantage is that every time a new block is added into a NoC system there is also a need for a wrapper (an adapter) that adapts the block interface into the used communication frame. If the communication frame is changed also a new wrapper has to be designed. There is a growing tendency to standardize communication frames.

3. Early characterization of functional blocks

In addition to modeling wires and different types of interconnect structures in SoC and NoC, we need to model functional blocks as well. Costs and performance of functional blocks define the requirements for the global interconnect structure, i.e. which type of interconnect scheme (see section 3.2.2) is chosen and how metal levels are utilized to different operational purposes (e.g. clock and power distribution and local/semilocal/global signaling). Additionally, cost functions of functional blocks, such as area and power consumption, affect directly properties like a maximum length or wire length distribution of global wires and also power supply voltage variation across the circuit.

In this section the focus is on modeling and estimating the characteristics of functional blocks which then act as cost producers and set constraints for global wiring.

3.1 Rent's rule for characterization of processors

In 1971 Landman and Russo published their well-known paper [16] on Rent's rule. Rent's rule describes a relationship between the number of signal I/O connections to and from a logic block and the number of logic gates it contains.

Different interpretations have been given for the rule afterwards but generally it has been applied to evaluate wire length distribution inside a block and thus to help in a placement and routing process.

Rent's rule can be described by equation 3.10 [16]

$$\#I/O = K_p N_{gates}^p, \tag{3.10}$$

where K_p is Rent's constant (number of average gate I/O), N_{gates} is the number of gates in a block, p is Rent's exponent and $\#I/O$ is the number of I/O pins in the external interface of a block. Bakoglu [10] has extracted some Rent's constants and exponents for the most general on-chip structures, such as static memory, microprocessor and gate array, and also for module, board and system level in a high-speed computer. The problem in using those values for Rent's parameters is that they match with the architectures that were used for about 15 years ago. Today processors and other digital circuits are mainly designed using logic synthesis tools which have developed very rapidly over the years and can optimize logic structures to meet various constraints set by an application. Synthesis-based Rent's exponents and constants are discussed in subsection 3.3.2. The way to apply Rent's rule in that subsection differs from the traditional approach; instead using the rule to *intrablock* wiring design we apply it to *inter-block* (i.e. global) wiring design.

Rent's rule has been used directly or indirectly in many earlier models. Bakoglu's SUSPENS model [10] is an obvious example; there Rent's rule is used to estimate an average interconnection length and finally clock frequency, power dissipation and chip size. This implies that the total number of logic gates is known beforehand. Donath's model [17] is used for estimating average wire length. The problem with SUSPENS is that it does not include an option for using modern metal level structures with variable wire pitch. Additionally, it lacks an option for on-chip memory that is used in all modern on-chip processors and System-on-Chip circuits. There are other chip performance models, which take modern interconnect structures into account better than SUSPENS. Those models have later been integrated under a single estimator environment in a design tool called GTX [18]. In GTX a user can additionally add own rules to take new physical effects into account.

Usually those models use either one or two values for the Rent's exponent. In the latter case the exponent is divided into an internal and an external exponent. The former exponent relates closely to wire length distribution and placement optimization in a block and the latter is used for external I/O connections from and to a chip. Because there is the limited amount of signal I/O pins in a chip package (although area-array bonding increases that number) the internal exponent tends to be higher than the external exponent. However, there is a need for a large set of functional blocks in modern reconfigurable processors depending on an application. This implies that multiple Rent's exponents and constants have to be extracted. Additionally, modern logic synthesis tools are given various constraints (e.g. the delay in a block) which affect logic organization inside the block and the selection of standard cell components to implement the required function.

3.2 Extraction of Rent's exponents for synthesized blocks

A designer has to be able to predict system level performance metrics already in the early phase of the design cycle thus avoiding many time-consuming synthesis runs. We have developed a method for linking different synthesis runs of given timing constraints to an early estimation model. The model uses only a few parameters with which one can estimate performance metrics of processor blocks accurately enough before the implementation of a design. The key component in early estimation analysis is Rent's rule [16] presented already in section 3.3.1.

First, one block is taken under examination. By using Rent's rule (3.10) different Rent's exponent values are calculated. Values for Rent's constant K_p (number of average gate I/O) and number of gates N_{gates} are extracted from synthesis reports. The number of block I/O connections ($\#I/O$) is defined in the specification of the block. There are two ways to evaluate the number of gates (N_{gates}), the first is an area-optimized case and the second is a power-optimized case. In the area-optimized case $N_{gates} = \frac{A_{tot}}{A_{gate}}$, where A_{tot} is the total area of the block and A_{gate} is the area of the gate type chosen to represent an average gate in the block. In the power-optimized case $N_{gates} = \frac{P_{tot}}{P_{gate}}$, where P_{tot} is the power consumption of the block and P_{gate} is the power consumption of the chosen average gate with a certain fanout. The block-level values for area and power consumption are extracted from synthesis results and the average gate values from datasheets of the standard cell library. N_{gates}, and thus also exponent p, is usually a bit different for the two cases.

A few words have to be said about the evaluation of the number of logic gates. First, by using power consumption values from the logic synthesis a relatively bad accuracy can be reached because switching activities of electrical nets vary. The use of area information can be better argumented because in modern technologies there is enough wiring capacity (6-8 metal layers) and thus the final area after placement and routing does not increase severely. The use of different metal layers to different operational purposes is discussed more in section 3.4.3. Second, we know the number of I/O connections to and from a block before the actual design implementation but not the number of gates. That's why we apply the rule in a contrary way as usually. Third, our model assumes that some estimate of the ratio of combinatorial and sequential logic in a block can be given before the actual implementation of the block. This should be roughly evaluated already when specifying the functionality of the block. That ratio is important when estimating the area of the block but especially when estimating power consumption. The different values of switching activity can be applied to different types of logic and thus to achieve a better estimate for power consumption.

The values of the Rent's exponent p found in the literature very often refer to a specific design case implemented with specific technology and specific logic design style. Hence, there is a need to derive Rent's exponent for individual blocks separately. Because in the logic synthesis process a CAD tool changes the organization and the type of standard cell components according to various delay, area and power consumption constraints, we need to define a separate Rent's exponent for each individual block as a function of a specific constraint.

In this way one gets varying values (e.g., when a delay constraint is changed) of both p and K_p for the block under examination. This result is still inadequate if one needs to change the delay figure more freely. Assume that one wants to use a delay constraint not matching with any of the reference points (used in exponent's calculation) but something in between the synthesized values. Then one has to plot Rent's exponent p as a function of delay and make a regression analysis for the "data points". In our case study (see section 3.3.4), we decided to use linear regression for curve fitting although it shows that some of the curves could obey a polynomial expression or a step function.

3.3 Regression analysis for predicting Rent's exponent

A linear regression analysis is performed to Rent's exponent variation as a function of delay constraint. An error function is given by equation (3.11) [19]

$$J = \frac{1}{2N} \sum_{i=1}^{N} \epsilon_i^2, \tag{3.11}$$

where ϵ_i is $d_i - (b + wx_i)$ and where d_i is the real exponent value and x_i is here the delay constraint. This is a mean square error (MSE) function. The goal is to minimize the error function by setting partial derivatives of J to zero with respect to an axis intercept value b and a slope w. After doing this the following equations are derived [19]

$$b = \frac{\sum_i x_i^2 \sum_i d_i - \sum_i x_i \sum_i x_i d_i}{N\left[\sum_i (x_i - x_{avg})^2\right]} \tag{3.12}$$

$$w = \frac{\sum_i (x_i - x_{avg})(d_i - d_{avg})}{\sum_i (x_i - x_{avg})^2}, \tag{3.13}$$

where x_i and d_i are as defined earlier and x_{avg} and d_{avg} describe average (mean) values of the delay constraint and the Rent's exponent, respectively.

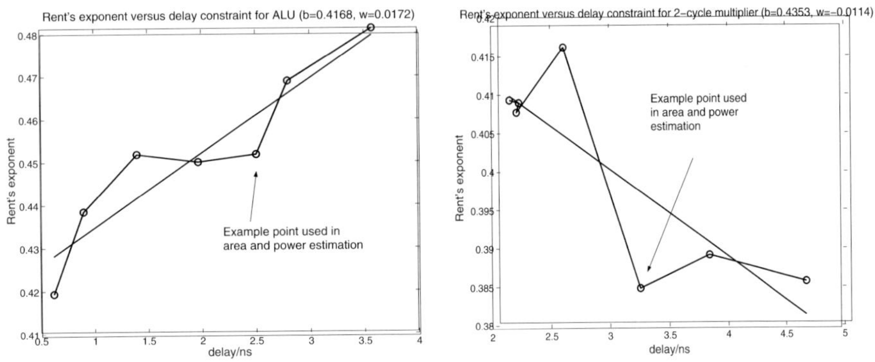

(a) ALU

(b) Mult(2)

Figure 3.2. Regression graphs (a) for an ALU (area-optimized case) and (b) for a 2-cycle multiplier (power-optimized case). ©2003 IEEE

3.4 Case study: Block-wise estimation of a XIRISC processor

In this case study we used seven different delay constraints. We used synthesis results to define a specific Rent's exponent p for each block used in our XIRISC processor case study as a function of delay constraint. There were two separately optimized cases for deriving Rent's exponent:

area-optimized case and power consumption-optimized case as explained in section 3.3.2. For both cases we used the information received from the synthesis report and assume that a standard 2-input NAND gate (with normal drive strength, fan-out of 2 and an input rise time of 17 ps) represents an average gate. We used 0.18 μm, 6-metal silicon technology.

Regression graphs are presented for an ALU (area-optimized case) and for a 2-cycle multiplier (power-optimized case) in Figure 3.2. The Rent's exponents as a function of the delay constraint for other processor blocks as well as more details of this case study can be found in [20]. One can see from Figure 3.2 that in the case of ALU Rent's exponent decreases gradually with tightening the delay constraint but in the case of 2-cycle multiplier an abrupt change in the exponent is noticed. Synthesis reports show that also the values of Rent's constant were simultaneously changed which we think is due to the change of the cell type used in the synthesis as the delay constraint gets tighter.

4. Optimize Global Interconnections for SoC/NoC

In subsection 3.4.1 we study different techniques to reduce the wire delay in a parallel wire structure. In subsection 3.4.2, strategies on repeater insertion under different constraints are investigated. The goal is to maximize bandwidth of the wire structure. Finally, in subsection 3.4.3 we present a case study where both signal distribution and power distribution have been optimized simultaneously. In the case study, a method called *interconnect partitioning* has been used and the maximum allowed power supply voltage variation in the power distribution network has been used as a design constraint.

4.1 Delay Reduction with Optimized Repeater Insertion

The most common method of reducing the delay over long interconnects is to insert repeaters (inverters) at appropriate points. We use the model derived in subsection 3.2.1.1 to show that both the number and size of repeaters can be optimized to compensate for dynamic effects [1].

In the delay analysis, the victim line is assumed to switch from zero to one, without loss of generality. When a line switches up(down) from zero(one) it is assumed to have been zero(one) for a long time. For simultaneously switching lines in the configuration of Fig. 3.1, six distinct switching patterns can be identified.

1) Both aggressors switch from one to zero. 2) One switches from one to zero, the other is quiet. 3) Both are quiet. 4) One switches from one

to zero, the other switches from zero to one. 5) One switches from zero to one, the other is quiet. 6) Both switch from zero to one.

Consider 3) above as the reference delay, where the driver of the victim line charges the entire capacitance. Cases 1) and 2) slow down the victim line, 4) is equivalent to 3), and 5) and 6) speed up the victim. In all cases except 5), the response of the distributed line for step inputs has a dominant pole nature. Since the time constants in question are linear combinations of R, C_s and C_c, changing coefficients are sufficient to distinguish between the different cases. The delay is as given in equation (3.14) where all λ_i take the values of Table 3.1. The coefficient μ_i in the same table is an empirical constant to model the Miller effect and will be used in the following equations.

$$t_{vic} = 0.4RC_s + \lambda_i RC_c \tag{3.14}$$

i	switching pattern	λ_i	μ_i
1	(a)	1.51	2.20
2	(b)	1.13	1.50
3	(c)	0.57	0.65
4	(d)	0.57	0.65
5	(e)	N/A	N/A
6	(f)	0	0

Table 3.1. Coefficients of the heuristic delay model for a distributed line with different switching patterns [1]. ©2003 IEEE

These constants were obtained by running sweeps with the circuit analyzer SPECTRE. Now the total delay of the line is affected by the driver strength, and the load at the end of the line. The simplest characterization of the driver is to consider it as a voltage source in series with an output resistance R_{drv}, with a capacitive load of C_{drv} at the input. The linear approximation of the buffers allows the use of superposition to find the delay, which is given by equation (3.15)

$$t_{T,vic} = 0.7R_{drv}(C_s + C_{drv} + \mu_i \times 2C_c) + R(0.4C_s + \lambda_i \times C_c + 0.7C_{drv}) \tag{3.15}$$

The lumped resistance R_{drv} combines with all the capacitances (both lumped and distributed) to produce delay terms with a coefficient of 0.7.

Similarly the distributed resistance of the line combines with various capacitances to produce different delay terms (it is assumed that the load at the end of the line is an inverter which is the same size as the driving inverter). The terms containing mutual capacitance C_c model crosstalk. The coefficients λ_i and μ_i make the expression for the total delay more accurate than using a single coefficient of 2 for the coupling capacitance to model the worst-case. For i=1, the above expression reduces to equation (3.16)

$$t_{T,vic} = 0.7R_{drv}(C_s + 4.4C_c + C_{drv}) + R(0.4C_s + 1.5C_c + 0.7C_{drv}) \quad (3.16)$$

If a universal factor of 2 is used for the coupling capacitance, the expression takes the form given in equation (3.17)

$$t_{T,vic} = 0.7R_{drv}(C_s + 4C_c + C_{drv}) + R(0.4C_s + 1.6C_c + 0.7C_{drv}) \quad (3.17)$$

Hence, with the empirical constants that we propose, factors of 4.4 and 1.5 appear before C_c, while in a conventional worst-case analysis they should be 4 and 1.6. If the driver impedance is set to zero, the difference between the two expressions is very small, but with non-zero driver impedances, the difference is significant. The accuracy of equations (3.16) and (3.17) was checked against simulated values. It shows that the empirical model contains the error to under 5 %, while the traditional method is more sensitive to the value of the driver impedance and has errors of up to 10 % for certain cases. To reduce delay, the long lines in Figure 3.1 are broken up into shorter sections, with a repeater (an inverter) driving each section as shown in Figure 3.3. Let the number of repeaters including the original driver be K, and the size of each repeater be H times a minimum sized inverter (all lines are assumed to be buffered in a similar fashion). The output impedance of a minimum sized inverter for the particular technology is $R_{drv,m}$ and the output capacitance $C_{drv,m}$ both of which are assumed to scale linearly with size. This arrangement is sketched out in Figure 3.3. In general, the line segments corresponding to the gain stages would not be equal in length, as repeaters are typically situated in "repeater stations", the locations of which are determined by overall layout considerations. Then the delay is given by equation (3.18)

$$t_{uneq} = \sum_{i=1}^{K} [A_i + B_i] + \frac{t_r}{2}, \quad (3.18)$$

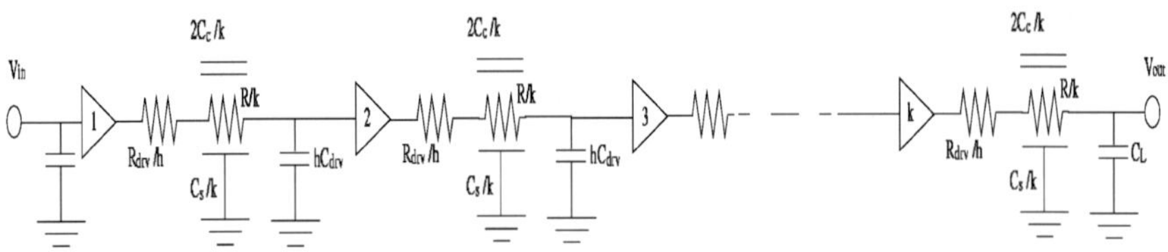

Figure 3.3. Repeaters inserted in long uniformly coupled nets to reduce delay [1]. ©2003 IEEE

where

$$A_i = 0.7(\frac{R_{drv,m}}{h_i} + R_{via})(c_s l_i + H_i C_{drv,m} + \mu_i \times 2c_c l_i) \tag{3.19}$$

and

$$B_i = r l_i (0.4 c_s l_i + \lambda_i c_c l_i + 0.7 H_i C_{drv,m}) \tag{3.20}$$

It is assumed that the load C_L is equal to the input capacitance of an H sized inverter. Also the signal rise time has been included here. For the long lossy lines that we consider here, usually the delay of the line is much greater than the rise time of the signal with which the driving inverter is gated, and the 50 % - 50 % delay from buffer input to output interconnect node is independent of rise time [21].

Now the minimum delay is obtained when the repeaters are equalized over the line, when the above expression reduces to equation (3.21)

$$t_{eq} = K\left[A_i' + B_i'\right] + \frac{t_r}{2}, \tag{3.21}$$

where

$$A_i' = 0.7\frac{R_{drv,m}}{H}(\frac{C_s}{K} + HC_{drv,m} + \mu_i \frac{2C_c}{K}) \tag{3.22}$$

and

$$B_i' = \frac{R}{K}(\frac{0.4C_s}{K} + \lambda_i \frac{C_c}{K} + 0.7HC_{drv,m}) \tag{3.23}$$

In order to find the optimum H and K for minimizing delay, the partial derivatives of (3.21) with respect to K and H are equated to zero, resulting in equations (3.24) and (3.25)

$$K_{i,opt} = \sqrt{\frac{0.4RC_s + \lambda_i RC_c}{0.7R_{drv,m}C_{drv,m}}} \tag{3.24}$$

$$H_{i,opt} = \sqrt{\frac{0.7R_{drv,m}C_s + 1.4\mu_i R_{drv,m}C_c}{0.7RC_{drv,m}}} \tag{3.25}$$

When a number corresponding to a certain case is substituted for i in the two equations, the number and size of repeaters to minimize the delay for that particular switching pattern (see also Table 3.1) results. Thus we have proposed a simple way to distribute the capacitance and take the effect of switching aggressors into account.

4.2 Optimal Signaling Over Parallel Wires

In the previous subsection, we have seen how repeater insertion can help to reduce delay over long interconnects. However, delay is not the only concern associated with the interconnect. Another major issue is the bandwidth supported by the interconnect under certain constraints, such as limited area, limited power consumption and limited freedom in choosing repeater insertion strategy. In the followin text we study how delay and bandwidth are related and derive an optimal bandwidth under different constraints.

For the wire arrangement shown in Figure 3.1, the worst-case delay of a line is defined as t_{WC}. Since in general it has to be assumed that the worst-case aggressor-victim switching pattern will occur on a given line, any calculation of bandwidth has to consider the worst-case delay as the *minimum* delay over a line. This minimum delay, as we shall show depends on the resources available for repeater insertion, but it shall always correspond to the switching pattern in case 1). Hence for all delay calculations, equation 3.21 with $i = 1$ is used. The line delay is matched to the minimum pulse width T, by allowing a sufficient margin of safety. The exact mapping depends on the type of line [22], but it is generally accepted that three 0-50% propagation delays are sufficient to let the signal cross the 90% threshold for RC lines [23]. Since we already consider the worst-case delay with good accuracy, a factor of 1.5 is deemed to be sufficient, resulting in equation (3.26)

$$T = 1.5t_{WC} \tag{3.26}$$

The total bandwidth in terms of bits per second is now given by equation (3.27)

$$BW = \frac{N}{T}, \tag{3.27}$$

where N is the number of signal wires that can be fitted into a given area. This expression changes if pipelining is carried out so that at any given time, more than 1 bit -up to a maximum of one bit per each gain section- is on the line. Since each repeater will refresh the signal and sharpen its rising or falling edge, the mapping between the propagation delay and the pulse width needs to be carried out for *each section.* Theoretically it is possible to gain an increase in bandwidth by introducing repeaters up to the limit where the bit width is determined by considerations other than the delay of a single stage, or where the delay of the composite net is greater than its constraint. In practice one rarely sees repeaters introduced merely for the sake of pipelining, when the total delay of the net, and power consumption increases as a result. If pipelining is carried out, it is a simple matter to multiply equation 3.27 by the appropriate factor.

The number of signal wires N that can be fitted into a given area depends on whether shielding is carried out or not. In general, shielding individual lines is only useful against capacitive crosstalk. The magnetic field will in all probability permeate the entire breadth and length of the bus, and can only be contained by very fat wires. Hence for the shielded case it is assumed that the shielding wires are the thinnest permitted by the technology, regardless of the size of the signal wires, as this serves the intended purpose while minimizing area for non-signal wires. From the geometry of Figure 3.1, we get the relation given in equation (3.28) for unshielded wires, and in equation (3.29) for shielded wires. Our problem definition is to maximize the bandwidth for a constant width W_T.

$$W_T = NW + (N-1)S \tag{3.28}$$

$$W_T = NW_{signal} + (N-1)(2S + W_{shield}) \tag{3.29}$$

Typically in a process the wires in a certain layer are limited to tracks determined by the minimum feature size of the technology. Within this frame, the designer has freedom to vary the spacing and the width of the wires. Now the problem definition can be stated as follows: for a constant width W_T, what are the N (number of conductors), s (spacing between conductors), and w (width of a conductor) values that give the optimum bandwidth? The variables are discrete as s and w are dictated by the process as well, and there are geometrical limits which cannot be exceeded. The optimal arrangement depends very much on the resources allocated for repeaters, and is investigated by simulations first. Then approximate analytic equations are developed that give close to optimal solutions, and can be used as guidelines to quickly obtain the true solution.

The simulations are carried out for a future technology with parameters estimated from guidelines laid out in [24]. The minimum feature size is 50 nm, and copper wires are assumed with the technology dependent constant b being 1.65, height above substrate h being 0.2 μm, and wire thickness t being 0.21 μm. The minimum wire width and spacing are each assumed to be 0.1 μm and the output impedance of a minimum sized inverter estimated to be 7kΩ and its input capacitance 1fF. In all cases the constraint for the wires is set to a total width of 15 μm. Of the three variables N, s and w, only two are linearly independent, as the third is defined by 3.28 or 3.29 for any values that the other two may take. We choose to vary N and s, and assume that w and s are variable in multiples of the minimum pitch. In the subsequent sections different constraints on the repeaters are considered.

4.2.1 Ideally Driven Line. Although ideal sources are never present in practice, the wire arrangement for the optimum bandwidth is interesting as it serves as a point of comparison for later results. Given in Figure 3.4 is the plot of how the bandwidth varies with N and s. It can be seen that there is a clear optimum of 16 conductors which is far from the maximum number of 150 conductors allowed by the technology constraints.

4.2.2 Unshielded Lines with Optimal Buffering. The bandwidth for changing N and s where the repeaters are optimally sized is plotted in Figure 3.5. It can be seen that the maximum bandwidth is obtained when the parallelism is the maximum allowed by the physical constraints of the technology, of $w = s = 0.1$ μm. This result is logical because the buffers which are optimally sized for each configuration compensate for the increased resistance and cross-talk effect. The values of H and K are 52 and 7 respectively, while the maximum bandwidth is 345.5 Gbits/s.

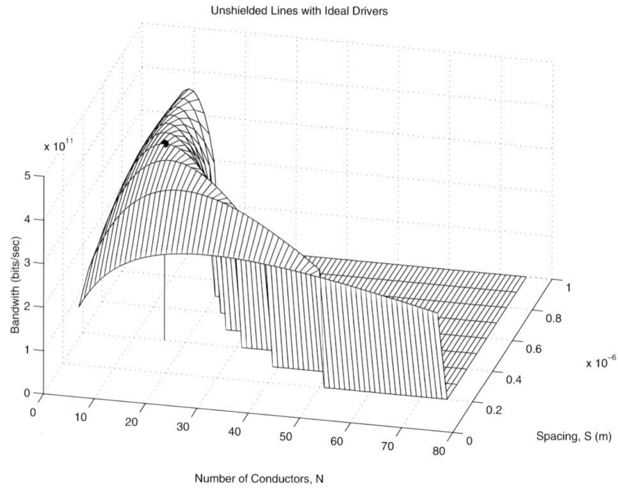

Figure 3.4. Bandwidth variation with number of conductors (N) and spacing between conductors (s) [1]. ©2003 IEEE

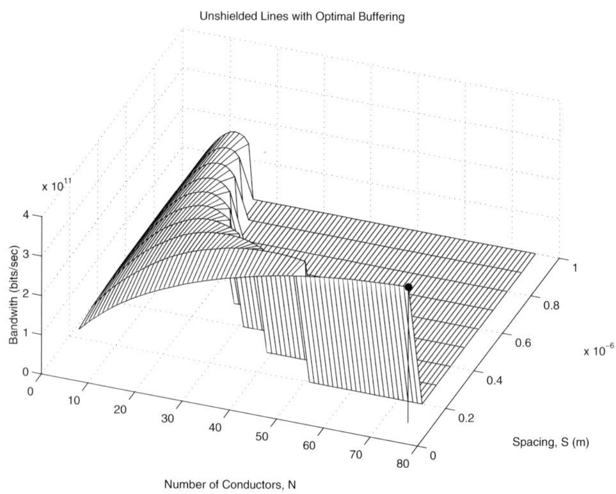

Figure 3.5. Bandwidth variation for unshielded lines with optimal repeater insertion [1]. ©2003 IEEE

4.2.3 Unshielded Lines with Constant Buffering.

Optimal repeater insertion results in a large number of huge buffers. Also, as is the case with optimal buffering in general whether the load is lumped or distributed, the delay curve is quite flat, and the sizes can be reduced with little increase in delay. Instead of optimal repeater insertion, if a constraint is imposed on the number and size of buffers for each line, the optimal configuration does *not* equate to the maximum number of wires.

If a constraint of $K = 1$ and $H = 20$ is laid down for each line the optimal configuration corresponds to $w = 0.16$ μm, $s = 0.2$ μm and $N = 42$, so that the $N \cdot H \cdot K$ product is 840. The maximum bandwidth is now 171.1 Gbits/s.

4.2.4 Unshielded Lines with Constrained Buffering. Typically the constraint would be on the total area occupied by the buffers, and hence K and H would be affected by N. If 3.30 describes the area constraint on the buffers, the optimum configuration is the solution to the constrained optimization problem of maximizing (3.27) subject to equations 3.28 and 3.30:

$$NKH \leq A_{max} \tag{3.30}$$

This adds a third independent variable to the objective function (3.27). The variable could be either K or H since N is constrained by equation (3.28) and A_{max} is a constant. It is a simple matter to incorporate all the relevant equations presented here into an iterative algorithm that can be used to obtain a computer-generated solution. As an example, assume that A_{max} is set to 500 for the same boundary conditions. It turns out that the optimal configuration is when $K = 1$, and shown in Figure 3.6 is a plot of the bandwidth where $K = 1$ and H changes according to N. The optimal wire arrangement turns out to be $w = 0.26$ μm, $s = 0.4$ μm and $N = 23$.

4.2.5 Shielded Lines with Optimal Buffering. In general, shielding each signal wire results in a *drop* in the overall bandwidth. The reason is that although shielding reduces the delay over each individual line, the reduction in the number of signal lines more than negates this effect. Shown in Figure 3.7 is a plot of the bandwidth where every other wire is a minimum sized shielding wire, and the signal wires are buffered optimally. The total bandwidth of 261.3 Gbits/s is less than in the unshielded case. This reduction is however accompanied with a saving in repeater size, and shielding can be considered as an option to reduce area and power consumption for repeaters.

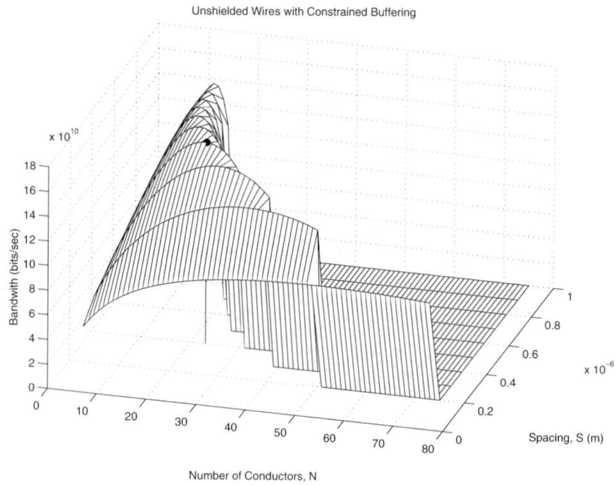

Figure 3.6. Bandwidth variation for unshielded lines with fixed total resource for repeater insertion [1]. ©2003 IEEE

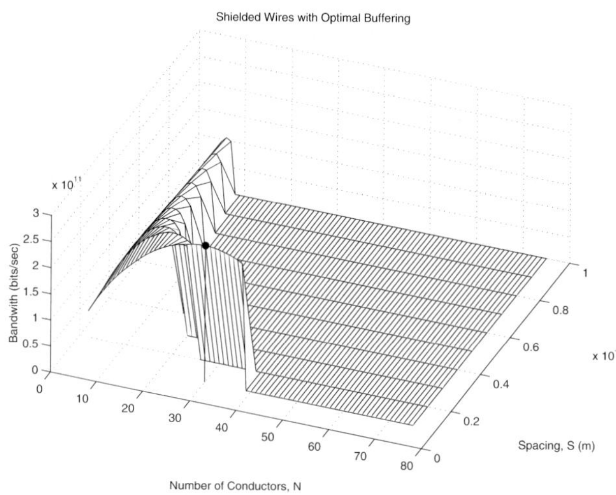

Figure 3.7. Bandwidth variation for shielded lines with optimal repeater insertion [1]. ©2003 IEEE

4.3 Case study: Budgeting wire capacity for robust power delivery and global signal integrity

In this case study, a dynamic interconnect library [7] was first employed to model the characteristics of system-level interconnections. This interconnect model consists of a network of 3D capacitance, resistance and inductance, and can accurately predict signal integrity and transmission properties at system level [7], based on the placement planning and wire planning of the system. Moreover, the model dynamically represents complex 3D multi-level interconnect structures, allows changing of wiring cross section, spacing, and usage of shielding and repeater insertion, to control the interconnect delay and crosstalk noise, and finally directs both placement and routing in the subsequent design stages. Details of the model were addressed in [7] and we will not discuss it in this context. In this case study, we used it for global interconnect modeling and checked the electrical performance of the longest wire.

4.3.1 Interconnect partitioning methodology. We present a design methodology for partitioning interconnects (power supply, clock and signal lines) on different metal levels. A group of metal levels that have the same cross-sectional dimensions is called a *tier*. Different tiers can be reserved for different operational purposes.

In our case study there were 6 metal levels available. The level 1 was used for intra-gate interconnections: a minimum pitch can be used for wiring on that level. Because intra-gate interconnections are usually inside a standard cell, this tier can be called a cell tier. Levels 2 and 3 are usually reserved for inter-gate interconnections inside a functional block (e.g. a decoder or an adder). In our case the interconnect layers 2 and 3 were identical in geometry. The tier consisting of these two metal levels can be called a local tier. Level 4 is called a semi-global tier. On that tier we assigned long-path signals such as control signals or bus signals. Levels 5 and 6 were used for power distribution and also for clocking hierarchy. Part of signal lines or even all of them can also be assigned on these levels depending on the maximum clock skew, power supply noise and supply voltage variation allowed between different parts of the system. These upmost levels form a global tier.

The first step in using the interconnect partitioning methodology was to find out the wiring capacity for each metal level. The equation for overall wiring capacity $WirCap_{total}$ is then a sum of wiring capacities over all levels:

$$WirCap_{total} = e_w \times \sum_{i=1}^{n_w} \left[(A_{logic} - A_i^{power} - A_i^{clk}) \times \frac{1}{p_i^w} \times \prod_{j=i}^{n_w - 1} (0.85)^{\frac{p_i^w}{p_{j+1}^w}} \right] \tag{3.31}$$

where A_{logic} is the area needed by the logic, A_i^{power} is the area of the power supply lines on a level i, A_i^{clk} the area of the clock tree interconnects on a level i ($A_i^{clk} = 0$ for levels 1..4), p_i^w is the wire pitch in a level i and e_w is routing efficiency of a routing tool. Finally, n_w is the number of metal levels (in this case it was 6). The factor 0.85 in equation 3.31 models the effect of via blockage on wiring capacity; the effect increases on lower metal levels. The value was suggested in [25].

After we calculated a wiring capacity for each level a constraint for the maximum allowable supply voltage variation was set. Then we can calculated power supply noise using our model for estimating core switching noise on distributed LRC power grid [26]. This implied that we first estimate area, power consumption, the used clock frequency and switching activity of electrical nodes. The last property is the most difficult to estimate in an early phase of the design flow.

After the constraint for the supply voltage variation was met we calculated how much there was space for signal wires after assigning wires for power distribution and subtracting the power wiring demand from the total wiring capacity. If there was enough space left on the upmost levels for all global signal wires the speed of the signals could be increased and thus a better performance was achieved. If not, we had to use lower metal levels to route global signals which degraded signal integrity.

The same methodology of calculating wire capacity for signals was applied also for local or semi-global signal wires (between logic gates) on lower levels.

4.3.2 Noise estimation.

The crosstalk noise arises from capacitive and inductive coupling of signal interconnections (e.g. in buses). It depends on many parameters such as wiring pitch, coupling length, driver/receiver impedance, and repeater insertion. Based on the dynamic interconnect model, we first check the maximum usable length for fast LC response. The usable length is given by equation (3.32) [7]

$$L_c = 2th\left[(1+\alpha)\rho\sqrt{\epsilon_k\epsilon_0}cK_c\right]^{-1} \tag{3.32}$$

where t is metal thickness, h inter-layer dielectric (ILD) thickness, α is the ratio of return-path and signal-path resistance, K_c a fringing factor for wire capacitance, ϵ_k the relative dielectric constant of ILD, ϵ_0 the dielectric constant of free space and c the light speed in free space. If a wire is longer than this, it is broken into shorter sections by repeater insertion. By such a procedure we can decrease the maximum wire delay and increase the system performance. The reason is that by breaking the wire with repeaters the wire delay is proportional linearly to wire length while using long wires without repeaters produces wire delays proportional to wire length squared. In section 3.4.1 the optimal repeater insertion to decrease the delay of a wire was discussed and analyzed.

To model crosstalk and signal propagation, we assumed that the effective driver impedance is the same as the wire characteristic impedance at high frequency. The receiver is a typical CMOS load that is modeled as a load capacitance. We used a matrix computation method to solve the coupled transmission line equations in frequency domain.

In addition, power supply noise, particularly the one caused by core logic switching, is critical for the performance of DSM CMOS [26]. Besides degrading the quality of power supply which in turn weakens the driving capability of the gates and hence increases the overall circuit delay, the noise is also coupled directly to signal lines in the circuit through the substrate, the wiring hierarchy, and radiation. In this case study, we assumed that the global power distribution was done on levels 5 and 6 with the topology of a meshed grid [27]. Two package structures, namely wire bonded pin-grid array and C4 bonded pin grid array were analyzed. Lead inductance for the packages was taken from [28]. We used the algorithm of [26] to compute the power supply noise. The load capacitance for each grid was calculated by $C_L = P/(V^2 \cdot f)$, where P is the estimated power consumption for this area, V is the power supply voltage and f is the clock frequency. The symbiotic bypass capacitance is estimated by $C_{sym} = [P/(V^2 \cdot f)] \cdot (1 - SF)/SF$, with SF the switching factor in this area. The symbiotic bypass capacitance refers to output capacitances of logic gates that are not switching; their output capacitances are connected to either positive supply or ground (except for tristate gates) [13].

4.3.3 Results for the case study. In our case study [29] we used two technologies ($0.18\mu m$ and $0.25\mu m$) and a case study architecture that has been presented in [30]. We assumed that metal level 1 belongs to a *cell* tier, levels 2-3 belong to a *local tier*, level 4 belongs to a *semi-global* tier and levels 5-6 to a *global* tier. Both technologies have 6 metal levels to offer.

First we estimated the maximum allowable length for a wire without repeaters on each level (see equation 3.32). We found that after the estimation of the processor core size and thus the global interconnection length, only levels 5 and 6 can offer a safe distribution for global signals. Thus we made an assumption that global signals are transported on those levels and all other signals are transported on levels 1-4. Then we had to evaluate if there was enough wiring space for signals. On the other hand, we had a constraint set to a power supply voltage variation. The constraint for the power distribution network was set so that the maximum allowable supply voltage variation was 10% of Vdd. The methodology how the power distribution network with a noise constraint is designed can be found in [26].

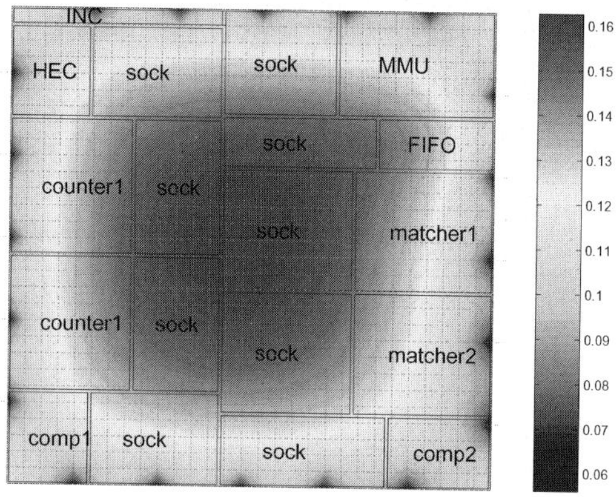

Figure 3.8. On-chip power supply noise (voltage) distribution if a 180-pin wire-bonded pin-grid-array package is used. It was found that 30 % of wiring resources for metal level 6 and metal level 5 need to be dedicated to on-chip power distribution ($0.18\mu m$).

In Figure 3.8, a 2-D power supply noise distribution is shown for a 180-pin wire-bonded PGA package. We used 30 by 30 meshed power grid, 20 I/O pairs for power/ground peripherally, lead inductance was 24nH per pin. The wire width is allowed to change and wire spacing is the minimum spacing allowed by the process on each level. We noticed that only 30% of the wiring space on metal levels 5 and 6 was needed for power distribution, the rest could thus be reserved for signaling (here signaling inludes also clock signal distribution).

In Figure 3.9, a 2-D power supply noise distribution for a C4-bonded multi-layered ceramic PGA package is shown. Pitch for area array connection is $200\mu m$, size is 13 by 13. Lead inductance is 8.2 nH. On-chip power grid size is 30 by 30. The wire width is changeable, wire spacing is the minimum spacing allowed by the process. Compared to the wire-bonded case, 5 % less wiring space is needed to power distribution on levels 5 and 6 which increases the wiring space for global signaling.

5. Conclusions

In this chapter we studied principles for system-level interconnect modeling. The main focus was to examine global wires that distribute

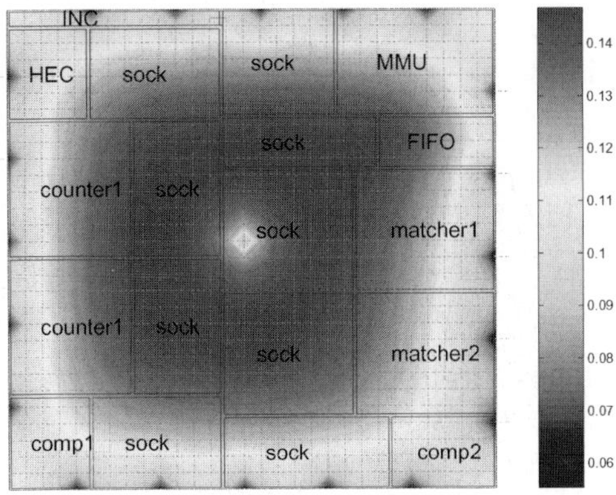

Figure 3.9. On-chip power supply noise (voltage) distribution if a C4 bonded multi-layered ceramic pin-grid-array package is used. For this case, 25% of wiring resources for M6 and M5 are dedicated to on-chip power distribution. The remaining 75% is left for signal distribution ($0.18\mu m$).

signals between different system blocks. Additionally, power distribution was dealt with in the end of the chapter by using the maximum allowed power supply voltage variation as a constraint to design power distribution network in a proper way. We examined electrical properties of on-chip wires and discussed shortly some possible interconnect schemes in SoC and NoC. Some interconnect schemes in SoC and NoC were shortly discussed. We used Rent's rule and multiple Rent's exponents to evaluate cost functions that different system blocks set for the global wiring (both signal and power distribution). We optimized our signaling to achieve the maximum bandwidth, the minimum delay by using properly sized and placed repeaters and finally presented a joint optimization case study in which both power distribution and signal distribution in global wires were simultaneously optimized. Our analysis revealed that both early cost/performance estimation of resources (i.e. functional blocks) and the joint optimization of global signal and power estimation are essential when designing future SoCs and NoCs.

References

[1] D. Pamunuwa, L. R. Zheng, and H. Tenhunen. Maximising throughput over parallel wire structures in the deep submicrometer regime. *IEEE Transactions on Very Large Scale Integration (VLSI) Systems*, 11(2):224–243, April 2003.

[2] E. Chiprout. Interconnect and substrate modeling and analysis: An overview. *IEEE Journal of Solid-State Circuits*, 33:1445–1452, September 1998.

[3] K. L. Shepard, D. Sitaram, and Y. Zheng. Full-chip, three-dimensional, shapes-based RLC extraction. In *Proceedings of ICCAD*, pages 142–149, November 2000.

[4] J. M. Rabaey. *Digital Integrated Circuits.* Prentice-Hall, Upper Saddle River, NJ, 1996.

[5] E. Barke. Line-to-ground capacitance calculation for VLSI: a comparison. *IEEE Transactions on Computer-Aided Design of Integrated Circuits and Systems*, 7(2):295–298, February 1988.

[6] C. P. Yuan and T. N. Trick. A simple formula for the estimation of the capacitance of two-dimensional interconnects in VLSI circuits. *IEEE Electron Device Letters*, EDL-3:391–393, 1982.

[7] L. R. Zheng, D. Pamunuwa, and H. Tenhunen. Accurate a priori signal integrity estimation using a multilevel dynamic interconnect model for deep sub-micron VLSI design. In *Proceedings of European Solid-State Circuit Conference*, pages 324–327, 2000.

[8] W. C. Elmore. The transient response of linear damped circuits. *Journal of Applied Physics*, 19:55–63, January 1948.

[9] J. Rubinstein, P. Penfield, and M. Horowitz. Signal delay in RC tree networks. *IEEE Transactions on Computer-Aided Design*, CAD-2(3):202–211, July 1983.

[10] H. B. Bakoglu. *Circuits, Interconnections, and Packaging for VLSI.* Addison-Wesley, Reading, MA, 1990.

[11] T. Sakurai. Approximation of wiring delay in MOSFET LSI. *IEEE Journal of Solid-State Circuits*, 18:418–426, August 1983.

[12] C. R. Paul. *Analysis of Multi-Conductor Transmission Lines.* John Wiley and Sons, New York, NY, 1994.

[13] W. J. Dally and J. W. Poulton. *Digital Systems Engineering.* Cambridge University Press, New York, NY, 1998.

[14] H. Kawaguchi and T. Sakurai. Delay and noise formulas for capacitively coupled distributed RC lines. In *Proceedings of Asian and South Pacific Design Automation Conference*, pages 35–43, June 1998.

[15] A. B. Kahng, S. Muddu, and E. Sarto. On switch factor based analysis of coupled RC interconnects. In *Proceedings of DAC*, pages 79–84, June 2000.

[16] B. S. Landman and R. L. Russo. On a pin versus block relationship for partitions of logic graphs. *IEEE Transactions on Computers*, C20(12):1469–1479, December 1971.

[17] W. Donath. Placement and average interconnection lengths of computer logic. *IEEE Transactions on Circuits and Systems*, CAS-26(4):272–277, April 1979.

[18] A. Caldwell, Y. Cao, A. B. Kahng, F. Koushanfar, H. Lu, I. L. Markov, M. Oliver, D. Stroobandt, and D. Sylvester. GTX: The MARCO GSRC technology exploration system. In *Proceedings in 37th IEEE/ACM Design Automation Conference*, pages 693–698, Los Angeles, 6 2000. ACM Press.

[19] J. C. Principe, N. R. Euliano, and W. C. Lefebvre. *Neural and Adaptive Systems: Fundamentals through Simulations.* Wiley&Sons, Inc, 2000.

[20] T. Ahonen, T. Nurmi, J. Nurmi, and J. Isoaho. Block-wise extraction of Rent's exponents for an extensible processor. In *IEEE Computer Society Annual Symposium on VLSI*, pages 193–199, February 2003.

[21] J. A. Davis and J. D. Meindl. Generic models for interconnect delay across arbitrary wire-tree networks. In *Proceedings of Interconnect Technology Conference*, pages 129–131, 2000.

[22] A. Deutsch, P. W. Coteus, G. V. Kopcsay, H. H. Smith, C. W. Surovic, B. L. Crauter, D. C. Edelstein, and P. J. Restle. On-chip wiring design challenges for gigahertz operation. *Proceedings of the IEEE*, 89(4):529–555, April 2001.

[23] R. Ho, K. W. Mai, and M. Horowitz. The future of wires. *Proceedings of the IEEE*, 89(4), April 2001.

[24] SEMATECH. International technology roadmap for semiconductors. http://public.itrs.net/files/1999_SIA_Roadmap/Home.htm, 1999.

[25] G. A. Sai-Halasz. Performance trends in high-end processors. *Proceedings of the IEEE*, 83(1), January 1995.

[26] L. R. Zheng and H. Tenhunen. Fast modeling of core switching noise on distributed LRC power grid in ULSI circuits. In *Proceedings of IEEE 9th Topical Meeting on Electrical Performance of Electronic Packaging*, pages 307–310, 2000.

[27] Y. L. Low, L. W. Schaper, and S. S. Ang. Modeling and experimental verification of the interconnected mesh power system (IMPS) MCM topology. *IEEE Transactions on Components, Packaging, and Manufacturing Technology part B*, 20:42–49, February 1997.

[28] R. R. Tummala. *Microelectronic Packaging Handbook, 2nd edition.* Chapman&Hall, 1997.

[29] T. Nurmi, L. R. Zheng, J. Isoaho, and H. Tenhunen. Early estimation of interconnect effects on the operation of System-on-Chip platforms. In *Proceedings of the 15th European Conference on Circuit Theory and Design (ECCTD'01)*, August 2001.

[30] S. Virtanen, J. Lilius, and T. Westerlun. A processor architecture for the TACO protocol processor development environment. In *Proceedings of the 18th IEEE NORCHIP Conference*, pages 204–211, November 2000.

Chapter 4

DESIGN METHODOLOGIES FOR ON-CHIP INDUCTIVE INTERCONNECT

Magdy A. El-Moursy
maelmou@ece.rochester.edu

Eby G. Friedman
friedman@ece.rochester.edu

1. Introduction

On-chip interconnects are now the primary bottleneck in the flow of signals through high complexity, high speed integrated circuits (ICs). Operating an IC at high frequency while dissipating low power is the primary objective for many modern circuit applications. The frequency at which ICs operate increases each year. In order to satisfy these performance objectives, the feature size of CMOS circuits is decreased for each advanced generation of technology. The reduction in feature size reduces the delay of the active devices. The effect on the delay due to the passive interconnects, however, has increased rapidly as described in the National Technology Roadmap [1].

Low power dissipation is another increasingly important design objective in current (and future) ICs. With shrinking feature size, the number of wires grows exponentially [1]. The interconnect capacitance often dominates the total gate load. On-chip interconnects therefore dissipate a large portion of the total power dissipation. Long interconnect that distribute clock signals and power can dissipate up to 40% to 50% of the total power of an IC [2]. Additional interconnect layers may enhance circuit speed while increasing the power dissipation. Interconnect design has, therefore, become a dominant issue in high speed ICs. Some insight into the complexity of modeling multilayer interconnects at high frequencies is presented in this chapter.

Interconnect wires can be classified into two categories, local wires and global wires. Local wires are those interconnects within logic units that connect the active devices. The delay of the local wires decreases with feature size since the distances among the devices decrease. Global wires are those interconnects that connect different logic units (*e.g.*, busses) or distribute signals across the die (*e.g.*, clock distribution networks). The delay of local wires decreases while the delay of global wires increases with advancing technology nodes [2]. Despite the reduction in feature size, the die size has tended to increase. The die size has historically doubled every ten years, as shown in Fig. 4.1. The length of the global wires does not scale with technology, while the cross section decreases. The reduction in crossection increases the line resistance and, consequently, the delay required for a signal to propagate along a line. The increase in die size further increases the length of the global lines which further increases the delay. Special attention should therefore be placed on the global lines, since these lines can limit the overall speed of a circuit [4].

In order to cope with these trends in advanced technologies, different design methodologies have been developed to decrease the time required for a signal to propagate through a long line. In this chapter, different techniques to drive long interconnect are reviewed.

In subsection 1.1, different models for on-chip interconnect are described. The increasing importance of considering line inductance in the interconnect model is discussed in subsection 1.2.

1.1 Interconnect Modeling

Modeling on-chip interconnect is important to determine the signal characteristics of a line. Accurate modeling enhances both the design and analysis processes. Local lines can often be modeled as a single lumped capacitor, as shown in Fig. 4.2a. If the line capacitance is much smaller than the load capacitance, local lines can be neglected in the delay analysis. Modeling local interconnect by a capacitive load becomes more important as the line capacitance becomes comparable to the load capacitance. Signal propagation through these lines is negligible as compared to the gate delay. Since these lines are short, the resistance is typically negligible. The contribution of the line resistance to the degradation in the signal propagation characteristics is therefore not significant.

The line resistance impedes the signal propagation in long lines. The delay through these lines can be comparable to or greater than the gate

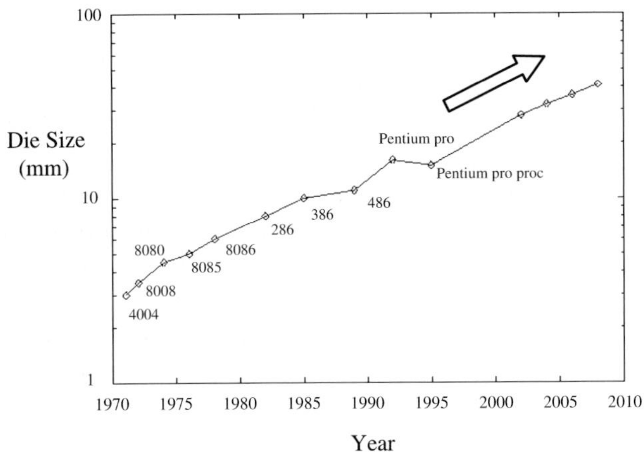

Figure 4.1. Die size for different generations of Intel microprocessors [3]

delay. Modeling a global line as a lumped capacitor is often highly inaccurate. Different models have been proposed to model *RC* lines. The simplest model is a lumped *RC* model as shown in Fig. 4.2b. In order to capture the distributed nature of the line impedance, *RC* lines are often divided into sections of distributed impedances [5]. Each section is modeled as an equivalent *RC* circuit. The T and Π circuits, shown in Figs. 4.3a and 4.3b, respectively, are widely used in modeling long interconnects. The accuracy of the model depends upon the number of sections used to model the interconnect. An *RC* model is adequate at low to medium frequencies (up to a few hundred MHz). However, at high frequencies (on the order of a GHz), the *RC* model is inadequate to accurately characterize the signal. An *RLC* model is necessary to accurately characterize these interconnects.

Global lines are usually wide, exhibiting low resistance. With the reduction in line resistance and the increase in clock frequencies, the line inductance has begun to affect the signal propagation characteristics [6]. The inductance should therefore also be considered in the interconnect model. A first order approximation of an inductive interconnect is shown in Fig. 4.2c. The T and Π equivalent distributed models for an *RLC* line are shown in Figs. 4.3c and 4.3d, respectively.

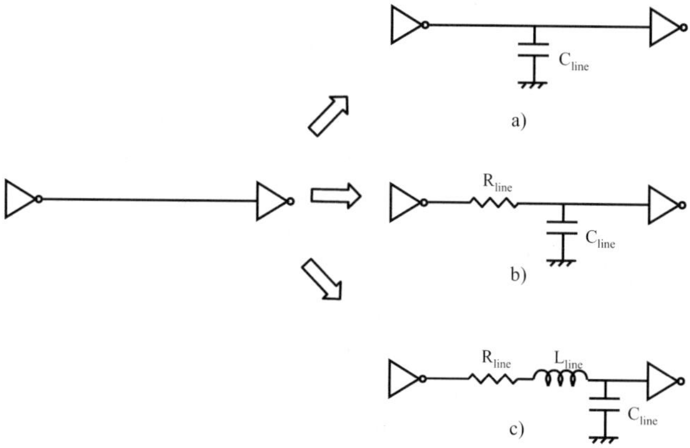

Figure 4.2. Different lumped interconnect models a) C model b) RC model c) RLC model

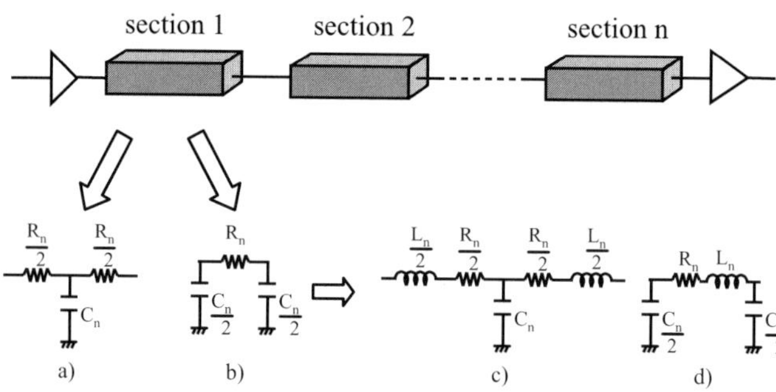

Figure 4.3. Distributed model for an interconnect a) RC T model b) RC Π model c) RLC T model d) RLC Π model

1.2 Importance of On-Chip Inductance

Many studies have been made to determine the conditions at which the line inductance should be considered in an interconnect model. The work described in [6] determines the limits on the line length and transition time at which the line inductance should be considered. As the equivalent output resistance of the gate that drives the interconnect decreases, the limits presented in [6] become more accurate. The lower limit on the line length beyond which the line inductance should be considered is shown in Fig. 4.4 for different line impedance parameters (L is the inductance and C is the capacitance per unit length, respectively). Increasing signal frequencies typically require faster signal transition times. As the signal transition time decreases, the lower limit on the line length also decreases, making shorter on-chip interconnects behave inductively. Medium length lines which do not behave inductively at low frequency, behave inductively as the frequency increases.

Alternatively, the number of long interconnect has increased rapidly. For example, an IC today may have only 10,000 to 20,000 global nets. This number, however, is expected to grow to more than 100,000. Considering the line inductance is, therefore, becoming more crucial in high speed, high complexity integrated circuits [7, 8].

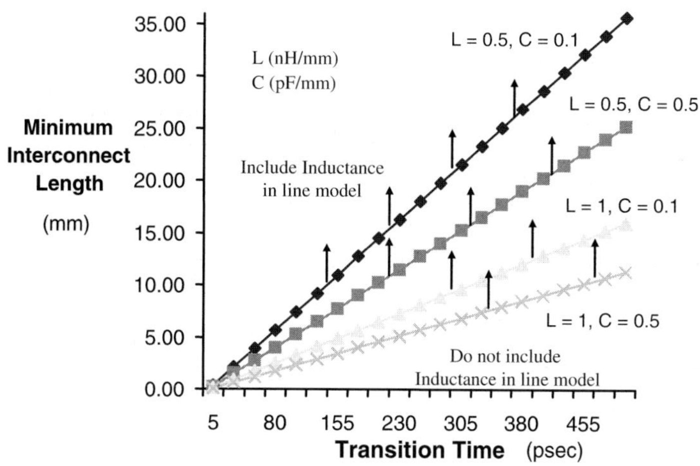

Figure 4.4. Lower limit of interconnect length above which the inductance should be considered in the line model

Another factor which increases the importance of line inductance is the introduction of new materials. New metal and dielectric materials have been introduced to reduce interconnect delay. Low-k dielectrics can decrease the line capacitance to half the capacitance of SiO_2. Furthermore, replacing aluminum lines with copper can also reduce the line resistance by a factor of two to three. These new materials increase the importance of considering the line inductance. As described in [6], the damping factor ζ can be used to characterize the significance of the inductance. For $\zeta < 1$, the line is underdamped, causing ringing in the signal. As shown in Fig. 4.5, when advanced materials are used, the damping factor decreases, increasing the importance of considering the line inductance.

In order to drive global interconnects, many design methodologies have been proposed. Different design techniques have been developed to reduce the propagation delay along a long resistive line. These techniques ignore the line inductance, which may lead to area and/or power inefficient circuits. In this chapter, efficient techniques to drive global interconnects are reviewed. The importance and difficulties in considering on-chip inductance in the design process are discussed. Trends in both the signal delay and power dissipation for inductive lines are described.

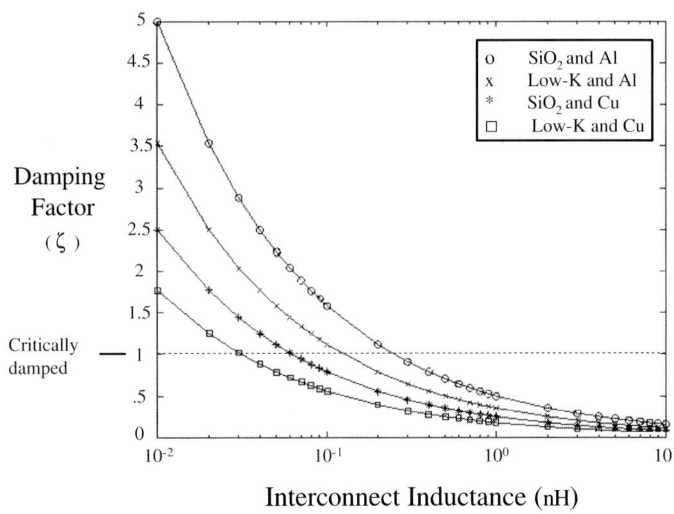

Figure 4.5. Damping factor as a function of the line inductance for different dielectric and metal materials

The chapter is organized as follows. In section 2, effective techniques to drive long resistive interconnect are described. In section 3, new trends in design methodologies to drive inductive interconnects that operate at high frequencies are presented. A discussion of future trends which may affect next generation circuit design methodologies is provided in section 4. In section 5, some comments and conclusions are summarized.

2. Design Methodologies to Drive RC Interconnects

As interconnect has become a dominant issue in high speed ICs, different design methodologies have been developed to improve the performance of long interconnects. These methodologies have historically concentrated on the distributed resistance of a long line. The most effective techniques used to drive long *RC* interconnect are discussed in the following subsections. In subsection 2.1, wire sizing is presented as an effective technique to increase circuit speed. Uniform repeater insertion is another effective technique as described in subsection 2.2. In subsection 2.3, optimum wire shaping for minimum signal propagation delay is discussed.

2.1 Wire Sizing

Interconnect widening decreases the interconnect resistance while increasing the capacitance. Many algorithms have been proposed to determine the optimum wire size that minimizes a target cost function. Some of these algorithms address reliability issues by reducing clock skew. Most of the previous work concentrate on minimizing delay [9]. The results described in consider simultaneous driver and wire sizing based on the Elmore delay model with simple capacitance, resistance, and power models. As the inductance becomes important, specific algorithms have been enhanced that consider *RLC* impedance models.

Previous studies in wire and driver sizing have not considered changes in the signal characteristics accompanied with changes in the characteristics of the line impedance. The interconnect impedance characteristics are more sensitive to the wire size in long, inductive interconnects. The work described in considers power dissipation while ignoring the inductive behavior of the interconnect and, therefore, the effect of line inductance on the power characteristics. In subsection 4.1, the power characteristics of an inductive interconnect are described. Changes in the matching characteristics are discussed in terms of sizing the inductive interconnect for minimum power and delay.

2.2 Repeater Insertion

Uniform repeater insertion is an effective technique for driving long interconnects. The objective of a uniform repeater insertion system is to minimize the time for a signal to propagate through a long interconnect. Based on a distributed *RC* interconnect model, a repeater insertion technique to minimize signal propagation delay is introduced in [10]. Uniform repeater insertion techniques divide the interconnect into equal sections, employing equal size repeaters to drive each section as shown in Fig. 4.6. Bakoglu and Meindl developed closed form expressions for the optimum number and size of the repeaters to achieve the minimum signal propagation delay in an *RC* interconnect [10]. A uniform repeater structure decreases the total delay as compared to a tapered buffer structure when driving long resistive interconnects while buffer tapering is more efficient for driving large capacitive loads [11]. Adler and Friedman developed a timing model of a CMOS inverter driving an *RC* load [12]. The authors used this model to enhance the repeater insertion process for *RC* interconnects. Alpert considered the interconnect width as a design parameter [13]. He showed that, for *RC* lines, repeater insertion outperforms wire sizing. As shown in subsection 4.2, this behavior is not the case for an *RLC* line. The minimum signal propagation delay always decreases with line width for *RLC* lines if a repeater system is used.

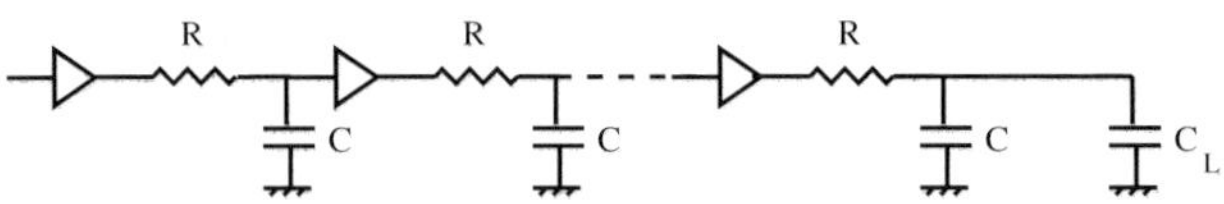

Figure 4.6. Uniform repeater system driving a distributed *RC* interconnect

The delay can be greatly affected by the line inductance, particularly for low resistance materials and fast signal transitions. Ismail and Friedman extended previous research in repeater insertion by considering the line inductance [14]. The authors showed that on-chip inductance can minimize the speed, area, and power of the repeater insertion process as compared to an *RC* line model. Banerjee and Mehrotra developed an analytic delay model and methodology for inserting repeaters into distributed *RLC* interconnect which demonstrated the importance of including line inductance as technologies advance [15].

With increasing demands for low power ICs, different strategies have been developed to minimize power in optimizing the repeater insertion process. Power dissipation and area have been considered in previous work. The line inductance, however, has yet to be considered in the optimization process of sizing a wire driven by a repeater system. Tradeoffs in repeater systems driving inductive interconnect are described in subsection 4.2.

2.3 Interconnect Shaping

Another technique to reduce the signal propagation delay is to shape the interconnect line. Interconnect shaping changes the interconnect width from the driver to the load as shown in Fig. 4.7. As described in [16], the optimum interconnect shape which minimizes the signal propagation delay in an *RC* interconnect is an exponential function. Different extensions to this work have been applied to consider other circuit parameters such as fringing capacitance.

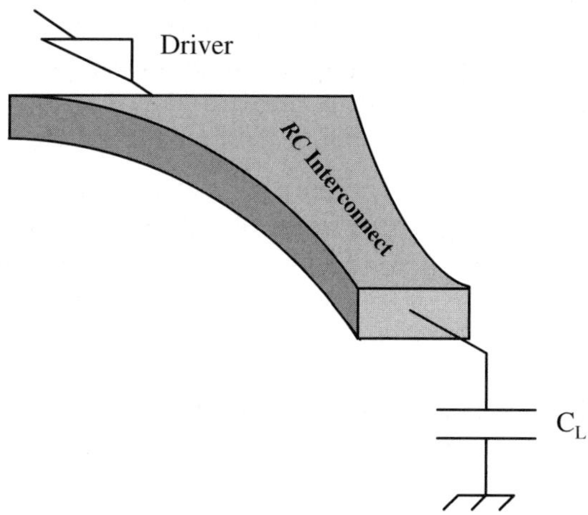

Figure 4.7. Tapered *RC* interconnect

The research described in [13] shows that wire tapering improves the speed by only 3.5% as compared to uniform wire sizing if an optimum repeater system is used to minimize the propagation delay of an *RC* line. The inductance, however, has not been considered in the line model described in [13]. The inserted repeaters increase the dynamic power due to the additional input capacitance of the repeaters. In subsection 4.4, the optimum shape of an *RLC* line that minimizes the delay is determined.

3. Integrating Inductive Effects into Existing Design Methodologies

Different design methodologies to drive long, inductive interconnect are discussed in this section. At high frequencies, long interconnects should be treated as lossy transmission lines. Transmission line properties affect the signal characteristics and change the nature of the circuit design methodologies. In the following subsections, different design methodologies are reevaluated assuming an inductive environment. In subsection 4.1, sizing a long inductive interconnect is discussed and the optimum width for minimum power dissipation is determined. Tradeoffs in sizing an inductive interconnect within a repeater system is presented in subsection 4.2. In subsection 4.3, the shielding effect of on-chip inductance is described. The optimum interconnect shape for minimum propagation delay is determined for an RLC line in subsection 4.4.

3.1 Wire Sizing for Inductive Interconnects

The width of an inductive interconnect affects the power characteristics of a circuit. A tradeoff exists between the dynamic and short-circuit power in inductive interconnects. The dependence of the power dissipation on the interconnect width is illustrated in Fig. 4.8. For the circuit shown in Fig. 4.9, a long interconnect line between two CMOS inverters can be modeled as a lossy transmission line. As the line inductance-to-resistance ratio increases with increasing wire width, the short-circuit power decreases (due to a reduction in the signal transition time). If the width of the interconnect exceeds a certain limit, the short-circuit power increases due to the change in the matching characteristics between the driver and the interconnect [17]. The dynamic power increases with line width since the line capacitance is greater. As shown in Fig. 4.8, an optimum interconnect width exists at which the total transient power is a minimum.

To better understand the signal behavior in terms of the interconnect width, an equivalent circuit of an inverter driving an inductive interconnect line is shown in Fig. 4.10a. The characteristic impedance of a lossy line can be described by the well known formula, $Z_{lossy} = \sqrt{\frac{R+jwL}{jwC}}$. Different approximations have been made to estimate Z_{lossy} in terms of per unit length parameters [18]. A general form of Z_{lossy} is $Z_0 + g\,R$ where g is a constant which depends on the line parameters.

At the end of the high-to-low input transition, the NMOS transistor is off. With the input low, the inverter can be modeled as an ideal voltage source with a variable output resistance R_{tr} as shown in Fig. 4.10b.

At small interconnect widths, the characteristic line impedance is large as compared to the equivalent output resistance of the transistor. The line is overdriven (the underdamped condition). Z_{lossy} decreases with increasing line width. The line remains underdamped until Z_{lossy} equals R_{tr}. A further increase in the line width underdrives the line as

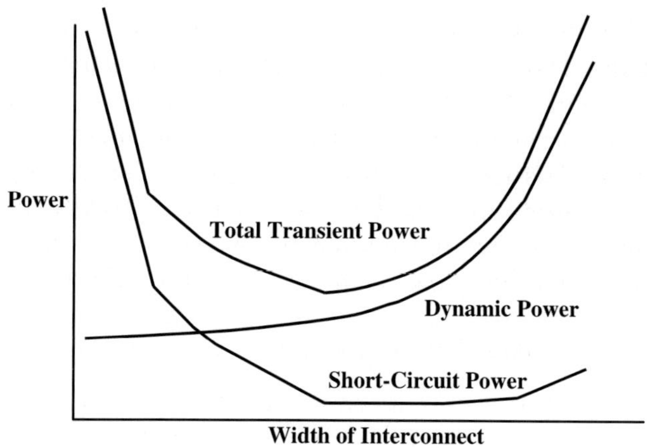

Figure 4.8. Dynamic, short-circuit, and total transient power as a function of the interconnect line width

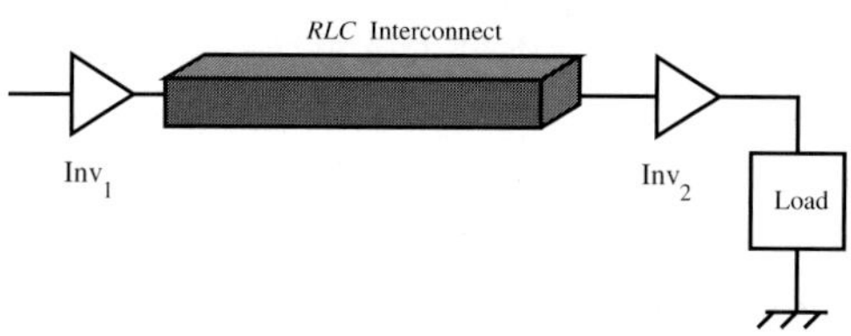

Figure 4.9. CMOS gates connected by an *RLC* interconnect line

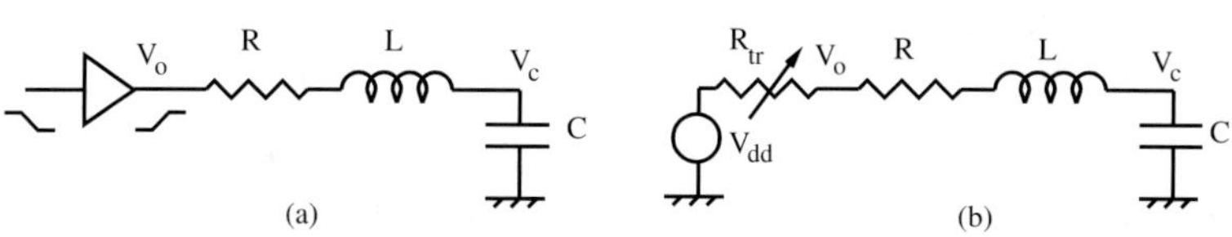

Figure 4.10. An inverter driving an *RLC* interconnect line (a) circuit diagram (b) equivalent circuit of inverter at the end of the high-to-low transition

Z_{lossy} becomes less than R_{tr}. As the line width is increased, the line driving condition changes from overdriven to matched to underdriven.

Increasing the line width makes an overdriven line behave more inductively. The resistance decreases linearly with a linear increase in width while the inductance decreases sublinearly [19]. As described in [20], the line approaches a lossless condition, where the attenuation constant approaches zero at large line widths. This effect further reduces the signal transition time. As the line width increases, Z_{lossy} decreases until the line impedance matches the driver impedance. A further increase in the width underdrives the line. At these widths, the capacitance begins to dominate the line impedance. With wider lines, the line becomes highly capacitive which increases the transition time, thereby increasing the short-circuit power dissipation in the load gate. For an overdriven line, the short-circuit power dissipation changes with line width as shown in Fig. 4.8. For an underdriven line, however, an increase in the line width increases the short-circuit power component. If the line is underdriven, the line should be as thin as possible to minimize the total transient power by decreasing the dynamic power.

A CMOS inverter driving a capacitive load of 250 fF through a 5 mm long interconnect line is used to demonstrate the signal behavior. Twenty RLC distributed impedance elements are used to model the interconnect line. The input signal V_{in} is a ramp signal with a 100 psec transition time. The signal V_c across the load capacitance is illustrated by the waveforms depicted in Fig. 4.11. In Fig. 4.11a, the line is thin. The line inductance does not affect the signal waveform since the resistance dominates the overall line impedance. As the line width increases, overshoots and undershoots appear in the waveform. As shown in Fig. 4.11b, the line inductance affects the signal characteristics and the signal transition time decreases (the overdriven condition). A further increase in the width matches the load with the driver and the overshoots disappear (see Fig. 4.11c). The signal transition time is minimum at this condition. As the wire is widened, some steps start to appear in the waveform (the underdriven condition) and the transition time increases (see Figs. 4.11d to 4.11f).

3.1.1 Optimizing Inductive interconnect Width for Minimum Power. A closed form solution has been developed in [21, 22] for the optimum interconnect width which minimizes the total transient power dissipation of a line. Proposed criteria for interconnect width optimization are applied to different target circuits. Using the optimum width rather than the minimum width, the total transient power is smaller since the short-circuit power is reduced.

As listed in Table 4.1, the optimum width of a copper line reduces the total transient power by 68.5% for $l = 5$ mm as compared to 28.6% for $l = 1$ mm. For aluminum, a reduction of 77.9% is achieved as compared to 37.8%. The more inductive the interconnect, the more sensitive the power dissipation is to a change in the line width (and the signal

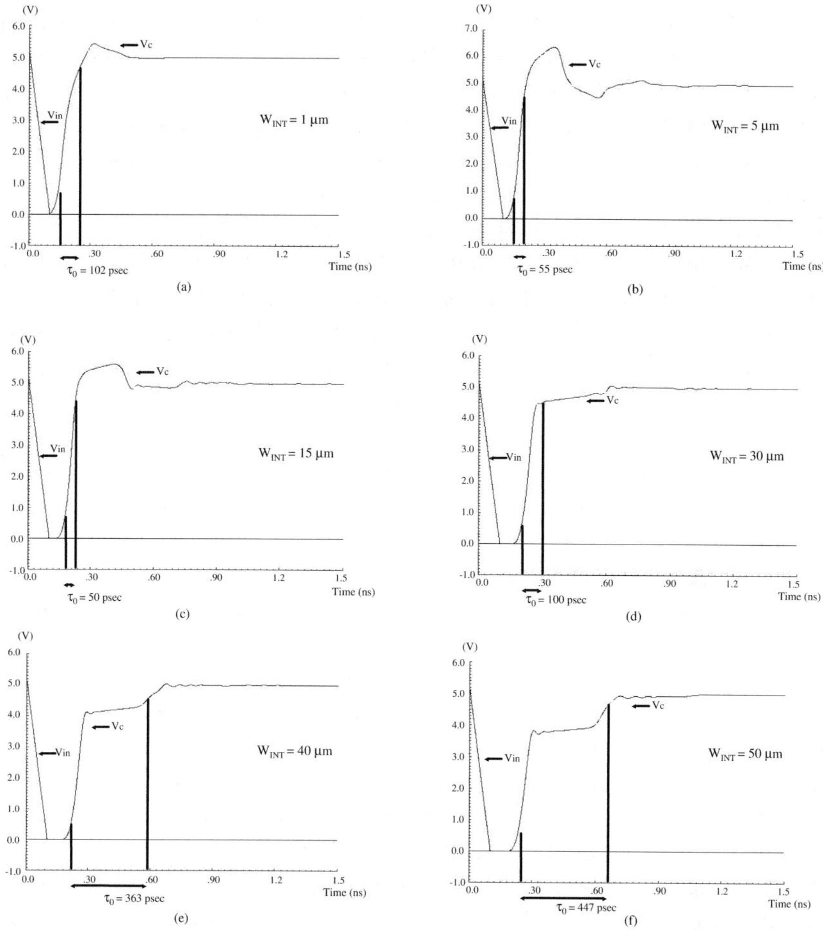

Figure 4.11. Output waveform at the far end of a long interconnect line driven by an inverter with different line widths. (a) Resistive, (b) Overdriven (inductive), (c) Matched, (d) Underdriven (inductive), (e) Underdriven, (f) Underdriven

characteristics). Wire width optimization is, therefore, more effective for longer, more inductive lines.

For l = 5 mm, the per cent reduction in power is 27.8% for copper as compared to 25.4% for aluminum. A reduction in copper of 41.9% is obtained versus 37.4% in aluminum for l = 1 mm. The inductance-to-resistance ratio of copper is higher, increasing the importance of the optimum width for less resistive (more highly inductive) lines. Alternatively, for thin lines, the line resistance has a greater effect on the signal characteristics. The reduction in power is higher for aluminum than for copper.

Table 4.1. Transient power dissipation for different line parameters

Resistivity ρ $\mu\Omega\, cm$	Total Transient Power Dissipation (μW)				
	Resistive Line (l = 1 mm)				
	Optimum	Thin	Reduction	Wide	Reduction
1.7 (Copper)	583	817	28.6%	808	27.8%
2.5 (Aluminum)	606	976	37.8%	813	25.4%
	Inductive Line (l = 5 mm)				
1.7 (Copper)	1121	3563	68.5%	1931	41.9%
2.5 (Aluminum)	1236	5592	77.9%	1973	37.4%

3.2 Wire Sizing Within a Repeater System

Uniform repeater insertion, as shown in Fig. 4.12, reduces the time for a signal to propagate through a long interconnect. Sizing an inductive interconnect driven by an optimum repeater system for minimum signal propagation delay is discussed in this subsection.

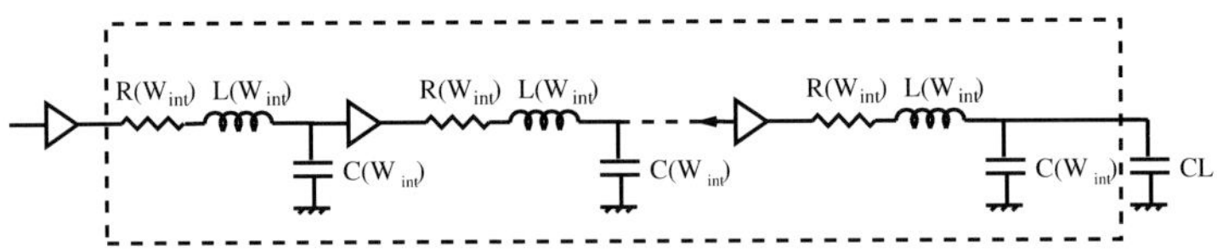

Figure 4.12. Uniform repeater system driving a distributed *RLC* interconnect

For an *RC* line, repeater insertion outperforms wire sizing [13]. Unlike an *RC* line, the minimum signal propagation delay always decreases with the line width for *RLC* lines if an optimum repeater system is used. In *RLC* lines, wire sizing outperforms repeater insertion as the minimum signal propagation delay with no repeaters is smaller than the minimum signal propagation delay using any number of repeaters. The minimum signal propagation delay always decreases with wider lines until the number of repeaters equals zero.

As shown in Fig. 4.13, the minimum propagation delay decreases while the power dissipation increases for wider interconnect. In the following subsections, the tradeoff between the minimum signal propagation delay and the total power dissipation is described in greater detail. In subsection 4.2.1, the effect of wire sizing on the minimum signal propagation delay is presented. The total power dissipation within the system is discussed in subsection 4.2.2. In subsection 4.2.3, the power delay product is used as a criterion to size a line within a repeater system.

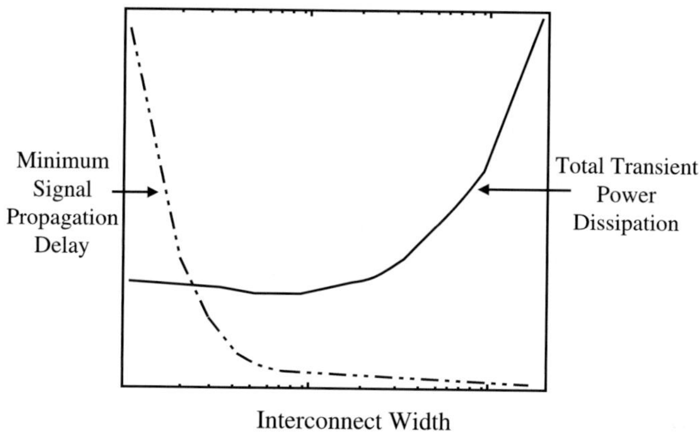

Figure 4.13. Minimum signal propagation delay and transient power dissipation as a function of line width for a repeater system

3.2.1 Propagation Delay in a Repeater System. The interconnect resistance decreases with wider lines, increasing $\frac{L}{R}$, the ratio between the line inductance and resistance, and decreasing the number of inserted repeaters to achieve a minimum propagation delay. For an RLC line, the minimum signal propagation delay decreases with wider wires until no repeaters should be used. Wire sizing outperforms repeater insertion in RLC lines.

For different line lengths l, the optimum number of repeaters $k_{opt-RLC}$ is shown in Fig. 4.14. As shown in the figure, for an RLC line the optimum number of repeaters which minimizes the signal propagation delay decreases with increasing line width for all line lengths. The number of repeaters reaches zero (or only one driver at the beginning of the line) for an interconnect width $= 3$ μm and 4 μm for $\text{l} = 5$ mm and 10 mm, respectively. Above a width of 4 μm, the wire should be treated as one segment. A repeater system should not be used above a certain width for each line length.

The line capacitance per unit length increases with line width. As the number of inserted repeaters decreases with wider lines, a longer line section is driven by each repeater. An increase in the section length and width increases the capacitance driven by each repeater. To drive a

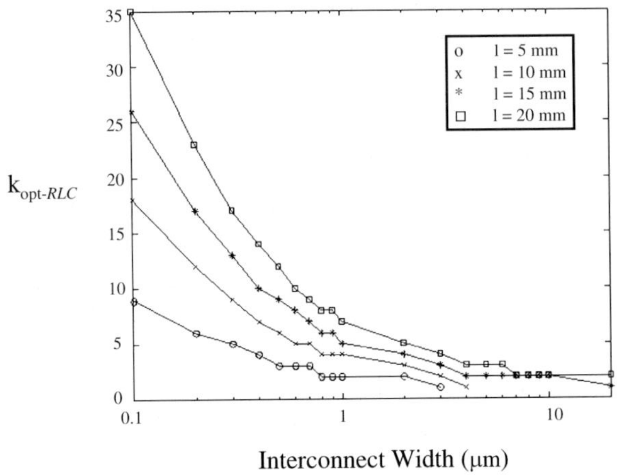

Figure 4.14. Optimum number of repeaters for minimum propagation delay for different line widths

large capacitive load, a wider repeater is required to minimize the overall delay. As shown in Fig. 4.15, the optimum repeater size $h_{opt-RLC}$ is an increasing function of line width.

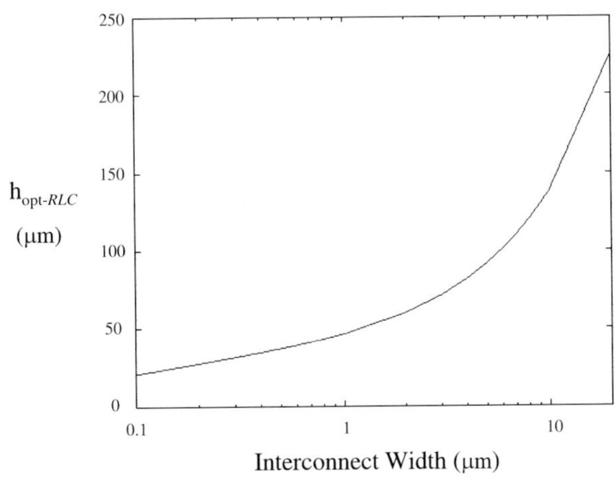

Figure 4.15. Optimum repeater size for minimum propagation delay for different line widths

The minimum signal propagation delay using an optimum repeater system decreases with increasing line width as the total gate delay decreases. For an inductive interconnect line, the total signal propagation delay is

$$t_{pd-total} = k_{opt-RLC}\, t_{pd-section}, \tag{3.1}$$

where $t_{pd-section}$ is the signal delay of each RLC section [14]. The minimum delay [obtained from (3.1)] is shown in Fig. 4.16. An increase in the inductive behavior of the line and a reduction in the number of repeaters decrease the minimum signal propagation delay that can be achieved for a particular repeater system.

The signal delay for different line lengths is shown in Fig. 4.16. The lower limit in the propagation delay decreases with increasing line width until the number of repeaters is zero. For a system of repeaters, there is no optimum width at which the total propagation delay is minimum. Rather, the delay is a continuously decreasing function of the line width. This characteristic is an important trend when developing a wire sizing methodology for a repeater system.

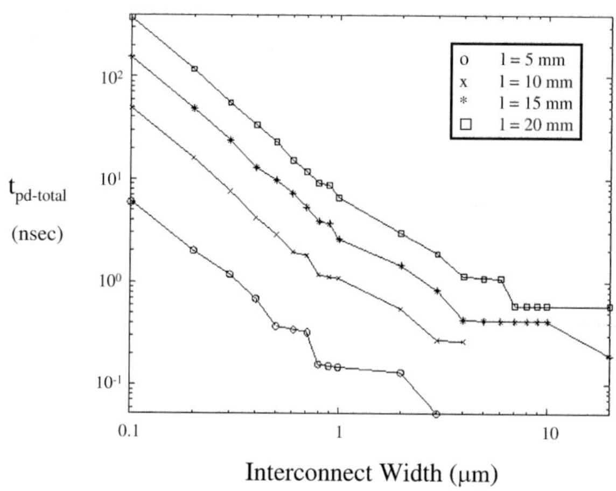

Figure 4.16. Minimum signal delay as a function of interconnect width for different line lengths

3.2.2 Power Dissipation within a Repeater System.

The dynamic power dissipated by a line increases with greater line capacitance (as the line width is increased). The dynamic power of the repeaters, however, decreases since fewer repeaters are used in RLC lines. Increasing the line width has two competing effects on the short-circuit power. As described in subsection 4.1, the short-circuit power decreases when a line is underdamped.

Alternatively, increasing the length of a line section by reducing the number of repeaters increases the short-circuit power of each section because of a higher section impedance. The minimum transient power, therefore, is dissipated with thin interconnect. The optimum line width for minimum transient power dissipation is obtained in [23, 24].

For short interconnects, few repeaters are required to produce the minimum propagation delay. For longer interconnect, an increase in the line capacitance rapidly increases the power dissipation, while the minimum propagation delay decreases more slowly.

In order to develop an appropriate criterion for determining the optimal interconnect width between repeaters, the total transient power dissipation of a system needs to be characterized. The total transient power can be described as

$$P_{total}(W_{int}) = V_{dd} f [k_{opt-RLC}(W_{int})(\frac{1}{2} I_{peak}(W_{int})\, t_{base}(W_{int}) \\ + h_{opt-RLC}(W_{int}) V_{dd}\, C_0) + V_{dd}\, C_{int}(W_{int})], \tag{3.2}$$

where W_{int} is the interconnect width, I_{peak} is the peak current that flows from V_{dd} to ground through an inverter driving an RLC load, t_{base} is the time period during which both transistors of the inverter are on, V_{dd} is the supply voltage, f is the switching frequency, and C_0 is the input gate capacitance of a minimum size repeater. All of the terms in (3.2) are functions of the line width except V_{dd}, C_0, and f. As described in subsections 3.1 and 3.2, both transient power components in repeaters decrease with increasing line width, thereby decreasing the total power until the line capacitance becomes dominant.

For an RLC interconnect, few repeaters are necessary to drive a line while achieving the minimum propagation delay [14]. For an inductive interconnect, the interconnect capacitance is often larger than the input capacitance of the repeaters. Increasing the width reduces the power dissipation of the repeaters and increases the power dissipation of the line. The reduction in power dissipated by the repeaters overcomes the increase in power due to the wider interconnect until the line capacitance dominates the line impedance. After exceeding a certain width, the total power increases with wider lines.

The total power dissipation as a function of line width for different interconnect lengths is shown in Fig. 4.17. As the line width increases from the minimum width (*i.e.*, 0.1 μm in this case), the total power dissipation is reduced. The minimum transient power dissipation therefore occurs with thin interconnect (see Fig. 4.17). The minimum transient power dissipation is obtained from

$$\frac{\partial P_{total}}{\partial W_{int}} = 0. \tag{3.3}$$

$\frac{\partial P_{total}}{\partial W_{int}}$ is a nonlinear function of W_{int}. Numerical methods are used to obtain values of W_{int} for specific interconnect and repeater parameters.

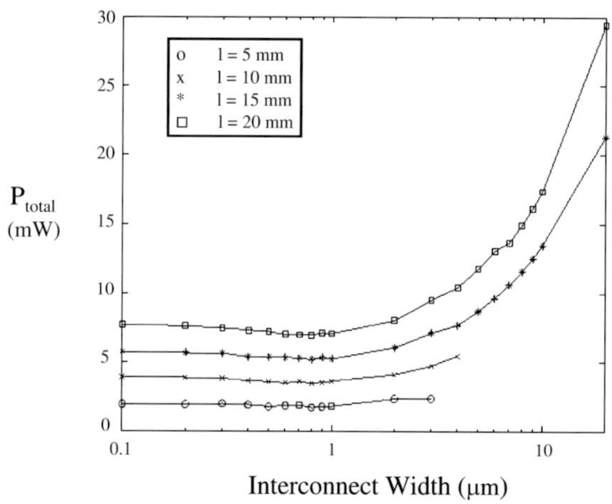

Figure 4.17. Total transient power dissipation as a function of interconnect width

For a range of reasonable interconnect width, the total transient power increases. As the line length increases, the total power dissipation increases rapidly with increasing line width as the interconnect capacitance becomes dominant. In subsection 4.2.3, the tradeoff between signal delay and power dissipation is considered in the development of a criterion for interconnect sizing.

3.2.3 Power Delay Product. From the discussions in subsections 4.2.1 and 4.2.2, the minimum signal propagation delay of an *RLC* interconnect driven by a repeater system decreases with increasing line width. Alternatively, the total transient power has a global minimum at narrow widths. Over the entire range of line width, the total transient power increases with increasing line width. At a line width smaller than the line width for minimum power, the power and delay both increase. An upper limit on the line width is reached where the minimum propagation delay of a repeater system is attained. Beyond that limit, a single segment sizing criterion should be used to optimize the width according to a cost function (*i.e.*, delay or power [17]). Between these two limits, a tradeoff exists between the power dissipation and signal propagation delay. An expression for the power delay product as a function of the interconnect width is

$$PDP(W_{int}) = P_{total}(W_{int})^{w_p}\, t_{pd-total}(W_{int})^{w_d}, \tag{3.4}$$

where w_p and w_d are the weights of the cost functions. A local minimum for the power delay product exists for each line length. The minimum power delay product is obtained by numerically solving the nonlinear equation,

$$\frac{\partial PDP}{\partial W_{int}} = 0. \tag{3.5}$$

In the following section, different criteria are applied to different systems to optimally size the interconnect within a repeater system. For an *RLC* line, there are three criteria to size interconnect in an unconstrained system. The first criterion is for minimum power while sacrificing speed. The optimum solution for this criterion is obtained from (3.3).

The second criterion is for minimum delay. As no optimum interconnect width exists for minimum propagation delay, the practical limit is either the maximum repeater size or no repeaters, whichever produces a tighter constraint. The criterion in this case is the maximum repeater size or line width. The optimum number of repeaters for a target line width is determined from [14]. Otherwise, no repeaters should be used and the design problem reduces to choosing the width of a single section of interconnect.

The third criterion is to satisfy both the power dissipation and speed. The weights w_p and w_d determine which design objective is more highly valued.

The three criteria are applied to a 0.24 μm CMOS technology to determine the optimum solution for different line lengths. No limit on the maximum buffer size is assumed. In order to characterize the line inductance in terms of the geometric dimensions, an interconnect line shielded by two ground lines is assumed. For a repeater system with the following characteristics, $C_0 = 1\ fF$ and $w_p = w_d = 1$, the optimum solution for each criterion is listed in Table 4.2.

The optimum line width using each design criterion is listed in the first row of each line length. The optimum number and size of the

Table 4.2. Uniform repeater system for different optimization criteria, $C_0 = 1\ fF$ and $w_p = w_d = 1$

$l = 5$ mm		Minimum Power	No Repeaters	Minimum PDP
W_{int} (μm)		0.8	2.1	2.1
# of Repeaters		1	0	0
Repeater Size (μm)		43.3	61.2	61.2
Minimum Delay (nsec)	Total	0.157	0.051	0.051
	Increase	208%	0%	0%
Power (μW)	Total	1730	1980	1980
	Increase	0%	14.5%	14.5%
$l = 15$ mm		Minimum Power	No Repeaters	Minimum PDP
W_{int} (μm)		0.8	20	3.9
# of Repeaters		5	0	1
Repeater Size (μm)		43.2	225.6	80.7
Minimum Delay (nsec)	Total	3.87	0.19	0.43
	Increase	1936%	0%	126.3%
Power (μW)	Total	5200	21,310	7580
	Increase	0%	310%	45.7%

repeaters for each line width is listed in the second and third row of each line length. The per cent increase in the minimum propagation delay based on the optimum power and power delay product as compared to no repeaters is also listed. The per cent increase in the total transient power dissipation is provided.

For an $l = 5$ mm line, the optimum interconnect width for both minimum power delay product and no repeaters is the same, producing a 14.5% increase in power as compared to the optimum width for minimum power and a reduction of 68% as compared to the optimum width for minimum signal propagation delay.

For short interconnects, few repeaters are required to produce the minimum propagation delay. For longer interconnect, an increase in the line capacitance rapidly increases the power dissipation, while the minimum propagation delay decreases more slowly.

For $l = 15$ mm, the optimum solution for the minimum power delay product increases the delay by 1.26 rather than 20 times for the solution for minimum power. The power increases by 45% rather than 3.1 times for the no repeater solution. Optimizing the interconnect to produce the minimum power delay product produces a smaller increase in both the power and delay as compared to separately optimizing either the power or delay. A reduction in the minimum propagation delay of 89% and in the power dissipation of 65% is achieved if the optimum width for the minimum power delay product is used rather than the optimum width for either minimum power or no repeaters.

3.3 Shielding Effect of On-Chip Interconnect Inductance

In this subsection, the effect of the line inductance on the effective impedance of a line is described. An important aspect of on-chip inductance is the shielding effect where the inductance hides (or shields) part of the load capacitance from the driver. The shielding effect can reduce the required size of a buffer to drive an *RLC* interconnect. The interconnect inductance introduces a shielding effect which decreases the effective capacitance seen by the driver of a circuit, reducing the gate delay. The effective capacitance of an *RLC* load driven by a CMOS inverter has been analytically modeled. The interconnect inductance increases the line propagation delay and decreases the gate delay, reducing the overall signal propagation delay to drive an *RLC* load as shown in Fig. 4.18 [25]. The minimum delay occurs when the load is matched with the driver. In subsection 4.3.1, the concept of an effective capacitance is extended to an *RLC* load. The effect of line inductance on the propagation delay is described in subsection 4.3.2. In subsection 4.3.3, sizing a driver for an *RLC* interconnect is discussed.

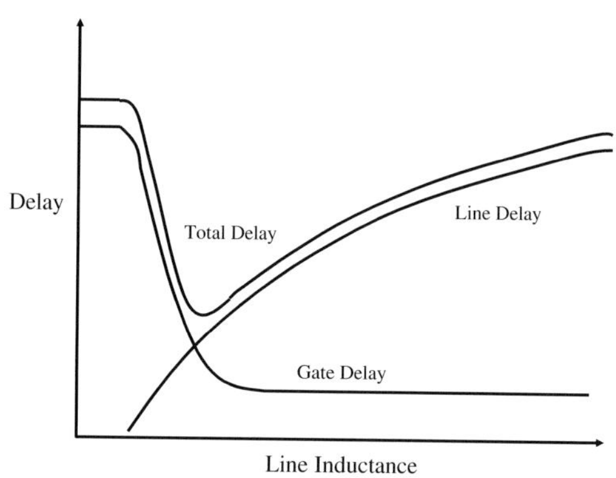

Figure 4.18. Propagation delay as a function of the line inductance

3.3.1 Effective Capacitance of an *RLC* Load. Reduced order models are used to increase the computation efficiency of the timing analysis process. A lumped model of an RLC load which uses the first two moments of the transfer function of a line is shown in Fig. 4.19a. A lumped RLC model can suffer from significant inaccuracy. Furthermore, the shielding effect of the load inductance is not considered. The circuit representation of a three moment reduced order model (π_{21} model) is shown in Fig. 4.19b [26].

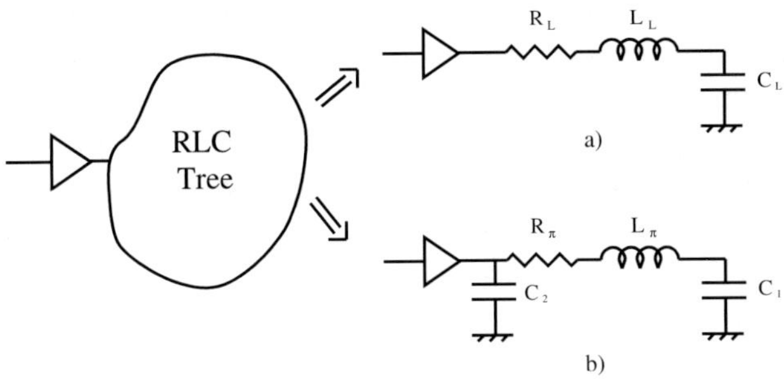

Figure 4.19. Reduced order model for a general RLC tree a) Lumped model b) π_{21} model

An efficient technique is presented in [27] to determine the values of R_π, L_π, C_1, and C_2 for a general RLC load. If the interconnect inductance is not considered, the RLC π_{21} model reduces to an RC π_{21} model with the same values of R_π, C_1, and C_2.

Intuitively, the effective capacitance is the equivalent capacitance which replaces the reduced order π_{21} model while producing the same delay at the load (as shown in Fig. 4.20). The effective capacitance of an RLC load is

$$C_{eff-RLC} = C_2 + C_{x-RLC}, \tag{3.6}$$

where C_{x-RLC} is characterized in [25]. C_{x-RLC} is less than C_1, reducing the total capacitance seen by the driver for the π_{21} model. C_{x-RC} for an RC model is determined for an RC load [28]. C_{x-RLC} is less than C_{x-RC} for an inductive load. C_{x-RLC} decreases with increasing load inductance as the inductive shielding effect increases. The gate delay is linearly proportional to the effective capacitance seen at the driving point. The gate delay becomes smaller since the effective capacitance decreases for larger values of inductance. The interconnect inductance shields part of the load capacitance, reducing the gate delay.

For a total load capacitance and resistance of 400 fF and 100 Ω, respectively, the impedance parameters of the π_{21} model are $R_\pi = 48$ Ω, $C_2 = 67$ fF, and $C_1 = 333$ fF. The ratio between the effective capacitance

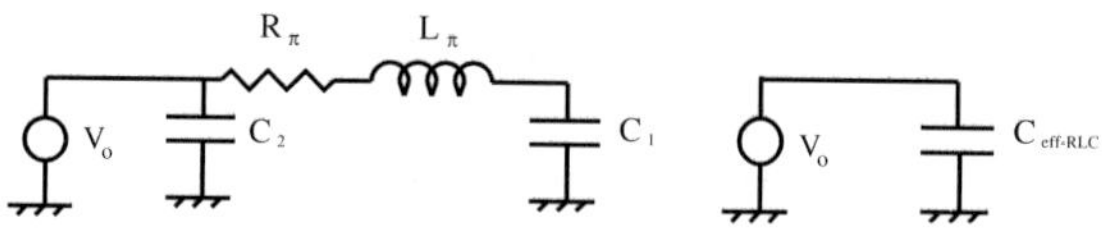

Figure 4.20. Effective capacitance of RLC π_{21} model

of the RLC and RC π_{21} models for different load inductances is shown in Fig. 4.21. The effective capacitance decreases as the load inductance increases. The curve illustrated in Fig. 4.21 is non-monotone due to the existence of the line inductance. The solution for the field equation results in a non-monotone shape.

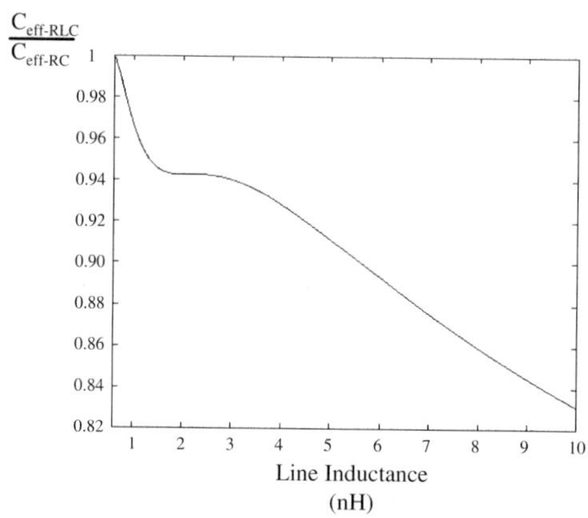

Figure 4.21. Ratio between the effective capacitance of an RLC and RC load

The shielding effect of the interconnect inductance increases the effect of the line inductance on the delay analysis. Ignoring the inductance overestimates the circuit delay, requiring an oversized buffer to drive the load. The effect of the interconnect inductance on the total signal propagation delay is discussed in section 4.3.2.

3.3.2 Effect of Line Inductance on the Delay Model.

The effective capacitance can be used to characterize the gate delay. The signal propagation delay depends on the active gate and passive interconnect components of the signal path. The gate delay is the time required to charge the capacitance seen by the driver through the equivalent resistance of the driver. The interconnect delay is the time required for the signal to propagate through the line. These two components cannot be separated as the driver and load represent a single system. The interconnect inductance reduces both the capacitance seen by the driver (as described in section 4.3.2) and the equivalent output resistance of the driver, reducing the overall gate delay.

For an ideal source driving an *RLC* line, the line delay can be modeled as [14]

$$t_{pd-RLC} = \frac{e^{-2.9\zeta^{1.35}}}{\omega_n} + 0.74R_{line}(C_L + 0.5C_{line}), \tag{3.7}$$

$$\zeta = \frac{R_{line}\omega_n}{2}(0.5C_{line} + C_L), \tag{3.8}$$

$$\omega_n = \frac{1}{\sqrt{L_{line}(C_{line} + C_L)}}, \tag{3.9}$$

where C_L is the load capacitance driven by the line and R_{int}, C_{int}, and L_{int} are the total line resistance, capacitance, and inductance, respectively.

The line delay increases with line inductance as shown in [25]. As the line inductance increases, two competing effects change the total delay of the signal. The delay due to the active transistor decreases while the delay due to the passive interconnect increases. A closed form solution characterizing the signal propagation delay of an inverter driving a reduced order π_{21} model of a distributed *RLC* line is presented in [25]. A comparison between this model and two related models is provided in subsection 4.3.3.

To exemplify the effect of the line inductance on the propagation delay, a CMOS inverter driving a long inductive interconnect with R_{line} = 50 Ω and C_{line} = 400 fF is considered. The total delay for different driver sizes based on a 0.24 μm CMOS technology is shown in Fig. 4.22. Different values of the line inductance with C_L = 50 fF are considered.

The propagation delay decreases with increasing line inductance until a minimum delay is reached. The total delay decreases with higher line inductance over a wide range of driver size (the NMOS transistor size W_n ranges from 10 μm to 50 μm). For small drivers (*i.e.*, $W_n < 5\,\mu$m), the line inductance has no effect on the propagation delay as the delay is dominated by the driver output resistance (and the line does not behave inductively). For large drivers (*i.e.*, $W_n > 50\,\mu$m), the line inductance increases the delay. The output resistance of these drivers is small, and the interconnect delay dominates the total delay. Large drivers are not preferred as the decrease in signal delay is not significant, while the

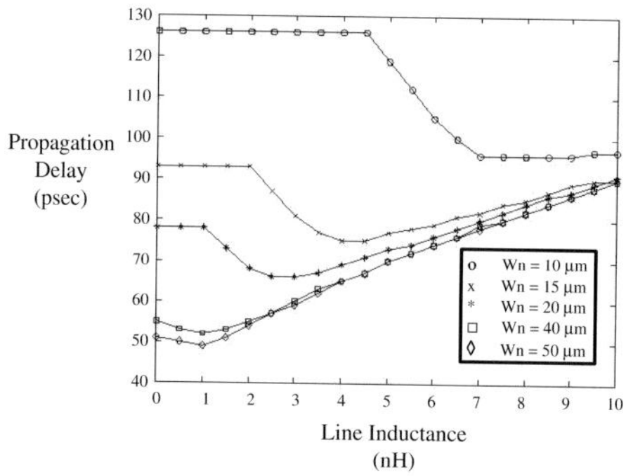

Figure 4.22. Total delay for different values of line inductance and driver size for a distributed *RLC* interconnect

required area and dissipated power are large. Furthermore, the input gate capacitance increases with larger drivers, increasing the delay of the previous logic stage. Buffer tapering can be used for large drivers, but the power dissipation increases with the addition of cascaded tapered inverters [11] to reduce the delay.

Curve fitting is employed to determine the optimum value of the line inductance to achieve the minimum propagation delay. The minimum delay is determined over a wide range of line inductance (from 0.1 nH to 10 nH), load capacitance (from 10 fF to 250 fF), inverter size (from 5 μm to 50 μm), line capacitance (from 100 fF to 1 pF), and line resistance (from 25 Ω to 100 Ω). The minimum delay occurs when the ratio between the equivalent output resistance of the driver R_{tr} equals the magnitude of the lossy characteristic impedance of the line $|Z_{line}|$, or $Z_T = 1$,

$$Z_T = \frac{R_{tr}}{|Z_{line}|}, \tag{3.10}$$

$$R_{tr} = \frac{V_{dd}}{k_n(V_{dd} - V_{tn})^{\alpha}} + \frac{V_{dd}}{k_n[2(V_{dd} - V_{tn})V_{dd} - \frac{V_{dd}^2}{2}]}, \tag{3.11}$$

$$|Z_{line}| = \sqrt{\frac{\sqrt{R_{line}^2 + (\omega L_{line})^2}}{\omega C_{line}}}, \tag{3.12}$$

$$\omega = \frac{2\pi}{t_r}, \tag{3.13}$$

$\alpha = 1.3$ and models the velocity saturation in a short-channel transistor, and t_r is the signal transition time at the output of the driving inverter.

The total propagation delay increases if the line inductance is less than the matched condition. Ignoring the line inductance overestimates the delay and the size of the driver. The line inductance is considered in subsection 4.3.3 in the design of an inverter driving a section of *RLC* interconnect. The savings in both power and area if the line inductance is considered is noted.

3.3.3 Driver Sizing under Inductive Environment. Three different models are used to illustrate the importance of an accurate model to represent both the driver and the interconnect. In Table 4.3, a comparison between the model provided in [14], a lumped *RLC* model, and the π_{21} model (which is described in [25]) is listed.

The line inductance reduces the total signal propagation delay as discussed in previous sections. Including the inductance in the interconnect model is important in the design of an appropriate driver. Excluding the inductance overestimates the delay of the circuit and underestimates the current sourced by the driver. Including the line inductance can reduce the driver size, saving area and power.

A 0.24 μm CMOS technology is used to demonstrate the effect of including line inductance in the design of a line driver. An interconnect line with $R_{line} = 10\,\Omega/mm$, $C_{line} = 105$ fF/mm, and $L_{line} = 650$ pH/mm is assumed to exemplify the reduction in the size of the line driver if inductance is considered. A symmetric CMOS inverter is used to drive a line loaded by a capacitive load of 50 fF to achieve a target delay. The target delay and driver size that achieves this delay are listed in Table 4.4. A reduction in power dissipation of 5% and gate area of 13% is achieved if line inductance is considered. Using low-k dielectric materials and copper interconnect will reduce both the line capacitance and resistance, increasing the effect of inductance on the signal behavior. A 17% reduction in power dissipation and 29% reduction in gate area are achieved for a low-k copper interconnect example.

3.4 Optimum Wire Shape of an *RLC* Interconnect

As described in subsection 4.2, repeaters can be used to minimize the signal propagation delay through a long interconnect. Alternatively, wire shaping can improve circuit speed. For *RC* lines, the optimum line shaping function that minimizes the signal propagation delay is an exponential function [16]. Wire tapering increases the interconnect width

Table 4.3. Propagation delay using different models for different line inductances

L_{line} nH	Cadence (psec)	Ismail [14] psec	Ismail [14] Err	Lumped psec	Lumped Err	$\pi 21$ psec	$\pi 21$ Err
	$W_n = 20$ μm, $R_{line} = 50$ Ω						
	$C_{line} = 400$ fF						
0.0	77	35.2	-54.2	59.5	-22.7	68.2	-11.3
1.0	74	35.7	-51.7	62.2	-15.8	70	-5.4
2.0	67	38.1	-43	66.6	-0.5	64.7	-3.4
3.0	68	41.4	-39	97.8	43.8	63.3	-6.8
4.0	70	45	-35.6	101.8	45.5	72.6	3.8
5.0	73	48.6	-33.4	105.2	44.1	80.3	10
Maximum		-54.27		-45.95		-11.35	
Average		41.51		30.75		6.11	
	$C_{line} = 1$ pF						
0.0	148	86.3	-41.6	97.8	-33.8	130.1	-12
1.0	147	86.4	-41.2	93.3	-36.5	129.6	-11.7
2.0	153	87.0	-43	86.5	-43.4	131.5	-14
3.0	145	88.7	-38.7	81.1	-44	134.2	-7.3
4.0	132	91.1	-30.9	76.6	-41.9	122.5	-7.1
5.0	118	94	-20.2	96.8	-17.8	117.1	-0.7
Maximum		-43		-45.64		-14	
Average		36.23		38.69		8.91	

Table 4.4. Reduction in area and power dissipation when considering line inductance for different dielectric and line materials

Dielectric Material	Resistivity	Target Delay (psec)	W_n (μm) *RC*	W_n (μm) *RLC*	Reduction in power	Reduction in area
SiO_2	Aluminum	100	19	16.5	5%	13%
	Copper	100	17.8	15.2	6%	15%
Low-K	Aluminum	60	23	19	9%	17%
	Copper	60	21	15	17%	29%

at the driver end of the line as shown in Fig. 4.23. Tapering reduces the total line resistance, increasing the inductive behavior of the line. In the following subsections, the efficiency of tapering an *RLC* interconnect is discussed. In subsection 4.4.1, the optimum wire shape that produces the minimum signal propagation delay of an *RLC* line is characterized. Different constraints on interconnect tapering are discussed in subsection 4.4.2. In section 4.4.3, a comparison between tapered RC and RLC lines is presented.

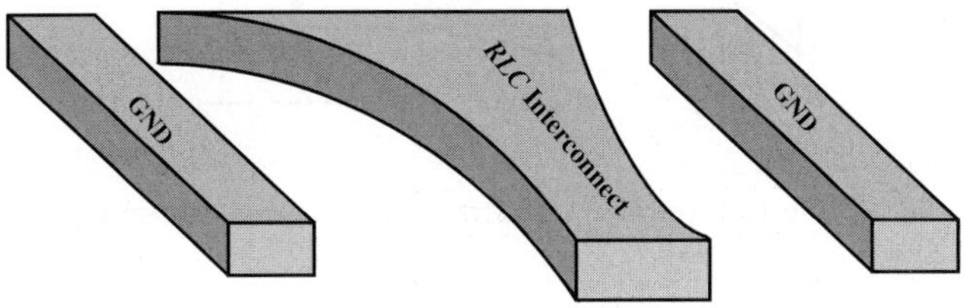

Figure 4.23. Coplanar tapered RLC interconnect

3.4.1 Optimum Wire Shape for Minimum Propagation Delay. The signal propagation delay of a distributed RLC interconnect is described in [7, 14]. Two time constants characterize the signal speed and shape of long interconnects, the resistive-capacitive (RC) time constant and the inductive-capacitive (LC) time constant (or the time of flight through the line $t_f = \sqrt{L_{int}C_{int}}$, where C_{int} and L_{int} are the line capacitance and inductance, respectively). For highly resistive (less inductive) lines, an RC delay model is adequate to characterize the signal delay. The optimum tapering factor for these lines is an exponential tapering factor [16]. If the inductive behavior of a line dominates the resistive behavior, the time-of-flight can dictate the time for the signal to propagate through the line. The optimum shape that minimizes the propagation delay of a line is the shape function that minimizes the time-of-flight.

The line inductance and capacitance per unit length, respectively, can be expressed in terms of the line width by the simple relations,

$$L_{int}(W) = \frac{L_0}{W(x)}, \tag{3.14}$$

$$C_{int}(W) = C_0 W(x) + C_f, \tag{3.15}$$

where L_0 is the line inductance per square, C_f is the fringing capacitance per unit length, and C_0 is the line capacitance per unit area. $W(x)$ is the line width as a function of x, the distance from the load as shown in Fig. 4.24.

The time-of-flight of the signal is

$$t_f = \sqrt{\int_0^l \frac{L_0}{W(x)} \int_0^x (C_0 W(y) + C_f) dy dx}, \tag{3.16}$$

where l is the line length. If functions F and $u(x)$ are defined as

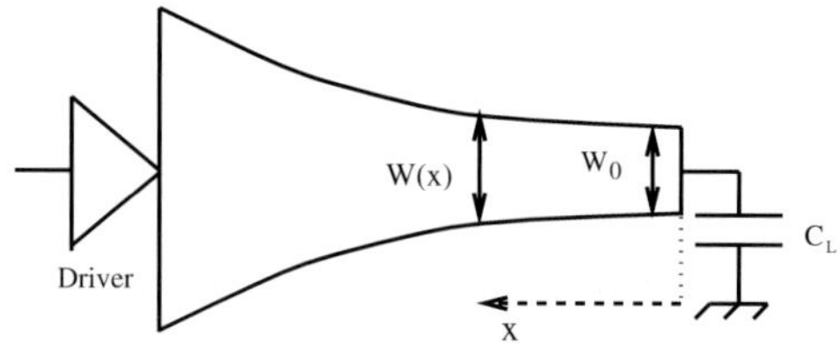

Figure 4.24. *RLC* line tapered by a general width tapering function $W(x)$

$$F \equiv \frac{L_0}{W(x)} \int_0^x (C_0 W(y) + C_f) dy,$$
$$u(x) \equiv \int_0^x W(y) dy,$$

respectively, and Euler's differential equation is used to minimize (3.16), as similarly described in [16], the optimum $u(x)$ should satisfy the differential equation,

$$u'(x) = \frac{2L_0C_0}{c} u(x) + \frac{2C_f L_0 l}{c}. \tag{3.17}$$

Thus,

$$W(x) = W_0 \, e^{\frac{2L_0C_0}{c} x}, \tag{3.18}$$

where $c = \frac{2C_f L_0 l}{W_0}$. W_0 is obtained by substituting (3.18) into (3.16) and differentiating (3.16) with respect to W_0. Setting the result to zero produces a nonlinear equation which can be solved numerically.

As shown in (3.18), the optimum tapering function of the width of an *LC* line is an exponential function. For either an *RC* or *LC* line, the general form of the optimum shaping function that minimizes the propagation delay is an exponential function. The *RC* and *LC* models are the two limiting cases of a general *RLC* interconnect. The optimum tapering function of an *RLC* line must satisfy the general exponential form $W(x) = qe^{px}$, where q is the line width at the load end, as shown in Fig. 4.24, and p is the tapering factor. The optimum value of q and p for an *RLC* line reduces to the value given in [16] if the line inductance is negligible (an *RC* line) and to the value given in (3.18) if the line resistance is negligible (an *LC* line). The optimum value of q and p for an *RLC* line is between these two limits.

As described in [25], for an *RLC* line, the signal propagation delay is minimum when the line is matched with the driver. The matching condition, from [25], is

$$R_{tr} = |Z_{line}(q,p)| \,. \tag{3.19}$$

$Z_{line}(q,p)$ is the lossy characteristic impedance of the line, given in (3.12), as a function of q and p. t_r is the signal transition time at the near end of the line which is determined from the reduced order model described in [25].

As there is one equation, (3.19), and two unknowns, q and p, there are two degrees of freedom to design an optimum RLC line tapered for minimum delay. For a width q, there is an optimum tapering factor p_{opt} which satisfies (3.19) and at which the propagation delay is minimum. Other design constraints, such as the minimum and maximum line width and power dissipation, are discussed in subsection 4.4.2 to determine a power efficient solution.

3.4.2 Constraints on Optimum Tapering for *RLC* Lines.

Tapering an interconnect assigns a small width for the line at the far end. The line width increases at the near end as shown in Fig. 4.24. As discussed in section 2, the width increases exponentially to obtain the minimum propagation delay. By choosing q and solving (3.19) as a nonlinear equation in one unknown, the optimum tapering factor p_{opt} can be determined. There are two practical limits for choosing q,

1 $q \geq W_{min}$, where W_{min} is the minimum wire width of a target technology.

2 $q \leq W_{max}e^{-pl}$, where W_{max} is the maximum wire width of a target technology.

These two constraints should be satisfied when designing a tapered line. q cannot be lower than the minimum wire width allowed by the technology. Alternatively, increasing q may result in a width at the near end (the largest width of the line) which may be greater than the maximum available wire width.

Another important design constraint is power dissipation. Wire sizing affects the two primary transient power components, the dynamic power dissipated in charging and discharging the line capacitance and the short-circuit power dissipated within the load gate. The short-circuit power is minimum when the line is matched with the driver [17, 21], which is also the optimum solution for minimum delay.

The dynamic power is directly proportional to the line capacitance. To decrease the line capacitance, the line width should be as narrow as possible, as the line capacitance increases superlinearly with the width

[29]. In order to satisfy both high speed and low power design objectives, q should be chosen equal to W_{min}. The optimum value for the tapering factor p_{opt} is obtained by solving (3.19) for $q = W_{min}$. Optimum wire tapering is compared in subsection 4.4.3 with uniform wire sizing for both RC and RLC lines.

3.4.3 Tapering versus Uniform Wire Sizing in *RC* and *RLC* Lines.

Interconnect tapering is more efficient in RLC lines than in RC lines. Two effects reduce the signal propagation delay of an exponentially tapered RLC line. The first effect is the shape of the line structure which minimizes both the RC and LC time constants.

The second effect is an increase in the inductive behavior of the line. Tapering an interconnect line decreases the line resistance, reducing the attenuation along the line. This effect increases the inductive behavior of the line. The inductive behavior of the line can be characterized by $\zeta = \frac{R_{line}}{2}\sqrt{\frac{C_{line}}{L_{line}}}$, the damping factor of a line [6]. As described in [6], when $\zeta < 1.0$, the inductive behavior of a line cannot be ignored. As shown in Fig. 4.25, the damping factor decreases as the line tapering factor increases, making the line behave more inductively. For $\zeta > 1.0$ (the dotted lines), the damping factor does not exhibit inductive behavior since the line is underdamped. The inductive effect of a line with $\zeta > 1.0$ is negligible.

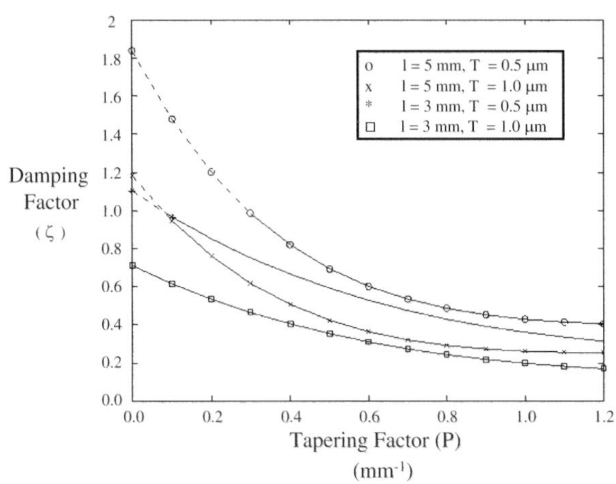

Figure 4.25. Interconnect damping factor as a function of tapering factor for different line parameters

The line inductance shields part of the line capacitance and decreases the equivalent output resistance of the gate that drives the line. The signal propagation delay decreases as the inductive behavior of the line becomes more pronounced [25]. This effect makes line tapering more attractive in long *RLC* lines.

Another criterion to optimize the interconnect width for minimum propagation delay is uniform wire sizing. A minimum width coplanar interconnect line is illustrated in Fig. 4.26 for two sizing criteria.

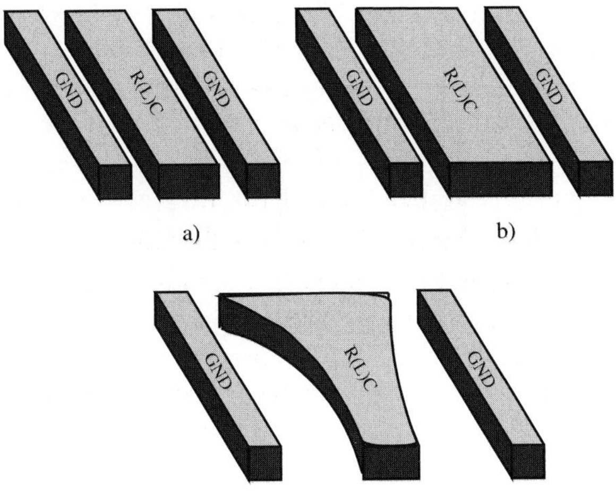

Figure 4.26. Coplanar interconnect a) minimum width b) uniform sizing c) exponential tapering

Exponential wire tapering outperforms uniform wire sizing as discussed in subsection 4.4.2. As with wire tapering, uniform wire sizing decreases the line resistance, making the inductive behavior more pronounced; however, the superlinear increase in the line capacitance limits the effect of the line inductance on reducing the signal propagation delay. Wire tapering, however, produces a smaller delay than the delay achieved from uniform wire sizing. Optimum wire tapering produces a greater delay reduction in *RLC* lines than in *RC* lines as the inductive behavior of the line further decreases the delay. The line inductance makes tapering more efficient than uniform wire sizing in *RLC* lines.

For an *RLC* line, tapering not only reduces the propagation delay, but also decreases the total power dissipation as compared to uniform wire sizing. An increase in the inductive behavior of the line reduces the signal transition time at the load, reducing the short-circuit current and, consequently, the total transient power dissipation [17, 20]. Simulation results are presented here to illustrate the efficiency of exponential wire tapering on both the propagation delay and power dissipation of *RLC* lines.

Different circuits have been investigated. The minimum delay is determined for both uniform wire sizing and exponential line tapering. As shown in Fig. 4.27, wire tapering outperforms uniform wire sizing for all circuits. For an *RLC* line, a greater reduction in the minimum delay is achieved as compared to an *RC* line.

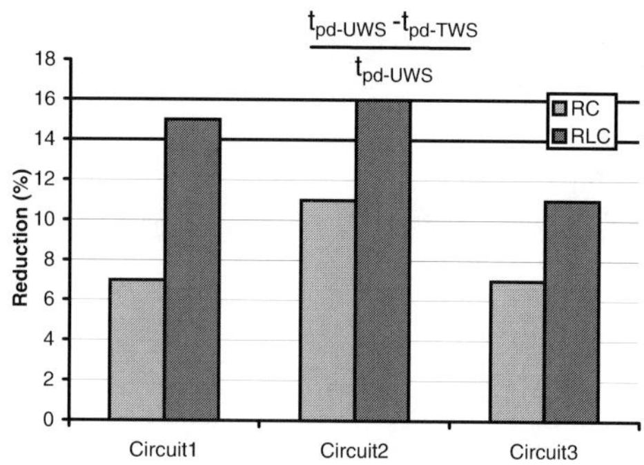

Figure 4.27. Reduction in propagation delay. UWS is Uniform Wire Sizing and TWS is Tapered Wire Sizing

In addition to a smaller propagation delay, the total transient power dissipation is less. A tapered line with q equal to the minimum width reduces the total line capacitance, decreasing the dynamic power (as compared to uniform wire sizing). Furthermore, the reduction in short-circuit power, due to the line inductance, increases the savings in total power dissipation. The relative reduction in power dissipation for an *RC* and *RLC* line is compared in Fig. 4.28.

A reduction in power dissipation of 16% and in propagation delay of 15% for an *RLC* line as compared to 11% and 7% for an *RC* line is achieved when optimum tapering is applied rather than uniform wire sizing.

4. Future Issues in Interconnect Design

With the increase in clock frequencies, different mechanisms affect the signal characteristics in long interconnects. These mechanisms alter the impedance characteristics of the interconnect. At high frequencies,

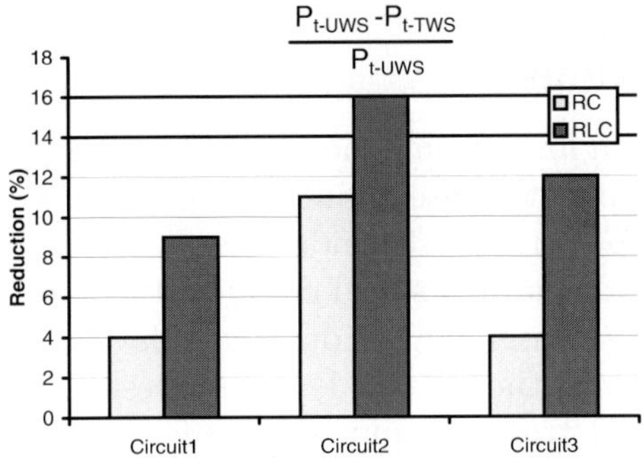

Figure 4.28. Reduction in total power dissipation

the current distribution along the interconnect cross section changes. The change in the current distribution affects the flow of current due to related phenomena such as skin and proximity effects. These effects cause a change in the line resistance and inductance. Advanced design methodologies are currently under development to consider the change in the line impedance parameters when the frequency exceeds the limit at which current interconnect models become inaccurate.

In this chapter, some of the most widely used criteria to drive long interconnects are considered. There are other criteria which use different techniques to enhance circuit speed. Current sensing and boosters [30, 31] are two promising techniques. Despite the limitations of these techniques, these approaches can be useful in certain applications. The line inductance should be considered if these techniques are applied in circuits that operate at high frequencies.

The leakage power in active elements which drive interconnect is often neglected. Leakage power is expected to become a significant portion of the total power dissipation in an integrated circuit. In future technology generations, leakage power should be considered in order to develop more efficient, high performance, low power design techniques.

From the discussion presented in subsection 4.4, tapering is an efficient technique to reduce signal delay. This technique could be expanded by tapering the line in three dimensions. Tapering the interconnect height can improve the performance of the circuit. However, implementing three-dimensional tapering would greatly complicate the manufacturing process.

5. Conclusions

Interconnect design has become more important as operating frequencies have increased since the interconnect affects the two primary design criteria, speed and power. Furthermore, line inductance cannot be neglected in next generation high speed circuits. Including line inductance in the design process can enhance both the delay and power as well as improve the accuracy of the overall design process. The line inductance introduces new circuit tradeoffs and design methodologies.

A tradeoff exists between dynamic and short-circuit power in inductive interconnect. This tradeoff is not significant in resistive lines as the signal characteristics are less sensitive to the line dimensions. The short-circuit power of an overdriven interconnect line decreases with line width, while the dynamic power increases. When the line exceeds the matched condition, not only the dynamic power but also the short-circuit power increases with increasing line width. The matched condition between the driver and the load has an important effect on the line impedance characteristics. If the line is overdriven, the short-circuit power decreases with increasing line width. When the line exceeds the matched condition, the short-circuit power increases with increasing line width (and signal transition time). To achieve lower transient power dissipation, the minimum line width should be used if the line is underdriven. For a long inductive interconnect line, an optimum interconnect width exists that minimizes the total transient power dissipation.

Repeater insertion outperforms wire sizing in RC lines. However, for RLC lines, the minimum signal propagation delay always decreases with increasing wire width if an optimum repeater system is used. In RLC lines, wire sizing outperforms repeater insertion as the minimum signal propagation delay with no repeaters is less than the minimum signal propagation delay using any number of repeaters. The minimum signal propagation delay always decreases with wider lines until the number of repeaters equals zero. In RLC lines, there is no optimum interconnect width for minimum signal propagation delay. The total transient power dissipation of a repeater system driving an RLC line is minimum at small line widths. Below the width for minimum power, both the signal delay and the power dissipation increase. Widening a line beyond the width for minimum power reduces the number of repeaters and the minimum signal propagation delay while increasing the total transient power dissipation. A tradeoff between the transient power dissipation and the signal propagation delay, therefore, exists in sizing the interconnect width. Optimizing the interconnect for minimum power delay product produces a much smaller increase in both the power and delay as compared to separately optimizing for either the power or delay. As interconnects becomes longer, the difference between the optimum width for minimum power and the optimum width for minimum delay increases, further enhancing the effectiveness of the proposed criterion.

On-chip inductance shields part of the interconnect capacitance. The effective capacitance of an *RLC* load decreases with increasing line inductance, reducing the gate delay of a driver. Furthermore, the line inductance reduces the equivalent output resistance of a driver, reducing the total propagation delay. A parameter Z_T, the ratio of the output driver resistance and the magnitude of the lossy characteristic impedance of the line, is introduced to characterize the signal propagation delay of a CMOS inverter driving an *RLC* interconnect. The minimum propagation delay is achieved when $Z_T = 1$ where the driver is matched with the lossy characteristic impedance of the line. A smaller buffer can be used to drive an interconnect line if the line inductance is considered, more accurately achieving the target delay than if the line inductance is ignored.

The optimum wire shape that produces the minimum signal propagation delay in a distributed *RLC* line is shown to be an exponential function. An exponentially tapered interconnect minimizes the time of flight of an *LC* line. The general form for the optimum shaping function of an *RLC* line is qe^{px}. The optimum wire width at the load end q and the optimum tapering factor p which achieve the minimum delay and low power are determined for the driver and load characteristics. Optimum wire tapering as compared to uniform wire sizing is more efficient in *RLC* lines than in *RC* lines. The line inductance makes tapering more attractive in *RLC* lines since tapering produces a greater reduction in delay as compared to uniform wire sizing. With a minimum wire width at the far end and an optimum tapering factor, both the propagation delay and the power dissipation are reduced. The line inductance increases the savings in power in an optimally tapered line as compared to uniform wire sizing.

On-chip inductance must be included in the design process in high frequency circuits. By including on-chip inductance, the efficiency of different circuit design techniques such as wire sizing, repeater insertion, line tapering, and driver sizing can be greatly enhanced.

References

[1] *National Technology Roadmap for Semiconductors*: Semiconductor Industry Association, 1997.

[2] *International Technology Roadmap for Semiconductors*: Semiconductor Industry Association, Edition 2001.

[3] S. Borkar, "Obeying Moore's Law beyond 0.18 Micron," *Proceedings of the ASIC/SOC Conference*, pp. 26-31, September 2000.

[4] A. V. Mezhiba and E. G. Friedman, "Trade-offs in CMOS VLSI Circuits," *Trade-offs in Analog Circuit Design The Designer's Companion*, C. Toumazou, G. Moschytz, and B. Gilbert (Eds.), Dordrecht, The Netherlands:Kluwer Academic Publishers, pp. 75-114, 2002.

[5] T. Sakurai, "Approximation of Wiring Delay in MOSFET LSI," *IEEE Journal of Solid-State Circuits*, Vol. SC-18, No. 4, pp. 418-426, August 1983.

[6] Y. I. Ismail, E. G. Friedman, and J. L. Neves, "Figure of Merit to Characterize the Importance of On-Chip Inductance," *IEEE Transactions on Very Large Scale Integration (VLSI) Systems*, Vol. 7, No. 4, pp. 442-449, December 1999.

[7] J. A. Davis and J. D. Meindl, "Compact Distributed *RLC* Interconnect Models–Part I: Single Line Transient, Time Delay, and Overshoot Expressions," *IEEE Transactions on Electron Devices*, Vol. 47, No. 11, pp. 2068-2077, November 2000.

[8] A. Deutsch *et al.*, "On-Chip Wiring Design Challenges for GHz Operation," *Proceedings of the IEEE*, Vol. 89, No. 4, pp. 529-555, April 2001.

[9] J. J. Cong and K. Leung "Optimal Wiresizing Under Elmore Delay Model," *IEEE Transactions on Computer-Aided Design of Integrated Circuits and Systems*, Vol. 14, No. 3, pp. 321-336, March 1995.

[10] H. B. Bakoglu and J. D. Meindl, "Optimal Interconnection Circuits for VLSI," *IEEE Transactions on Electron Devices*, Vol. ED-32, No. 5, pp. 903-909, May 1985.

[11] B. S. Cherkauer and E. G. Friedman, "A Unified Design Methodology for CMOS Tapered Buffers," *IEEE Transactions on Very Large Scale Integration (VLSI) Systems*, Vol. VLSI-3, No. 1, pp. 99-111, March 1995.

[12] V. Adler and E. G. Friedman, "Repeater Design to Reduce Delay and Power in Resistive Interconnect," *IEEE Transactions on Cir-*

cuits and Systems II: Analog and Digital Signal Processing, Vol. 45, No. 5, pp. 607-616, May 1998.

[13] C. J. Alpert, A. Devgan, J. P. Fishburn, and S. T. Quay, "Interconnect Synthesis Without Wire Tapering," *IEEE Transactions on Computer-Aided Design of Integrated Circuits and Systems*, Vol. 20, No. 1, pp. 90-104, January 2001.

[14] Y. I. Ismail and E. G. Friedman, "Effects of Inductance on the Propagation Delay and Repeater Insertion in VLSI Circuits," *IEEE Transactions on Very Large Scale Integration (VLSI) Systems*, Vol. 8, No. 2, pp. 195-206, April 2000.

[15] K. Banerjee and A. Mehrotra, "Analysis of On-Chip Inductance Effects for Distributed *RLC* Interconnects," *IEEE Transactions on Computer-Aided Design of Integrated Circuits and Systems*, Vol. 21, No. 8, pp. 904-915, August 2002.

[16] J. P. Fishburn and C. A. Schevon, "Shaping A Distributed-*RC* Line to Minimize Elmore Delay," *IEEE Transactions on Circuits and Systems I: Fundamental Theory and Applications*, Vol. 42, No. 12, pp. 1020-1022, December 1995.

[17] M. A. El-Moursy and E. G. Friedman, "Optimizing Inductive Interconnect for Low Power," *System-on-Chip for Real-Time Applications*, W. Badawy and G. A. Jullien (Eds.), Kluwer Academic Publishers, pp. 380-391, 2003.

[18] G. Cappuccino and G. Cocorullo, "A Time-domain Model Power Dissipation of CMOS Buffer Driving Lossy Lines," *Electronics Letters*, Vol. 35, No. 12, pp. 959-960, June 1999.

[19] E. B. Rosa, "The Self and Mutual Inductances of Linear Conductors," *Bulletin of the National Bureau of Standards*, Vol. 4, No. 2, pp. 301-344. Government Printing Office, Washington, January 1908.

[20] Y. I. Ismail, E. G. Friedman, and J. L. Neves, "Exploiting On-Chip Inductance in High Speed Clock Distribution Networks," *IEEE Transactions on Very Large Scale Integration (VLSI) Systems*, Vol. 9, No. 6, pp. 963 - 973, December 2001.

[21] M. A. El-Moursy and E. G. Friedman, "Inductive Interconnect Width Optimization For Low Power," *Proceedings of the IEEE International Symposium on Circuits and Systems*, Vol. 5, pp. 273-276, May 2003.

[22] M. A. El-Moursy and E. G. Friedman, "Optimizing Inductive Interconnect for Low Power," *Canadian Journal of Electrical and Computer Engineering*, pp. 183-187, Vol. 27, No. 4, October 2002.

[23] M. A. El-Moursy and E. G. Friedman, "Optimum Wire Sizing and Repeater Insertion in Distributed *RLC* Interconnect," *Proceedings of the 26th Annual IEEE EDS/CAS Activities in Western New York Conference*, p. 6, November 2002.

[24] M. A. El-Moursy and E. G. Friedman, "Optimum Wire Sizing of *RLC* Interconnect With Repeaters," *Proceedings of the IEEE Great Lakes Symposium on VLSI*, pp. 27-32, April 2003.

[25] M. A. El-Moursy and E. G. Friedman, "Shielding Effect of On-Chip Interconnect Inductance," *Proceedings of the IEEE Great Lakes Symposium on VLSI*, pp. 165-170, April 2003.

[26] P. R. O'Brien and T. L. Savarino, "Efficient On-Chip Delay Estimation for Leaky Models of Multiple-Source Nets," *Proceedings of the IEEE Custom Integrated Circuits Conference*, pp. 9.6.1-9.6.4, May 1990.

[27] X. Yang, C-K. Chang, W. H. Ku, and R. J. Carragher, "Hurwitz Stable Reduced Order Modeling for *RLC* Interconnect Trees," *Proceedings of the IEEE/ACM International Conference on Computer-Aided Design*, pp. 222-228, November 2000.

[28] J. Qian, S. Pullela, and L. Pillage, "Modeling the 'Effective Capacitance' for the *RC* Interconnect of CMOS Gates," *IEEE Transactions on Computer-Aided Design of Integrated Circuits and Systems*, Vol. 13, No. 12, pp. 1526-1535, December 1994.

[29] N. Delorme, M. Belleville, and J. Chilo, "Inductance and Capacitance Analytic Formulas for VLSI Interconnects," *Electronics Letters*, Vol. 32, No. 11, pp. 996-997, May 1996.

[30] A. Nalamalpu, S. Srinivasan, and W. Burleson, "Boosters for Driving Long On-Chip Interconnects– Design Issues, Interconnect Synthesis and Comparison with Repeaters," *IEEE Transactions on Computer-Aided Design of Integrated Circuits and Systems*, Vol. 21, No. 1, pp. 50-62, January 2002.

[31] R. M. Secareanu and E. G. Friedman, "Transparent Repeaters," *Proceedings of the IEEE Great Lakes Symposium on VLSI*, pp. 63-66, March 2000.

Chapter 5

CLOCK DISTRIBUTION FOR HIGH PERFORMANCE DESIGNS

Stefan Rusu
Intel Corporation, Santa Clara, CA

1. INTRODUCTION

The clock is a periodic synchronization signal used as a time reference for data transfers in synchronous digital systems. Since the clock plays a central role in the operation of a synchronous system, significant effort is invested in the design, optimization and verification of high performance clock distribution schemes. Since the clock edges determine the state updates in a synchronous system, higher clock frequencies are generally (but not always) associated with a higher system performance. As Figure 6-1 shows, clock frequency for desktop microprocessors has increased significantly over time, with the latest microprocessors running at above 3GHz in 0.13μm technology. This increase is driven by the process technology scaling, aggressive circuit design techniques and deeper pipelines. The clock distribution is particularly affected by process scaling that increases the resistance of long interconnect lines. This trend is aggravated by the increase in die size. Indeed, even as transistor geometries shrink, high-performance digital systems have an increasing number of execution units and larger caches. The clock needs to be distributed to all the circuits on the die, so the clock lines are getting longer and require more buffering levels.

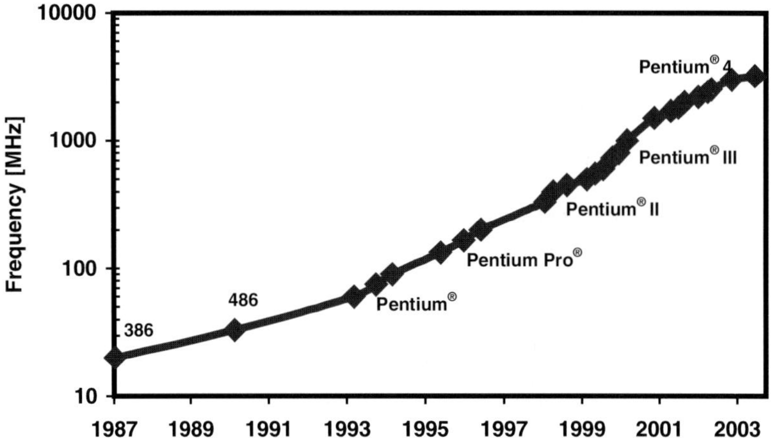

Figure 5-1. Processor Frequency Trend

This chapter is organized as follows: Section 2 defines the main characteristics of a clock distribution network and how they have evolved over time. Section 3 reviews several clock distribution schemes, with specific examples from high-performance microprocessor designs. Section 4 presents several deskew circuits, while Section 5 describes jitter reduction circuit techniques. Since power is a limiting factor in most digital designs today, we review several low-power clock distribution ideas in Section 6. Finally, Section 7 discusses several future directions in clock distribution, including distributed VCOs and PLLs, as well as rotary and standing wave clock distribution.

2. CLOCK PARAMETERS AND TRENDS

Figure 6-2 shows the main parameters of a clock distribution network. The *clock skew* is the spatial variation of the clock signal as distributed through the chip. We distinguish between global (chip-level) and local (block level) skew. A proper clock distribution design typically attempts to reduce the clock skew to zero across the entire chip. However, as we will show in Section 4, intentional skew insertion is sometimes used to relieve timing critical paths, which allows a chip to run at a higher frequency.

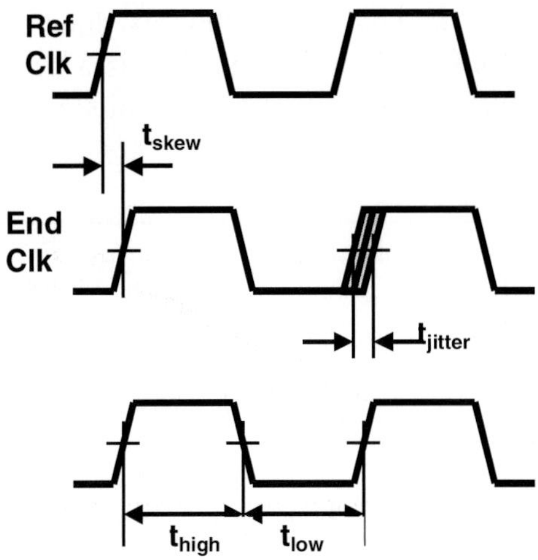

Figure 5-2. Characteristic parameters for a clock system

Figure 6-3 shows the clock skew as a function of the cycle time. The points in the chart represent the skew vs. operating frequency for various large processors, as reported in ISSCC or JSSC papers by major industry players. As expected, the skew decreases as the frequencies are increased.

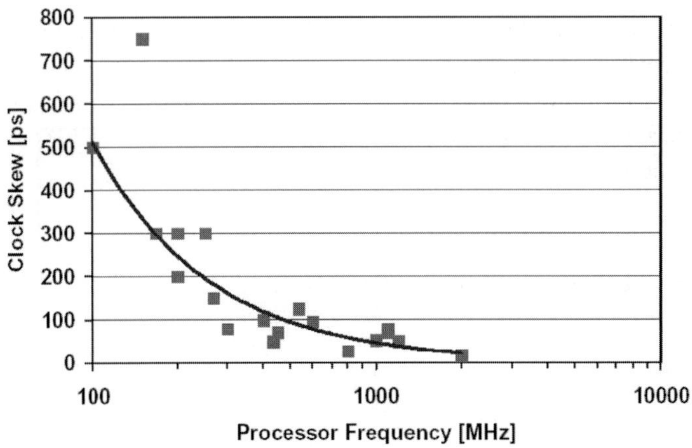

Figure 5-3. Clock skew trends

A better way to measure the clock skew is to look at it as a percentage of the cycle time, as shown in Figure 6-4. Notice that the skew has been averaging about 5% of the cycle time, although a wide variability exists between different designs. As the frequencies cross the 1GHz border, it becomes more difficult to hold on to this 5% limit. Going forward, we expect the clock skews to continue to increase as a percentage of the cycle time, even with the extensive use of deskew circuits, which will be described in Section 4.

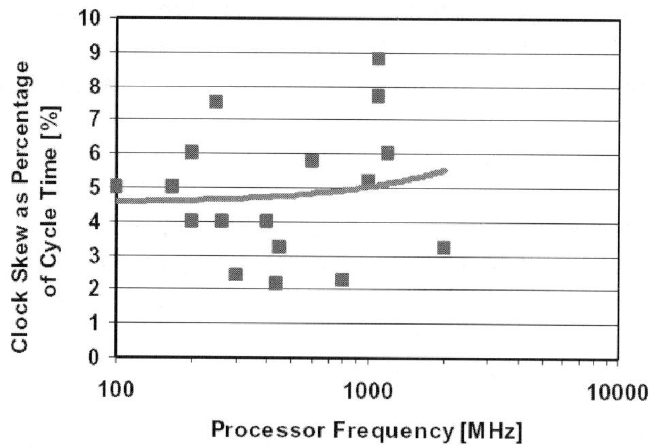

Figure 5-4. Clock skew as percentage of the cycle time

The main sources of clock skew are shown in Figure 6-5 [1]. Notice that more than half of the skew is caused by device mismatches. The difference in the local supply levels for the intermediate clock buffers accounts for about a quarter of the skew.

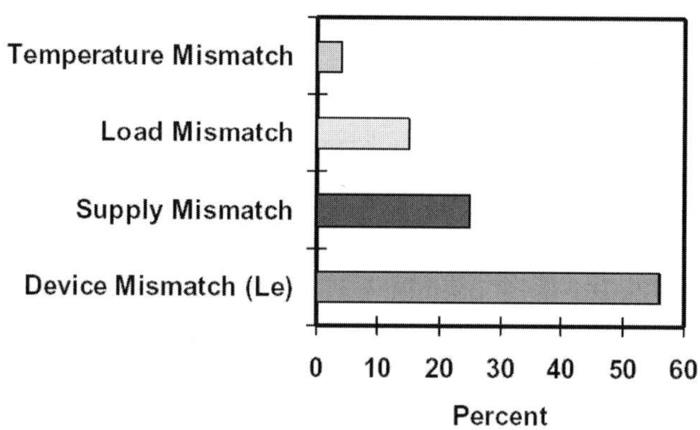

Figure 5-5. Clock skew sources [1]

The load mismatch is due to imbalances in the clock distribution trees and can be corrected by investing additional design effort. However, most of this extra effort is manual, so typically design teams accept the residual skew that remains after balancing out the clock trees using automatic tools. Another approach is to use a clock grid that shorts all the clock end-points. As we will see in Section 3, clock grids have lower skews but consume significantly more power. Finally, the temperature mismatch has a small impact on the clock skew. This is good news for the clock designers, since large temperature gradients exist in modern high performance designs.

The *clock jitter* is the temporal variation of the clock signal with respect to a reference edge. We differentiate between long-term jitter (that accumulates over a long period of time) and cycle-to-cycle jitter, which is measured between adjacent clock cycles. Long-term jitter accumulates as a phase error relative to the reference clock and degrades setup and hold timings for the entire circuit. The cycle-to-cycle short-term jitter is seen by the logic circuits as a frequency shift, similar to the clock skew.

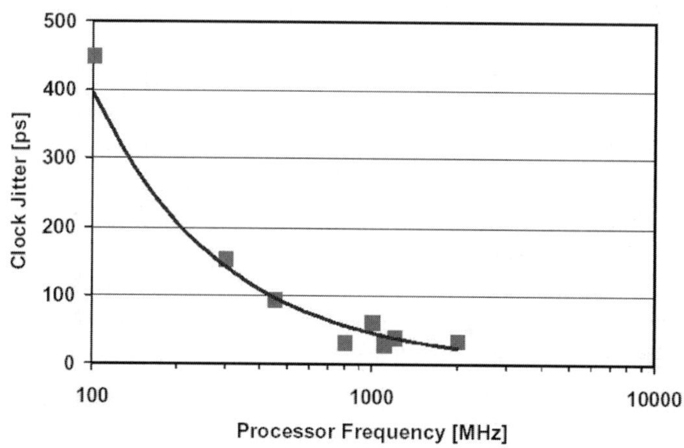

Figure 5-6. Clock jitter trend

Figure 6-6 shows the clock jitter trend as a function of the processor frequency as reported in major processor papers at ISSCC or JSSC. Notice that this graph has fewer data points, since not all papers report on the clock jitter results.

The main sources of clock jitter are the power supply noise coupling into the voltage-controlled oscillator (long term jitter) and the supply noise modulation of the clock network buffer delay (short term cycle-to-cycle or multi-cycle jitter). Jitter reduction techniques will be discussed in Section 5.

The clock *duty cycle* is the ratio of the clock high and low times. Ideally we would like the duty cycle to be 50/50, although small intentional deviations may enable higher operating frequencies in phase-based designs. The most common way for achieving a 50/50 duty cycle is to synthesize a double clock frequency and divide it by two before the distribution tree.

3. CLOCK DISTRIBUTION NETWORKS

Figure 6-7 shows several clock distribution options. The most common distribution network is the *tree*, where buffers are inserted along the clock distribution path forming a tree structure. All paths from the root of the tree to all the branches have an identical number of buffers, although their sizes may be adjusted to match the different loads. The number of buffer stages between the tree root and the clocked registers depends on the total capacitive loading, metal resistance and allowed skew. To further reduce skew, we can short the outputs of the intermediate buffers, creating a *mesh* clock tree structure.

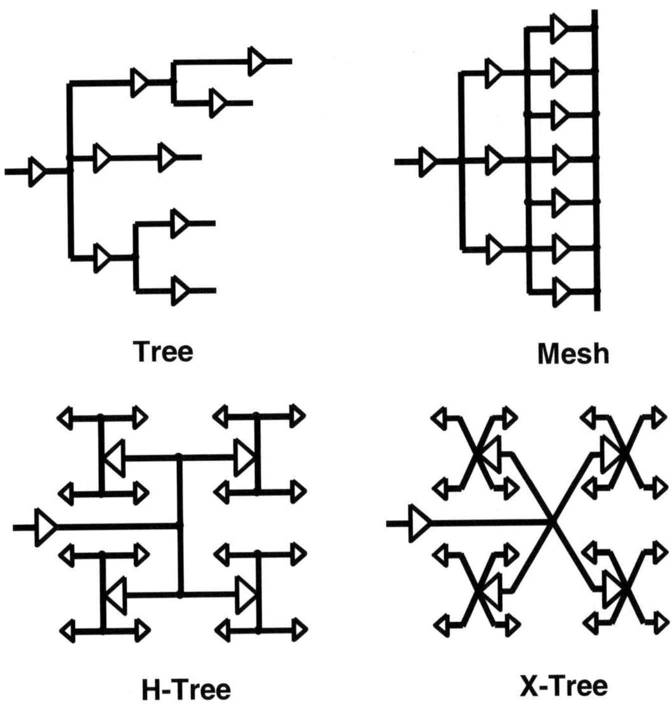

Figure 5-7. Common clock distribution structures

Another approach to ensure zero clock skew uses hierarchical *H* or *X-tree* structures. In this approach, the clock is driven from the center of an H structure to its four corners. Each corner drives another, smaller H structure. This distribution process is repeated through progressively smaller, hierarchical H structures. The end-points of the smallest H shape drive the clocked registers. Due to the symmetry of the H or X-trees, the path from the clock source to each end-point has the same delay. As the interconnect resistance increases with the technology scaling, adequate intermediate buffers are required along the H-tree distribution.

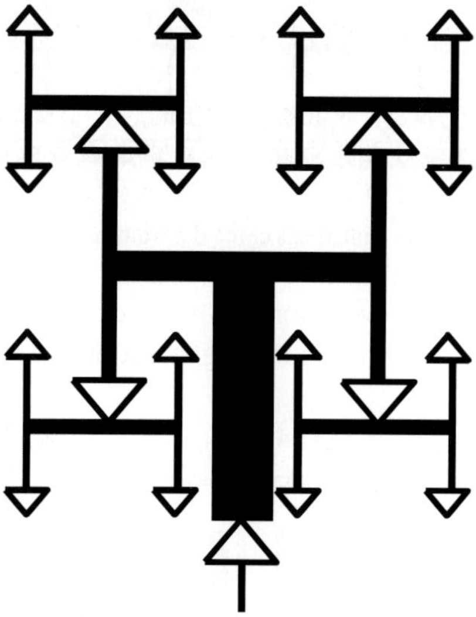

Figure 5-8. Tapered H-tree

A *tapered H-tree*, shown in Figure 6-8, matches the clock line impedance to minimize reflections at branching points. The width of the clock trunk decreases as the signal propagates through the tree. The impedance of the line exiting each branch point is designed to be twice the impedance of the line feeding the branch point. Notice that perfectly symmetrical H-trees are difficult to implement in actual designs due to floorplan constraints. Actual clock trees require careful extraction and characterization to achieve a balanced design.

Figure 6-9 shows the clock distribution of the Pentium® 4 Processor [2]. The clock is distributed using a triple spine approach to cover the large die. Each spine contains a binary distribution tree, with each of the 47 leaf nodes providing an independent domain clock. Local clock drivers are used to buffer the clock load as well as produce the proper frequency and clock type for each particular block. The drivers are connected to the appropriate domain clocks through delay-matched taps. The maximum RC delay from the output of the local drivers to the input of a latch is restricted to less than 38ps to minimize the local clock skew.

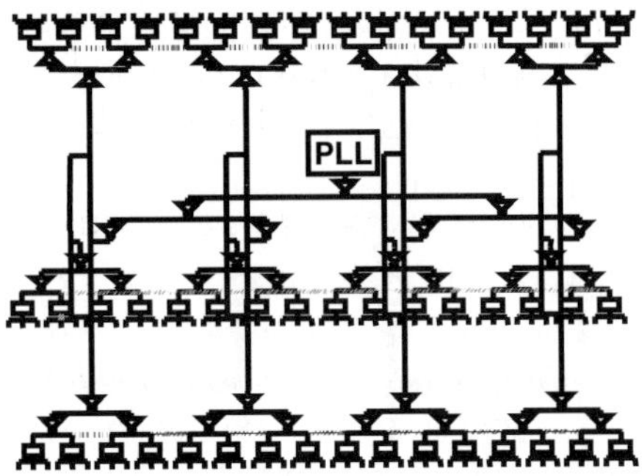

Figure 5-9. Clock distribution for the Pentium® 4 Processor [2]

Figure 6-10 shows the evolution of the clock distribution network for the Alpha microprocessor [3] through three generations of the design.

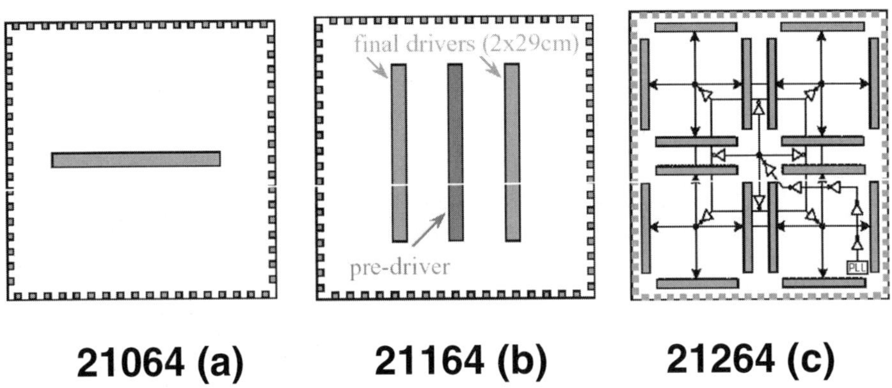

Figure 5-10. Evolution of the clock distribution network for the Alpha processors [3]

The 21064 had a two-phase single-wire clocking scheme. The driver was located in the center of the die as shown in Figure 6-10 (a). The final clock load was 3.5nF, and it required a final driver with a gate length of 35 cm. To handle the large transient currents in the power grid when the clock driver switched, on-chip decoupling structures were placed around the clock driver. Roughly 10% of the chip area was allocated to decoupling capacitance.

Figure 6-10 (b) illustrates the clock distribution on the 21164. The main clock driver was split into two banks and was placed midway between the center of the die and the edges. The pre-driver was located in the center of the die to distribute the clock to the two main drivers. The clock skew was reduced by a factor of 2 using this approach. In addition, by distributing the main clock driver over a larger area, the localized heating seen on the 21064 was reduced.

In the 21264, the power consumption became a major concern in designing the clocking system. To reduce it, a single wire global clock (GCLK) was routed over the entire chip as a global timing reference. The GCLK drivers were distributed around the four quadrants as shown in Figure 6-10 (c), to reduce clock grid delay and distribute clock power. The GCLK drives a hierarchy of thousands of buffered and conditioned local clocks used across the chip. There are several advantages to this clocking scheme. First, conditioning the local clocks saves power. Second, circuit designers can take advantage of multiple clocks to add local skew that benefits timing critical paths. Finally, using local buffering significantly lowers the GCLK load, which reduces GCLK skew to less than 75ps.

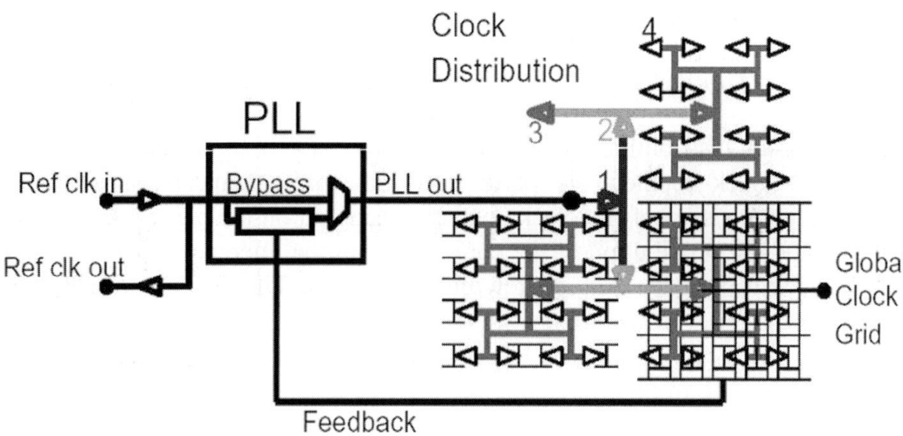

Figure 5-11. Power4 microprocessor clock distribution [4]

The Power4 microprocessor clock distribution is shown in Figure 6-11 [4]. Considering the complexity of this 174M-transistor dual-processor chip, the clock distribution is relatively simple. A single chip-wide clock domain is used, with no active or programmable skew-reduction circuitry. A single PLL is used near the center of the chip to minimize the global clock distribution delay. The clock is distributed by 64 tuned trees driving a single full-chip clock grid at 1024 points. The grid smoothes out clock skew caused by across-chip process variations, but it does consume more power than a balanced tree structure. Experimental measurements using pico-probes at 19 locations across the die showed a maximum skew of 25ps. Optical probing on 9 of the 64 sector buffers confirmed less than 18ps skew at the leading edge of the photon pulses.

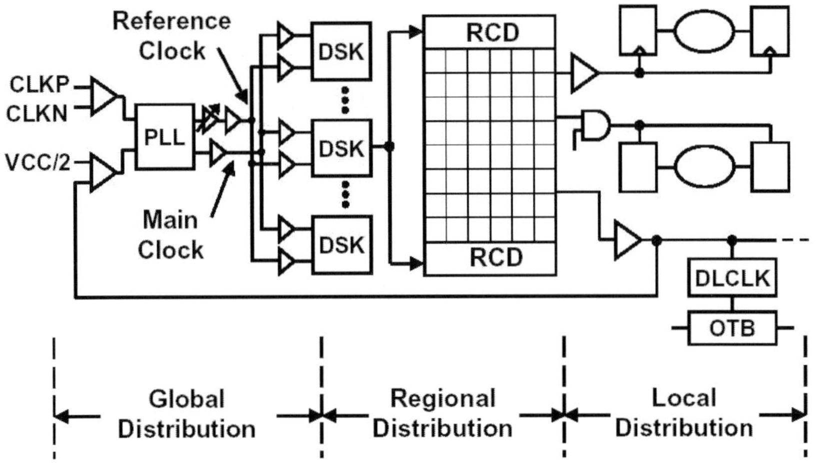

Figure 5-12. Clock distribution for the Itanium® Processor [5]

The clock distribution of the Itanium® Processor is shown in Figure 6-12 [5]. The clock topology is partitioned into three segments. The global distribution consists of the clock synthesis using an on-die phase-locked loop (PLL) and the distribution of the core clock and the reference clock from the PLL clock generator to the deskew buffers (DSK). The regional distribution includes the clock distribution from the DSKs to the 30 regional clock grids. Finally, the local distribution consists of the local clock buffers (LCBs) taking the input from the regional clock grid and the local interconnect to support the clocked elements. The clock deskew function will be discussed in detail in Section 4. The regional clock grid is implemented using metal 4 horizontal and metal 5 in the vertical direction. As with the global clock network, the regional clock grid contains full lateral shielding to ensure low capacitance coupling and good inductive return paths. The regional clock grid utilizes up to 3.5% of the available metal 5 and up to 4.1% of the available metal 4 routing over a region.

The last topic in this section is the inductive effect in clock distribution networks and the need to carefully model these effects when designing high-frequency clock trees.

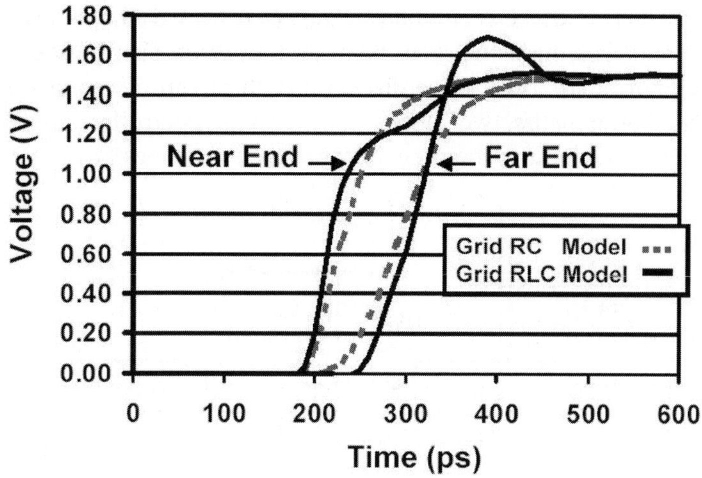

Figure 5-13. Inductive effects on clock distribution grids [6]

Low resistance copper interconnects together with fast clock edge rates result in inductive wire impedance comparable to the pure resistive impedance. As a result, high frequency clock grids are modeled as two-dimensional distributed RLC transmission line structures, as shown in [6]. Detailed 3D field simulations are used to extract accurate per unit length RLC clock grid values in the presence of finite coplanar current return paths. Transmission line effects, like signal reflections near the clock drivers as well as overshoot and undershoot at the far end of the grid, are clearly observed in Figure 6-13. Distributed wire inductance increases the delay between the early clock (near the driver) and the late clock (at the end of the grid). Clock loads near the driver are inductively shielded from remote loads during the clock transition and observe a faster clock. On the other hand, remote loads receive the clock later due to additional signal phase imposed by wire inductance. Optimal clock driver sizing requires that the equivalent output impedance be smaller than the grid characteristic impedance to guarantee a high amplitude incident wave, and yet that the clock rise and fall times at the drivers are sufficiently slow to ensure that voltage overshoot and undershoot are not excessive at the far ends.

4. DESKEW CIRCUITS

The first active deskew circuit was reported by Geannopoulos, *et al.* in [1]. The circuit equalizes the skew between two clock distribution spines, as shown in Figure 6-14. The phase detection (PD) circuit determines the phase relationship between the two clocks. The deskew controller adjusts one of the delay lines to minimize the skew between the two spines.

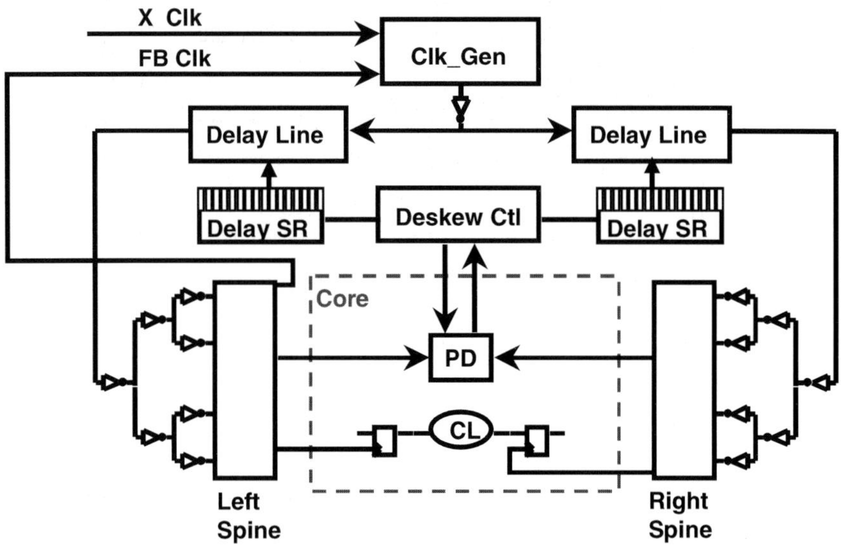

Figure 5-14. Block diagram of a two-spine clock deskewing circuit [1]

A more refined scheme was introduced in [5], where the entire chip area was split in 30 skew zones, as shown in Figure 6-15. Each zone has an independent deskew circuit that adjusts the regional clock delay to match it with the reference clock. This adjustment is repeated until a minimum phase error is achieved. Therefore, any load mismatches and within-die variations in the core clock distribution are automatically compensated. Since all the clock regions use the same reference clock, the residual skew of the reference clock, the uncertainty of the phase detector and the mismatches of the feedback clocks determine the overall skew across these regions.

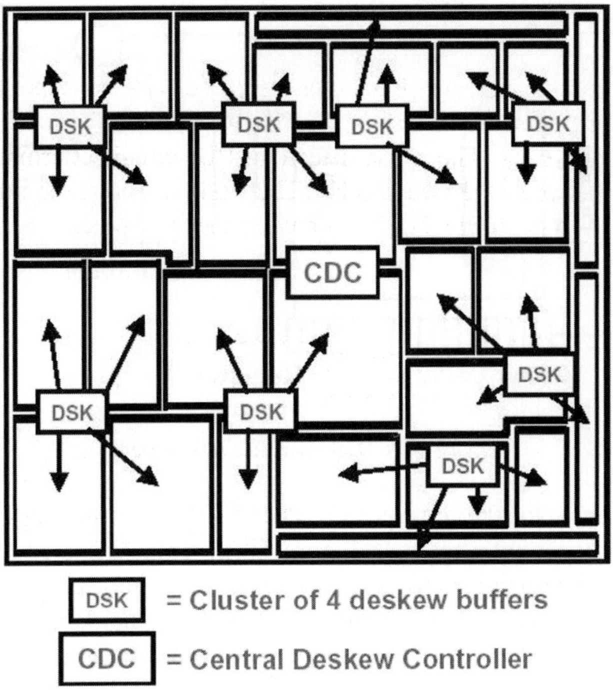

Figure 5-15. Itanium® Processor clock deskew scheme [5]

The deskew operation is executed in parallel for the 30 clock regions during the initial microprocessor reset. The global deskew controller monitors the progress and signals the deskew completion. Once this occurs, the DSK delay register settings are fixed until the next power up sequence. This mode compensates for the process variations and most of the voltage and temperature variations. An alternative operating mode is to allow the deskew operation to continue during normal microprocessor operation. This mode compensates for the dynamic effects such as temperature variations and supply voltage drift over time, but has the additional risk of creating new timing critical paths.

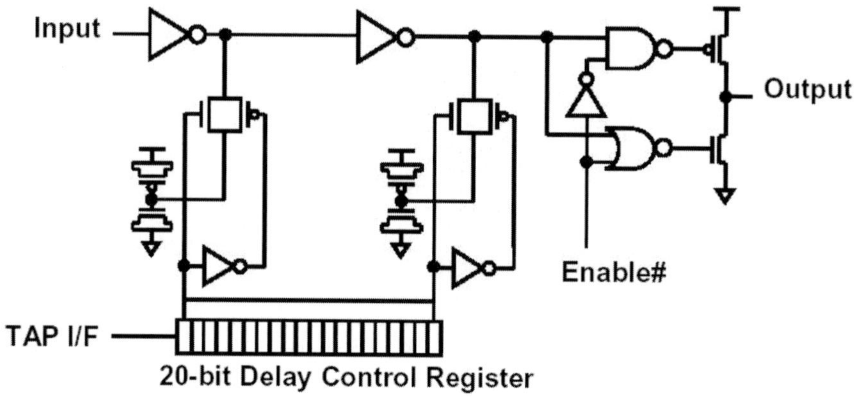

Figure 5-16. Deskew buffer schematic [5]

Adjustments to the core clock delay are accomplished through the variable delay circuit shown in Figure 6-16. This is a digitally controlled analog delay line using a 20-bit delay control register, a two-stage variable delay circuit and a push-pull style output buffer. The delay control register forms a 20-steps linear delay coding that provides a good balance between the delay step-size resolution and the total buffer delay range. Delay adjustment can be accomplished by shifting a "1" from one end of the register to decrease its delay or by shifting a "0" from the opposite end to increase its delay. In addition to the input derived from the local deskew controller, the delay control register also accepts input from the test access port (TAP) interface. This feature permits a manual adjustment of the deskew buffer delay through the TAP interface, which can be used for post-silicon timing optimization. The variable delay circuit is constructed of CMOS inverters and two arrays of passive loads. The delay across the inverters varies in accordance to the setting stored in the delay control register. Advantages of this design over a starving inverter approach are linear delay steps and more symmetric layout. The push-pull style output stage consists of twelve parallel drivers that can be enabled individually via mask options to match the extracted loading of each region. This allows one standard design to accommodate a wide range of regional clock loads. The measured delay range of the deskew buffer is 170ps with a step size of 8.5ps.

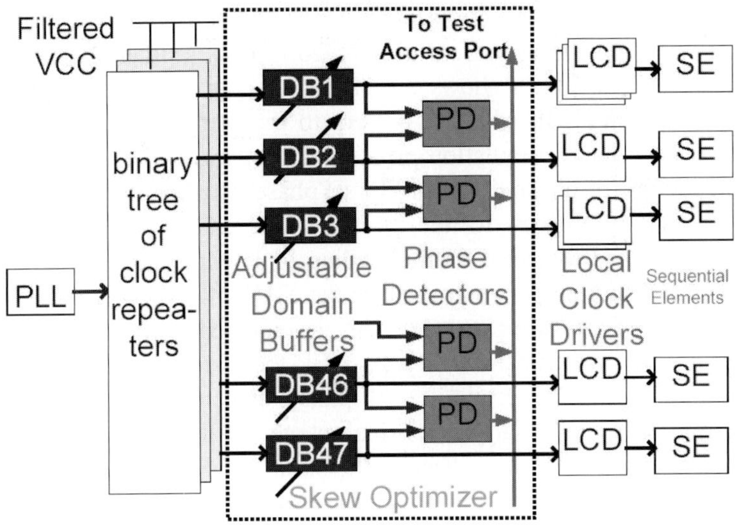

Figure 5-17. Logical diagram of the skew optimization circuit in the Pentium® 4 Processor

Figure 6-17 presents the schematic of the deskew circuit implemented in the Pentium® 4 Processor [2]. The main components of the skew optimization circuit are 47 adjustable delay domain buffers and a phase-detector network of 46 phase detectors. The delay adjustment control for the domain buffers and the output of the phase detectors are accessible from the test access port (TAP).

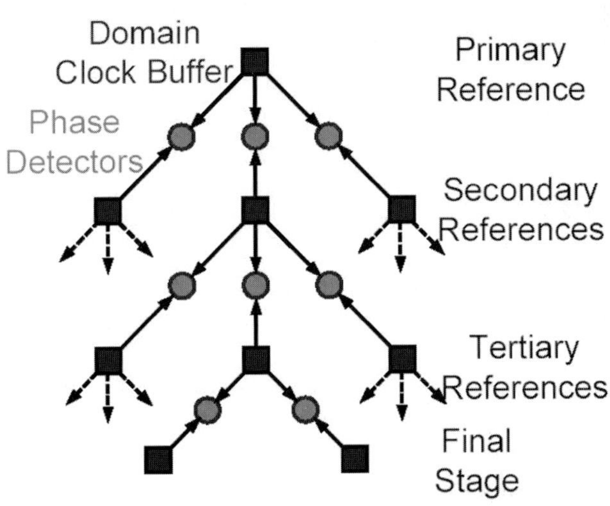

Figure 5-18. Phase detector network in the Pentium® 4 Processor

One domain buffer at the center of the die is chosen as the primary reference. The remaining buffers are categorized as secondary, tertiary, and final buffers. Figure 6-18 shows the phase-detector network, which coupled with the TAP, aligns the domain buffers to the primary reference. To limit the phase detector accumulation errors, the domain buffers go through at most three levels of phase detection. First, phase detectors adjust the delay of the secondary references to the primary reference. The phase detector outputs a high or low based on the leading or lagging inputs. The output is read out into a scan chain controlled by the TAP. Based on the outcome, the clock domain buffers are adjusted. This is repeated until all the secondary reference clocks are deskewed. Then, after the secondary reference delays have been adjusted, a second set of phase detectors adjust the delay of tertiary references. Similarly, the final stage buffers are adjusted to the tertiary references. With this scheme, the skew is adjusted to within an error of about 8 ps, limited mainly by the resolution of the adjustable delay elements.

Table 5-1. Summary of clock deskew techniques

Author	Source	Zones	Skew Before	Skew After	Step Size
Geannopoulos	ISSCC 1998	2	60ps	15ps	12ps
Rusu	ISSCC 2000	30	110ps	28ps	8ps
Kurd	ISSCC 2001	47	64ps	16ps	8ps
Stinson	ISSCC 2003	23	60ps	7ps	7ps

Table 6-1 summarizes all published clock deskew designs. Notice that all designs manage to reduce the clock skew to less than a quarter of the value measured without the deskew mechanism. As the process technology shrinks, the step size is reduced, without requiring any additional control bits.

5. JITTER REDUCTION TECHNIQUES

The most common jitter reduction technique is filtering the supply voltage for the clock buffering stages. Figure 6-19 shows a simple low-pass RC filter designed to reduce the core supply noise for the clock buffers [2]. The resistance is implemented using PMOS devices. The optimal design has an IR drop of 70mV with a RC constant of 2.5ns, while minimizing the layout area required by the capacitor. The actual component values for the RC filter were adjusted for the different clock buffer types to a fixed IR drop and RC time constant. The filter circuit model simulations with typical supply noise waveforms show up to 5X noise amplitude reduction on the filtered supply.

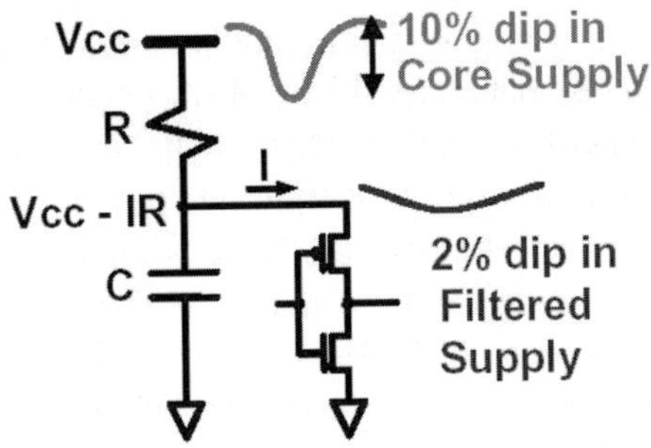

Figure 5-19. Low jitter, RC-filtered power supply for clock drivers [2]

Another technique is to use an active voltage regulator as described in [6] and shown in Figure 6-20. The core voltage of 1.5V is used as a reference after passing it through a low-pass filter (LPF). The I/O voltage of 2.5V is used to generate a clean, local supply for the delay-lock loop (DLL) circuits. The regulator attenuates supply noise frequencies in excess of 1MHz by more than 15dB, while lower supply noise frequencies are easily tracked by the DLLs themselves.

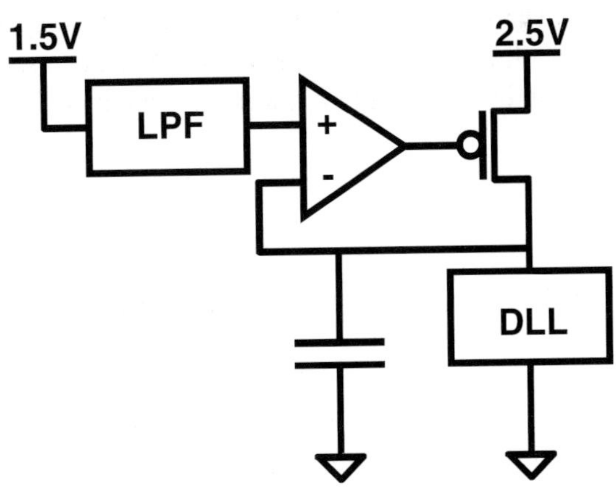

Figure 5-20. Active clock supply regulator [6]

6. LOW POWER CLOCK DISTRIBUTION

In modern VLSI devices, the clock distribution network and the clocked sequential elements (latches and flip-flops) are the largest components of the total power dissipation, accounting for 20% to 45% of the total power. The flip-flops and the last branches of the clock distribution network that drive them consume 90% of the total clock power. The reason for this large power consumption of the clock system is that the transition probability of the clock is 100% while that of the ordinary logic is about one-tenth to one-third in average. It is therefore desirable to minimize the power consumed by the clock distribution and sequential elements.

To accomplish this, we start from the dynamic power equation $P = CfV^2$, where C is the total switched capacitance, f is the operating frequency and V is the supply voltage. Lowering the clock frequency contradicts the basic trend outlined in Section 1 of pursuing higher operating frequencies through design and process technology scaling. This is only feasible on a part-time basis, like lowering the clock frequency during idle periods of time in mobile computing devices. On a practical basis, lower active power is best achieved by reducing the voltage and switched capacitance of the design. Since the voltage is squared in the power equation, it has a larger impact. Kojima *et al.* [7] proposed a half-swing flip-flop (HSFF) design, where the voltage swing of the clock is reduced to half the operating voltage. The HSFF requires four clock signals as shown in Figure 6-21. Two clock phases with a swing between *Vdd* and *Vdd*/2 drive the PMOS devices, while the other two phases with a swing between Gnd and *Vdd*/2 drive the NMOS transistors. A theoretical analysis of this scheme shows that the clocking power is reduced by 75% compared to the full clock swing distribution.

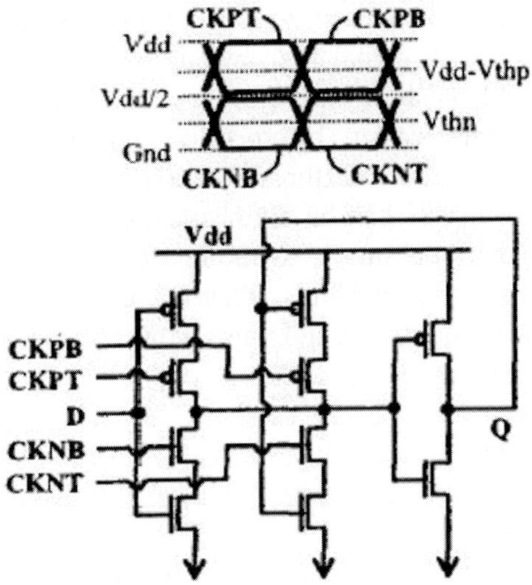

Figure 5-21. Flip-flop design using half-swing clock [7]

Experimental savings of 67% were demonstrated on a 0.5μm CMOS test chip with only 0.5ns degradation in speed. However, this scheme requires additional area for the special clock drivers and suffers from skew problems between the four clock phases.

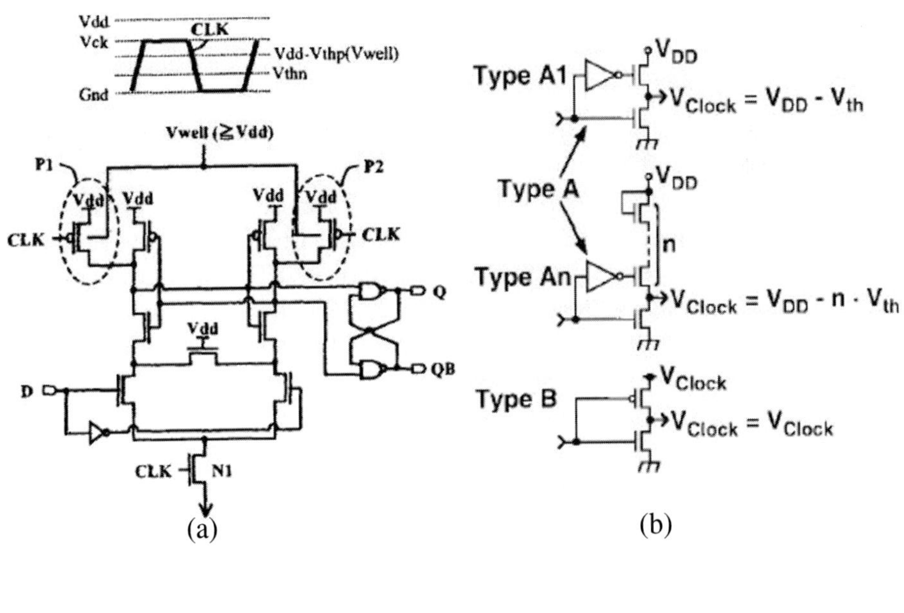

Figure 5-22. Reduced clock swing flip-flop [8]

Kawaguchi *et al.* [8] proposed a reduced clock swing flip-flop (RCSFF) that needs only one clock signal that swings between *Gnd* and *Vck*, where *Vck* is lower than *Vdd*, as shown in Figure 6-22 (a). To control the leakage of the pullup P-transistors driven by this low swing clock, a well-biasing technique is used to increase the threshold voltage of these devices. Several reduced swing clock drivers can be used, as shown in Figure 6-22 (b). Type A drivers use the same *Vdd* supply as the rest of the core, while Type B use an external V_{clock} supply. While this scheme can achieve up to 63% clock power reduction, it requires additional layout area for the well biasing scheme.

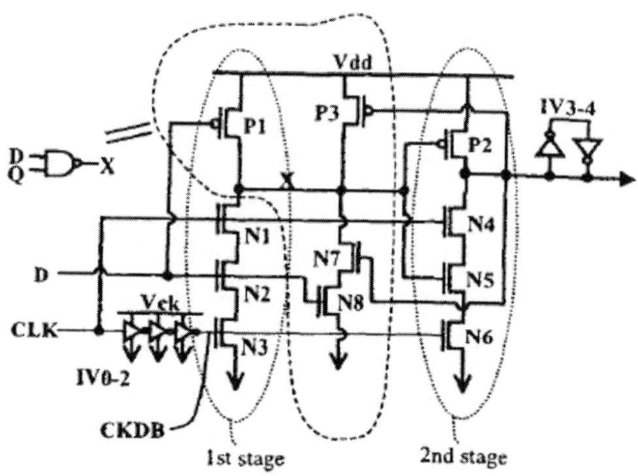

Figure 5-23. NAND-type Keeper Flip Flop [9]

A NAND-type Keeper Flip-Flop (NDKFF) proposed by Tokumasu *et al.* [9] is shown in Figure 6-23. Notice that all transistors driven by the clock signal are NMOS devices, which enables the NDKFF to operate with a reduced clock swing without concern over PMOS pullup leakage. The NAND-type keeper eliminates unnecessary transitions at the internal node X, further reducing the power consumption.

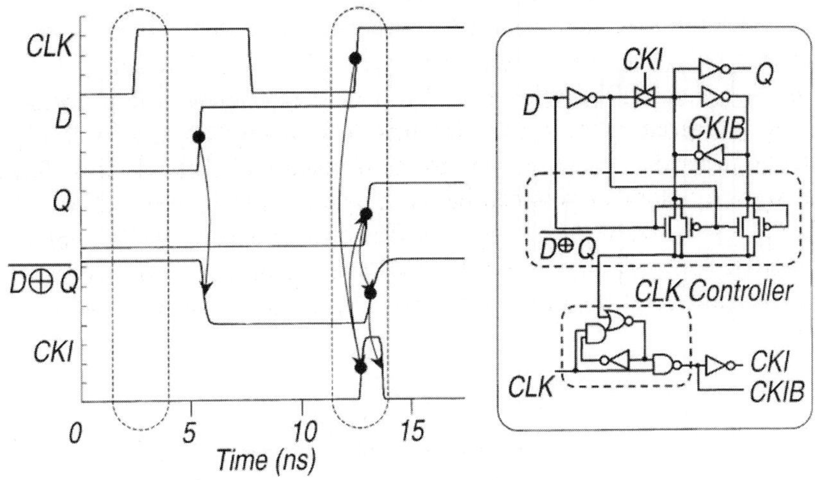

Figure 5-24. Clock-on-demand flip-flop [10]

Hamada [10] proposes a clock-on-demand flip-flop shown in Figure 6-24. In this design, the internal clock CKI is activated only when the input data D will change the output Q. This amounts to the finest granularity (single bit level) of clock gating. The clock-on-demand flip-flop generates a self-aligned pulsed clock internally, that enables the latch circuit to operate like an edge-triggered flip-flop. For a data transition probability of 0.95, the clock-on-demand flip-flop dissipates the same power as a conventional flip-flop. Since in practical applications the data transition probability is between 0.1 and 0.3 on average, the clock-on-demand flip-flop has a power advantage over the conventional design. However, the clock-on-demand flip-flop has a longer setup time and is prone to hold time violations like all other pulsed-latch designs.

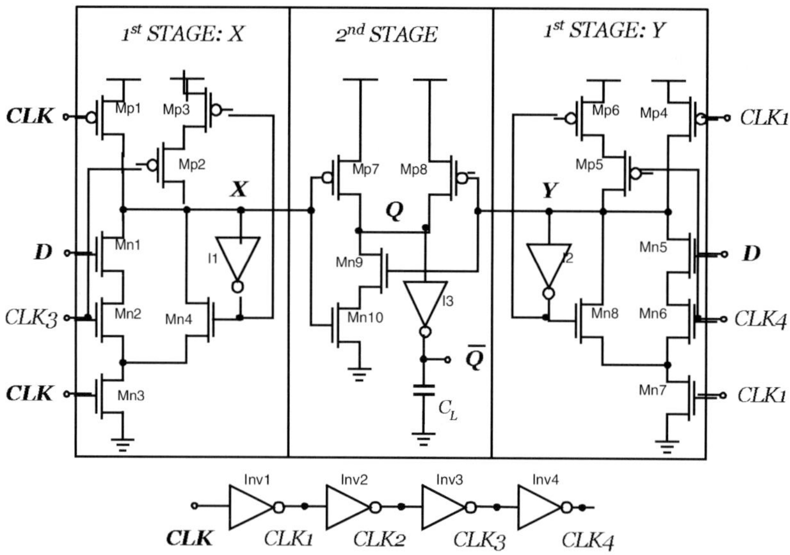

Figure 5-25. Dual edge flip-flop circuit [11]

Another approach to reduce the clock power is to use dual-edge triggered storage elements that can achieve the same throughput as single-edge clocked flip-flops at half the clock frequency and half the clock-related power consumption. Nedovic *et al.* [11] proposed the design shown in Figure 6-25. The first stage consists of two symmetric pulse-generating latches that create data conditioned clock pulses on each edge of the clock. The second stage is a 2-input NAND gate, effectively used as a multiplexer. Simulation results in 0.11μm technology show an energy-delay product and delay comparable to the best single-edge designs. The clock load of this design is similar to the clock load of single-edge flip-flops used in high-performance processor designs, allowing a power savings of about 50%. The main drawback of dual-edge clocking is that it requires tight control of the clock duty cycle and uncertainties of both clock edges.

7. FUTURE DIRECTIONS IN CLOCK DISTRIBUTION

Mizuno *et al.*, [12] proposed using distributed voltage-controlled oscillators (VCO) to generate local clocks as shown in Figure 6-26. Metal lines of equal length *l* short the outputs of the VCOs to minimize the skew between the multiple oscillators. The voltage V_c is the frequency control signal for the VCOs that is distributed across the chip instead of the global clock signal. Careful shielding and filtering are required to insure the noise immunity for this analog voltage level.

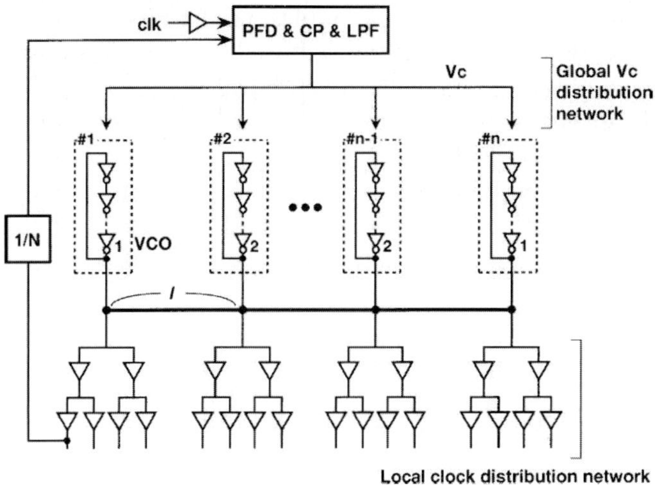

Figure 5-26. Clock distribution using synchronous distributed oscillators [12]

A two-dimensional matrix configuration distributes the VCOs over the entire chip area, as shown in Figure 6-27. Using this scheme, each VCO is placed close to the local clock distribution network. A test chip fabricated in 0.25μm CMOS technology achieved a mean skew of 17ps.

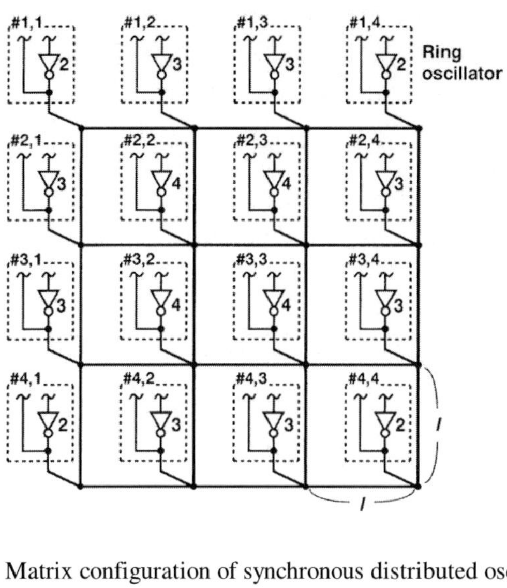

Figure 5-27. Matrix configuration of synchronous distributed oscillators [12]

Gutnik *et al.*, [13] propose distributing phase-locked loops (PLLs) at multiple points across the chip, as shown in Figure 6-28. Each locally generated clock is distributed only to a small section of the chip (tile). Phase detectors at the boundaries between tiles produce error signals that are summed by an amplifier in each tile and used to adjust the frequency of the node oscillator. Since this technique requires many nodes (16 in the example shown in Figure 6-28), the total area and power consumption of all PLLs is higher than the single PLL conventional approach. The voltage-controlled oscillator uses an NMOS-loaded ring oscillator to minimize the power supply noise. A 4x4 test chip in a standard 0.35μm CMOS technology demonstrated a long-term jitter between neighboring tiles of less than 30ps and cycle-to-cycle jitter of less than 10ps.

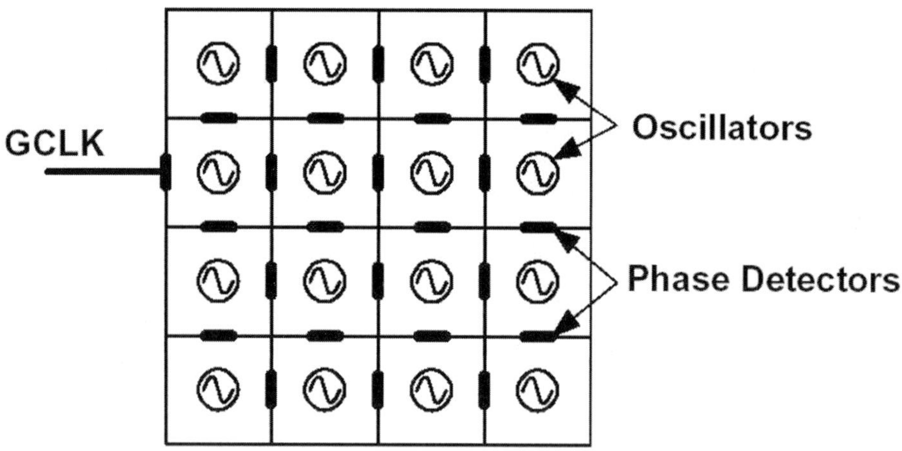

Figure 5-28. Distributed phase locked loops [13]

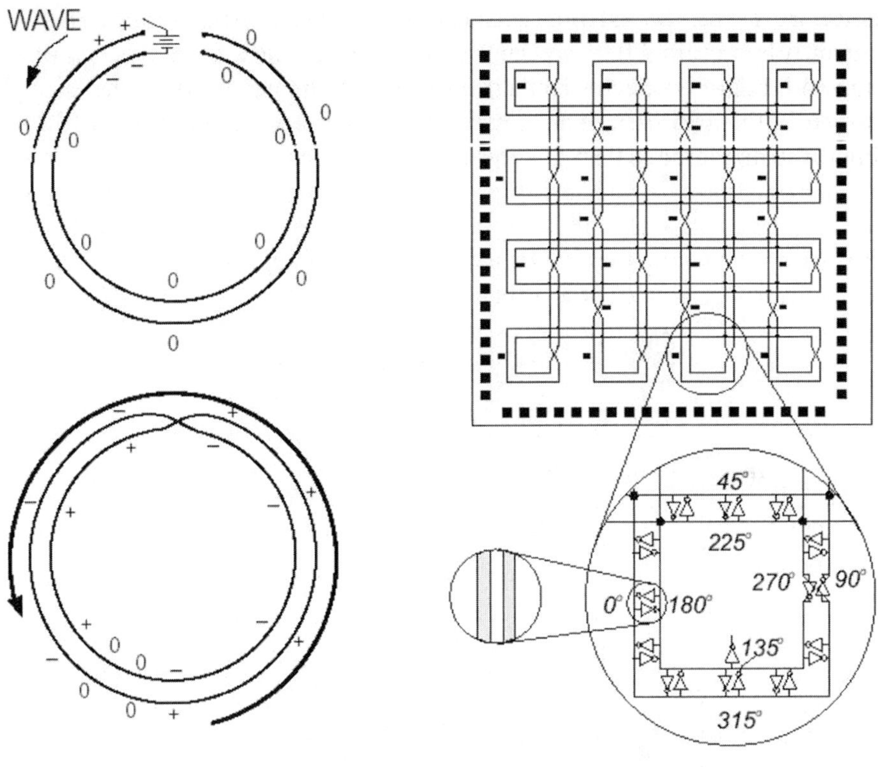

(a) (b)

Figure 5-29. Rotary clock distribution architecture [14]

Wood, *et al.*, [14] propose a rotary clock distribution architecture to achieve a low skew and jitter, gigahertz rate clock distribution. Figure 6-29 (a) illustrates the theory behind the rotary clock architecture, using a simple, open loop of differential conductors connected to a battery through an ideal switch. When the switch is closed, a voltage wave starts to move counter-clockwise around the loop. Once the wave is started, it can be maintained through a logical inversion by crossing over the wires instead of the battery supply. To overcome losses, anti-parallel inverter pairs are used. The energy is recirculated in the closed electromagnetic path, providing a significant power savings, since losses are due only to I^2R dissipation in the wires and not CfV^2 dissipation as in the conventional clock distribution. Figure 6-29 (b) shows the layout of a rotary clock with 25 interconnected rings. Each ring consists of a differential line driven by anti-parallel inverters distributed around the ring. The clock wave frequency depends on the electrical length of the ring and the inductance and capacitance of the lines. A prototype circuit was built in a 0.25μm 2.5V CMOS process and has a 12000μm long ring with 60μm conductors on a 120μm pitch. Simulations predicted a clock frequency of 925MHz, while measured waveforms clocked at 965MHz. Jitter was measured to be 5.5ps rms.

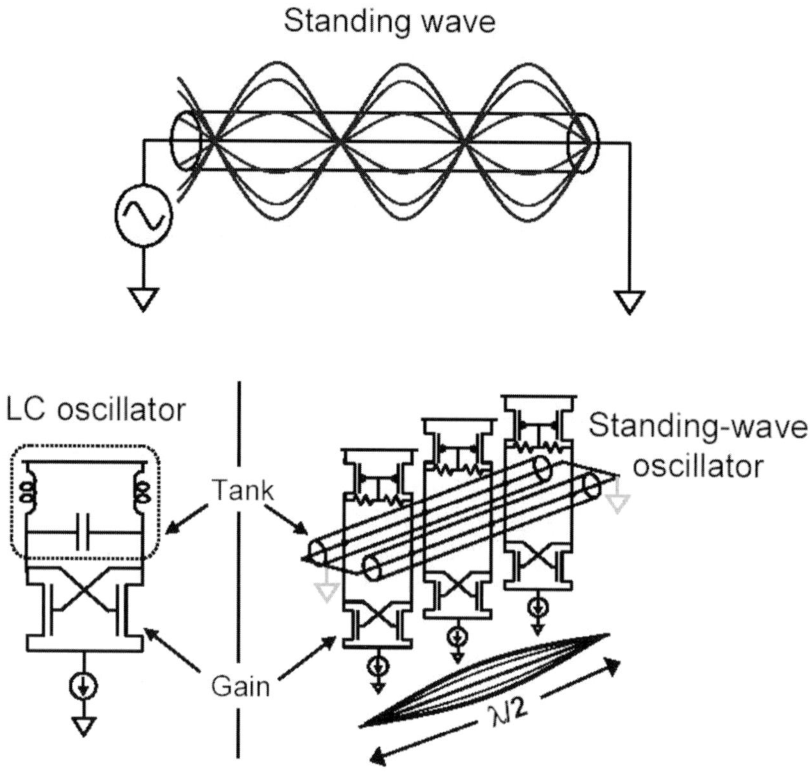

Figure 5-30. LC vs. standing wave oscillator [15]

O'Mahony [15] describes a 10GHz standing-wave clock distribution system that achieves sub-picosecond skew and jitter using on-chip interconnects and distributed buffers to create a network of coupled oscillators. Standing waves have the same phase at all points, as opposed to the rotary clock scheme discussed earlier that generates traveling waves. A standing-wave oscillator is similar to a differential LC oscillator (both shown in Figure 6-30) where the gain and tank are distributed. The NMOS cross-coupled pairs provide enough gain to compensate for wire losses. The PMOS diode-connected loads set the common mode voltage and allow injection of an external clock reference.

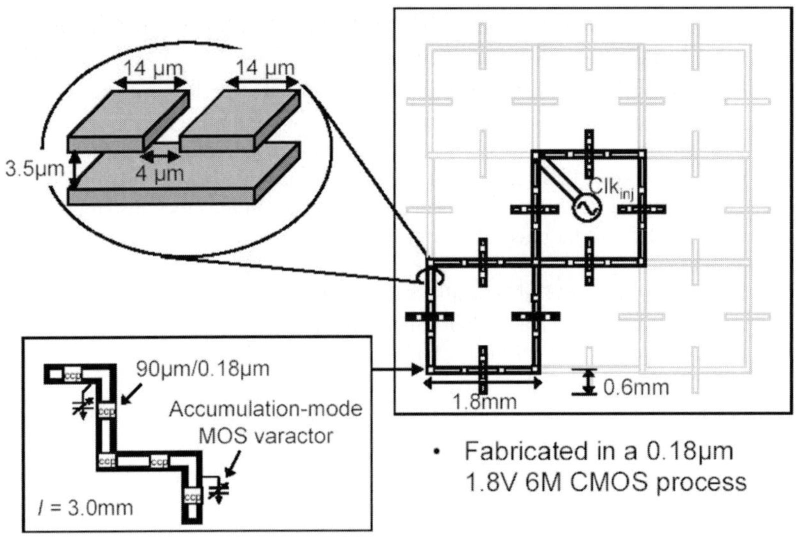

Figure 5-31. 10GHz standing wave clock distribution test chip [15]

Figure 6-31 shows a prototype standing wave oscillator clock network implemented in a 0.18μm, 1.8V CMOS process with six AlCu metal layers. The differential λ/2 lines are 3mm long, 14μm wide and are 4μm apart in metal six. The clock jitter added by the clock grid is below 0.5ps. The measured skew is 0.6ps (0.6%) when the grid is tuned to 10GHz with a single control voltage for all varactors and 3.2ps when half of the grid is de-tuned by 1%. The worst-case skew between any two adjacent points is 1.4ps for the de-tuned grid.

REFERENCES

1. G. Geannopoulos, X. Dai - "An adaptive digital deskewing circuit for clock distribution networks", ISSCC Digest of Technical Papers, 1998, Pg 400 –401
2. N. Kurd, *et al*, - "A multi-GHz clocking scheme for the Pentium® 4 microprocessor", IEEE Journal of Solid-State Circuits, Nov. 2001, Pg 1647-1653
3. P. Gronowski, *et al*, "High-performance microprocessor design", IEEE Journal of Solid-State Circuits, May 1998, Pg 676 – 686
4. P. Restle, *et al*, - "The clock distribution of the Power4 microprocessor", ISSCC Digest of Technical Papers, 2002, Pg 144 –145
5. S. Tam, *et al*, - "Clock generation and distribution for the first IA-64 microprocessor", IEEE Journal of Solid-State Circuits, Nov. 2000, Pg 1545-1552
6. T. Xanthopoulos, *et al*, - "The design and analysis of the clock distribution network for a 1.2GHz Alpha microprocessor", ISSCC Digest of Technical Papers, 2001, Pg 402-403
7. H. Kojima, *et al*, - "Half-swing clocking scheme for 75% power saving in clocking circuitry", IEEE Journal of Solid-State Circuits, April 1995, Pg 432 –435
8. H. Kawaguchi, T. Sakurai, - "A reduced clock-swing flip-flop (RCSFF) for 63% power reduction", IEEE Journal of Solid-State Circuits, May 1998, Pg 807 -811
9. M. Tokumasu, *et al*, - "A new reduced clock-swing flip-flop: NAND-type keeper flip-flop (NDKFF)", Custom Integrated Circuits Conference, 2002, Pg 129 -132
10. M. Hamada, *et al*, - "Flip-flop selection technique for power-delay trade-off", ISSCC Digest of Technical Papers, 1999, Pg 270 -271
11. N. Nedovic, *et al*, "A low power symmetrically pulsed dual edge-triggered flip-flop", ESSCIRC Proceedings, 2002
12. H. Mizuno, *et al*, - "A noise-immune GHz-clock distribution scheme using synchronous distributed oscillators", ISSCC Digest of Technical Papers, 1998, Pg 404-405
13. V. Gutnik, *et al*, - "Active GHz clock network using distributed PLLs", ISSCC Digest of Technical Papers, 2000, Pg 174-175
14. J. Wood, *et al*, - "Multi-gigahertz low-power low-skew rotary clock scheme", ISSCC Digest of Technical Papers, 2001, Pg 400 -401
15. F. O'Mahony, *et al*., - "10GHz clock distribution using coupled standing-wave oscillators", ISSCC Digest of Technical Papers, 2003, Pg 428-429

II

LOGICAL AND ARCHITECTURAL ISSUES

Chapter 6

ERROR-TOLERANT INTERCONNECT SCHEMES

Heiko Zimmer
Institute of Microelectronic Systems
Darmstadt University of Technology, Darmstadt, Germany
heiko@mes.tu-darmstadt.de

Axel Jantsch
Institute of Microelectronics and Information Technology
Royal Institute of Technology (KTH), Stockholm, Sweden
axel@imit.kth.se

1. Introduction

The Network-on-Chip paradigm targets future Systems-on-Chip built of a large number of reusable, independent Intellectual-Property-blocks (IPs). A typical approach is to align these IPs as tiles in a regular manner, each associated with a wrapper providing access to the on-chip network. The network itself is a regular structure composed of switches/routers and the interconnecting links.

The objective of implementing a Network-on-Chip is to decouple computation from communication by offering a uniform, reliable, and versatile communication platform for all the inter-IP communication required by a typical SoC application. Thus, the need for custom wiring to build an application-specific communication infrastructure is overcome. Furthermore, placement and routing are simplified for the whole NoC because both the IPs and the network components are encapsulated from one another except for a defined network interface providing network access in terms of services usable by the IP for all communication it requires with its surroundings.

To fully exploit the advantages this approach offers, the network must provide defined, reliable communication services to the resources attached to it. This problem is far from trivial since the network links will most likely not be error-free in future deep submicron technology generations.

While much work is done to develop robust transmission schemes, it is expected that especially crosstalk will seriously affect interconnect reliability. When the physical layer of an on-chip network fails despite all preventive measures taken, the effect should be controlled. This is what the error-tolerant interconnect schemes provide: They increase the network reliability and hide imperfections from the applications which use the communication services offered.

Reliable network services are of great importance since applications have demands on communication that must be fulfilled to achieve correct application behaviour. In a classical approach, the communication infrastructure of a SoC is a combination of shared buses and custom designed interconnect to meet specific requirements. However, when all communication should be transported over a common medium, the on-chip network, it must be taken into account that the demands are different for different applications and may include bandwidth guarantees, integrity requirements or deadlines for completion of a specific task in real-time applications.

Typically, an application running on a SoC is comprised of multiple processes associated to different NoC-resources. Naturally, the characteristics of data transport within one application are not uniform since different traffic types such as control messages, audio signals and video streams have to coexist. Even seemingly regular data streams become irregular during processing. For instance video streams are usually encoded such that only the delta between frames are transmitted, which makes the communicated data volume higly dependent on the video content.

Closing the gap between the hardware platform's possibilities and the applications' requirements is the demanding task of error-tolerant interconnect schemes. Their aim is to provide a network with defined properties to the application. In the ideal case, what the applications see is an error-free communication medium fulfilling all their communication needs.

Furthermore, this idea hides the physical implementation details of a specific technology. By providing defined services, the border between platform design (technology, layout, error-tolerant interconnect scheme) and application design (using communication at defined QoS-levels) is clear.

Which measures are taken to implement an error-tolerant interconnect scheme depends on the specific applications' requirements and the constraints imposed by the selected platform/architecture. Therefore, we give a general overview before focusing on one specific example.

2. Architectural Considerations

One main objective of networks on chip is the idea that they are formed by a regular structure. This helps to overcome the problems associated with custom and global interconnects: Designing and verifying specialized custom interconnects takes considerable amounts of time, global interconnects traversing the whole chip will be inefficient from an energy viewpoint and very susceptible to crosstalk-related errors. By building a regular interconnection architecture that is defined by the platform, the electrical parameters of the interconnects can be tightly controlled. For instance, in a regular mesh all links have the same length. Obviously, much effort can be put into fine-tuning the electrical parameters of these links (in order to optimize reliability, energy consumption etc.) and an efficient implementation will be very rewarding. This example shows that the NoC-approach enables designers to fully take advantage of a given technology to implement the low-level details of transmission in a place- and energy-efficient manner. At the same time, the technological details influencing interconnect design are hidden from the application.

In the past, various NoC-architectures have been proposed, among them tori, hypercubes, meshes and trees. In defining the architecture, many characteristics are decided that have an impact on the possibilities to realize fault-tolerant (i.e., reliable) communication.

2.1 Architecture-imposed Constraints

Among the most important parameters for data protection schemes are (a) the width of the links between resources and switches, (b) if parallel or serial transmission is employed on these links, and (c) how much delay, area and power overhead can be tolerated in the switches.

Considering for instance a $n \times m$ mesh it is obvious that an area-efficient switch design will be very rewarding: Since a total of $n \cdot m$ switches form the NoC, even small savings can have a significant effect on the overall area overhead introduced by the on-chip network. This clearly motivates carefully balanced design decisions when implementing network components, especially those dealing with error-tolerance for they can be arbitrarily complex.

2.2 Errors in DSM

Compared to today's circuits, the number of transient errors is expected to increase significantly in future technology generations [1]. The reliability of long interconnects, especially buses, will suffer from crosstalk faults. At the same time, technology scaling gives rise to new error sources. Among them is radiation which has previously only been a concern with memories, but will affect logic and low capacitance/low power buses as well. Generally speaking, the amount of transient faults increases with the decrease of feature size.

Crosstalk. The main source of errors on buses will be crosstalk, which is strongly influenced by the coupling capacitance between wires and can lead to increased delay or cause voltage glitches on wires [2]. Furthermore, crosstalk can inherently cause multi-bit and bidirectional errors [3] (i.e., disturbing multiple wires, causing both $0 \rightarrow 1$ and $1 \rightarrow 0$ errors).

Crosstalk depends not only on the physical and electrical parameters but also on the dynamic switching behaviour. Thus, the system has either to be designed for the worst case or significant information about the transmitted data has to be known at design time. One example is an algorithm for wire placement in address buses [4]: A lot of information about the switching activity (data patterns) on the bus is needed to achieve good results. For data buses carrying random data, however, this approach is not useful.

Many other approaches aim at reducing the bus crosstalk energy by minimizing the number of simultaneous transitions on adjacent bit-lines. A good summary of approaches to reduce the number of transitions can be found in [5] which describes a dictionary-based encoding scheme for data buses. The authors investigated various data flows and found correlation between adjacent bus lines. At the expense of additional hardware (so that encoding/decoding can take place within one cycle), a dictionary-based compression method can be used to transfer less data, thereby causing less signal transitions.

Another possibility to reduce crosstalk energy is to use time redundancy. It was shown in [6] that it is very energy efficient to detect errors and retransmit erroneous data. By comparing various codes, the authors found that more powerful codes imply higher power consumption due to more complex encoders and decoders. The best energy efficiency has been achieved using a detection scheme with retransmit.

The work presented in [7] is targeted towards encoding the bus in a way that minimizes crosstalk. The authors present a special encoding, which is shown to reduce the crosstalk delay variations by a factor of two. The main drawback is that an area overhead of more than 60% is required, and an important finding is that encoding for low crosstalk and low energy may not be compatible since worst-case crosstalk patterns can occur among the patterns that have very low energy consumption.

Electro-Magnetic Interference. Electro-magnetic interference clearly is a mechanism that can induce transient errors. And on-chip interconnections are very likely to be the victims of such influence because of their relatively long wires. Errors due to electro-magnetic interference can thus be expected on the buses between the switches of the Network-on-Chip.

With higher integration of more complex building blocks on one chip, electrical noise from the environment will no longer be the only error source. Especially mixed-signal ICs will also be susceptible to distortions generated on-chip by HF components or by the large number of devices switching simultaneously at high clock frequencies.

Radiation. Transient faults caused by alpha particles (emitted by the package) or cosmic radiation have first appeared in semiconductor memories, which are typically the most highly integrated devices fully exploiting a given technology's capabilities. Error correcting codes have been successfully applied to these memories in order to minimize the effect of both transient errors and permanent errors introduced during the manufacturing process (resulting in an improved yield).

Due to decreased node capacitances, logic [8] and low capacitance/low power buses [9] are likely be affected by radiation-induced errors as well.

Process Variations. At decreasing feature size, unavoidable process variations may lead to a greater variance of circuit parameters. While transistors are the devices whose parameters are easily changed by process variations, the effect on the high number of interconnects on an always-increasing number of layers has also to be taken into account. Even a small change in the cross-coupling capacitances may lead to additional delays on several or all lines of a bus, thereby increasing the delay variation.

Other Error Sources. Given the high costs of microchip production (masks, manufacturing, testing etc.) that increase with each technology generation, it may become necessary to relax the current, inherent requirement of 100% correctness for devices and interconnects on purpose. According to the International Technology Roadmap for Semiconductors, this paradigm shift might be forced during the time-frame the NoC-architecture is developed for, since it may be the only means to reduce production, verification and test costs per chip by increasing the yield significantly [1].

Fault-tolerance will then not only be implemented to cope with runtime errors but also to mask out a certain amount of permanent errors introduced during manufacturing. This could lead to a characterization of acceptable errors, which will not render the chip unusable but can be tolerated by the implemented error-control scheme.

Furthermore, it is expected that the number of soft errors only occurring under specific conditions of voltage, timing and temperature will increase. Since these types of errors are difficult or impossible to test for, they will lead to an higher overall rate of transient errors during circuit operation.

While interconnect reliability has been tackled successfully on the physical layer in the past, this will not necessarily be possible in future deep submicron NoC-applications. Measures like shielding (adding a ground wire between two signal wires) might be too expensive in a highly integrated network environment built of massively parallel buses. Furthermore, Worm et. al. raise the question whether maintaining all design margins for on-chip buses to cope with worst-case manufacturing variations will be too expensive in terms of area consumption. [10]

Taking all these arguments into consideration, it becomes clear that guaranteeing completely reliable links between the NoC-switches by manufacturing – while desirable – will perhaps not be the most efficient solution when both technical and economical arguments are taken into account. When the physical layer of an on-chip network is not able to hide all errors that occur from the upper protocol layers, it is important to understand what the error sources are and which error characteristics are expected in order to design suitable error-tolerance schemes.

3. Application Demand and Quality of Service

An implementation objective of the on-chip network is the principle of communication between resources. This communication is not as tightly coupled as it could be if custom interconnect (along with a custom protocol) was used. This leads to a new way of looking at SoC design with a strong emphasis on communication. After all, a future NoC-application will rather resemble a distributed system than a traditional SoC.

When an application is mapped onto a NoC-platform, it is split into multiple blocks which need to communicate. Likewise, when multiple applications are running on one NoC, they must be able to interact. While this is not new and requirements for specific communication have formerly been implemented by buses or custom designed connections, the NoC-approach aims at transferring all inter-block communication over a chip-wide uniform network infrastructure.

This in turn means that the network must be so versatile that it can handle (adapt to) a large number of communication streams of which each one can have its own characteristics. These characteristics will certainly not be uniform. Therefore, a closer look at parameters for traffic description is appropriate. Since communication aspects will play a major role in future NoC-based SoC design, applications will have to explicitly describe their communication streams.

Especially in a large-scale NoC that handles a significant number of concurrent communication streams, the traffic will not be uniform. Supporting different types of traffic can be rewarding and efficient in terms of resource usage and overall transmission times. An example for this are the advantages of combining guaranteed-throughput and best-effort traffic shown in [11] for the Æthereal network.

An overview of parameters that will be helpful in describing on-chip communication streams is listed in Table 6.1.

Parameter	**Description**
latency	maximum latency/delay allowed
bandwidth	minimum bandwidth required
completion time	maximum duration of a certain transmission
traffic distribution	e.g., burstiness
resilient error rate	error rate that is tolerable

Table 6.1. Quality-of-Service Parameters to describe communication streams

A set of parameters like these enables applications to describe their communication connections in a form that is advantageous for the NoC. Based on these parameters, the network determines whether and how a connection request can be served.

4. Error-Tolerance Schemes

Combining the contradicting requirements presented in the previous sections leads to a trade-off in terms of error-tolerance: Which error characteristics have to be coped with? Which QoS guarantees have to be realized? This gap must be closed by an error-tolerant interconnect scheme.

Obviously, a NoC can be customized and optimized to be a platform for a class of applications. On the one hand is the implementation of error-tolerance schemes guided by constraints specific for an application-class (e.g., real-time applications). On the other hand impose architecture- and technology-dependent constraints limits on the design freedom.

Error-tolerance can only be achieved by adding some form of redundancy so that errors can be compensated. As the following list suggests, there are many approaches possible, each with its unique advantages and shortcomings.

1 information redundancy (additional information is transported)

   (a) additional information on one link (e.g., a checksum)

   (b) identical information sent over multiple links (broadcast to increase the probability that this information will reach its intended recipient)

2 time redundancy (e.g., repeated transmission or retransmission on request)

3 space redundancy (multiple components working in parallel, the output is typically determined by a majority vote)

The important questions that have to be answered when deciding on an error-tolerance scheme are not only which of the above approaches to achieve redundancy is chosen but also where the error-tolerance scheme is implemented.

In a NoC, error-tolerance can be implemented at various levels. One option is to implement error-control measures on the switch-to-switch links, another option is to define a protocol between connection end-points to cope with packets which were lost or garbled on their way through the network. Regardless of the type of redundancy chosen, implementing switch-to-switch error control usually has the advantage that errors cannot accumulate when the data is transported over multiple hops but are noted as soon as they appear. On the other hand is the design simplified when only end-to-end error-tolerance measures are used.

Typically, a good solution will be comprised of both switch-to-switch and end-to-end error-tolerance schemes, fine-tuned to complement one another.

When making design decisions, it is important to keep in mind that a typical NoC is comprised of a large number of identical network components, namely switches and network interfaces. This leads to energy and area constraints which influence both how much computational complexity can be included in network components and where which amount of buffers can be used. For instance has the number of buffers or pipeline-stages in switches a direct influence on the latency of a communication stream.

When designing a NoC, many parameters can be controlled directly, for sure an advantage in comparison to traditional network concepts. Additionally, in comparison to larger-scale networks new parameters become critical and must be considered during network design in order to achieve a well-balanced trade-off. The most obvious one is energy consumption, a major concern in SoC design.

This motivates a look at error-tolerance schemes from an energy-efficiency viewpoint. In contrast to traditional networks is the energy consumed by encoding/decoding logic in the same order of magnitude as the energy needed for transmitting data. While low-power transmission schemes are susceptible to transmission errors, they will nevertheless prevail future NoC-designs because communication energy minimization will be the only means to meet global energy consumption limits. The result of [6] where the energy-efficiency of transmission schemes was examined indicates that a combination of error-detection and retransmission (if necessary) can be very efficient. Approaches incorporating more complex codes to protect the transmitted data needed too much energy for their encoders and decoders in comparison to the retransmissions they could avoid.

In order to be able to retransmit erroneous packets, buffering has to be implemented into the on-chip network. Additionally, communication between network components becomes necessary so that retransmissions can be requested. Typically, positive or negative acknowledgements along with some sort of sequence number are employed.

The most simple retransmission scheme expects an acknowledgement for each sent packet. This can either be a positive acknowledgement if the packet was successfully received or a negative acknowledgement indicating a transmission error. Optimizations of this basic protocol incorporate sequence numbers to acknowledge a whole group of packets or selectively request retransmission of erroneous packets. These actions aim at reducing the number of acknowledgement packets that have to be sent and result in both a lower network load and higher throughput. Problems with this kind of protocols usually arise when acknowledgement packets are scrambled or lost. Therefore, timers are implemented and the resulting protocols can become very complex due to the large number of possible states.

An approach different to costly error- and flow-control are broadcast-based transmission schemes. For instance, [12] proposes a so-called "gossip protocol" which transmits data from sender to receiver over multiple routes in a broadcast fashion. Since every node broadcasts received packets, the data will eventually reach its intended destination. However, the additional network load this scheme causes in comparison to unicast transmission is hardly compensated by its conceptual simplicity.

As this short overview illustrates, there are as many potential sources of faults and errors threatening future systems on chip as there are strategies and techniques to protect data from corruption and loss. In the remaining sections of this chapter we become more concrete and focused and discuss a two level error protection scheme for a particular NoC architecture, the Nostrum. Even more specifically we focus mostly on the data protection from errors occurring on the wires between the switches of the communication network.

5. Error-Tolerant Interconnects in the Nostrum NoC

5.1 The Nostrum Architecture

The Nostrum NoC architecture [13] is based on a regular $n \times m$ mesh of (network) switches and resources. Every resource is connected to one switch and the switch can be connected to up to four neighbors. The network is based on packet switching with packets of a fixed size determined by the width of the connection between adjacent switches. This is a massively parallel bus structure that allows for bidirectional transmission of two network packets each transfer cycle.

In the current implementation, a packet is 128 bit wide and comprised of both a header and application payload. The Nostrum NoC transports a whole network packet over the wide links between the network switches each cycle. Every switch can send and receive up to five packets at the same time. Since the switches use deflective routing [14, 15] and neither pipelining nor buffering takes place in the switches, no retransmission of scrambled packets between adjacent switches is possible.

This buffer-less design style is motivated by area and power consumption. Introducing buffers into switches would mean to introduce buffers to every input or output connection. Since each switch has five unidirectional inputs and five outputs, each 128 bit wide, the area overhead introduced by the network components would drastically increase. Without buffers, however, tight speed-constraints are introduced: The switch must be designed in a way that allows for fast analysis of the packets. All packets have to be analyzed before the switching decision can be made.

5.2 Connection Types

Nostrum offers best effort (BE) and guaranteed latency (GL) communication services. BE communication is based on individual packets that are communicated and routed independently. Hence, each of these packets contains a relatively long header with the full address information. Since BE packets can deviate from the shortest path between source and destination when traffic load is high [14, 15], no maximum delay can be guaranteed. However, packets that have been longer in the network increase their priority and thus their chance to reach the destination. In contrast, GL communication is based on a *virtual circuit* that is a direct stable connection between a source and a destination. A virtual circuit has to be opened and closed. A virtual circuit is opened by reserving particular time slots in every switch from source to destination. These time slots cannot be used by other packets. Consequently, packets traversing an open virtual circuit cannot be disturbed by other BE or GL packets, thus they experience a deterministic latency from source to destination. The GL packets have a much shorter header without address information, because the switches know where GL packets are to be sent.

The discussion below about error protection is valid for both BE and GL packets. Thus the different reliability services are available for both kinds of traffics. The only difference is that GL packets have a much shorter header and therefore allow for a simplified protection scheme in some of the services described below. For the sake of simplicity we therefore discuss only BE traffic in the following sections.

5.3 Quality of Service for Data Integrity

As discussed above, implementing an on-chip network to handle all inter-IP traffic benefits from offering different QoS-characteristics to its applications.

End-to-End Services. Nostrum offers a two level data protection scheme: end-to-end and link layer data protection. The end-to-end service denotes a technique implemented in the network interfaces between the resources and the communication network. We distinguish two modes: the *send-and-forget* (SaF) service and the *acknowledge-and-retransmit* (AaR) service. We only sketch a possible end-to-end protection mechanism very briefly here and concentrate then on the link layer data protection in the following sections.

In SaF mode no additional action at the network interface is taken, based either on the assumption that the underlying communication is lossless and fully reliable, or because the application can tolerate loss and data corruption at the level provided by the lower layers.

AaR mode defines a window of size $N, 1 \leq N \leq 64$. The receiver, i.e. the network interface at the receiver side, sends an acknowledgement packet to the sender for each N received packets. The acknowledgement packet has one bit for each of the N packets indicating if that packet should be retransmitted by the sender. There are three possible reasons why the receiver requests a retransmission of a packet. (a) The packet has been flagged "corrupted" by the link level data protection mechanism (see below). (b) An end-to-end error detection code, again implemented in the network interface, has detected a corrupted packet. (c) A packet has not arrived after a maximum elapsed time. This maximum time can be determined easily for guaranteed latency packets but has to be set heuristically for best effort traffic.

The sender in an AaR communication buffers $2N$ packets. After having sent N packets it expects the corresponding acknowledgement packet while keeping sending packets. If $2N$ packets have been sent and the first N packets have not been acknowledged, the sender stops and waits for the acknowledgement packet. When it arrives, the sender retransmits requested packets and removes the positively acknowledged packets from its buffer. Then the normal transmission of packets can be resumed.

As mentioned this is only a sketch of a possible data protection mechanism. For a complete AaR protocol all time-out periods and the end-to-end data encoding have to be decided. Also, the actions have to be defined for the case when data corruption and loss increase beyond an acceptable level and jeopardize for example maximum latency guarantees.

In the following we elaborate in more detail the lower level protection mechanism and define four reliability levels for the link layer. It should be noted that all four levels can be combined arbitrarily with both the SaF and the AaR mechanisms leading to eight distinct data protection classes. All of them can be applied to both best effort and guaranteed latency services. Thus, an application can choose from a variety of communication services with different performance and reliability properties.

Link Level Services. For the Nostrum link layer we focus on protecting data from errors occuring on the wires between the switches and not in the switches themself. This is motivated by the fact that the area of the switch to switch interconnect is significantly larger than the area of the switch. In Nostrum the resources form a two dimensional mesh. Making the assumption that the resources have a footprint of $2\text{mm} \times 2\text{mm}$ and the switches have a footprint of $100\mu\text{m} \times 100\mu\text{m} = 10^4\mu\text{m}^2$, the wires between two switches will occupy an area of $2\text{mm} \times 100\mu\text{m} = 2 \cdot 10^5\mu\text{m}^2$, i.e. 20 times the area of the switches. Under the simplifying assumption that the the number of faults is proportional to the area, we expect many more faults to affect the inter-switch wires than the switch logic. Also, we hope to capture the faults in the switches with the end-to-end data protection scheme. However, these assumptions need yet to be substantiated with realistic fault models of future technology generations and quantitative experiments.

In Nostrum, link level error protection is achieved by bus encoding. We propose four QoS-classes describing the characteristics of the switch-to-switch transport of a packet. The characteristics of these service classes range from offering maximum bandwidth at the expense of error tolerance to reduced application bandwidth allowing for more redundancy so that tighter integrity or latency bounds can be met: While the transport of multimedia data may require maximum bandwidth, data integrity is the main concern for transfers from and to memories.

Maximum Bandwidth. Since the amount of wires available for routing the inter-switch buses is limited, the redundancy introduced by coding will directly influence the application payload that can be transported in a packet. Consequently, maximum bandwidth is available to the application when no encoding is applied.

Guaranteed Integrity. To ensure data integrity as far as possible, error detection methods can be used. Data correction is not attempted since it might result in miscorrection. Instead, if erroneous data is detected, a flag can be set or the whole packet can be dropped. Due to the lack of packet buffering in Nostrum's switches, however, no immediate retransmission can be performed. Thus, the main property of this mode is that if the data arrives and no error has been detected, it can be assumed to be correct.

Minimum Latency. To achieve minimum latency, packets must always be forwarded. Only error correction is performed, the underlying assumption is that all errors can be corrected by the code used. This mode is well suited for applications that can tolerate rare errors (due to miscorrection) but depend on receiving data at a constant rate.

It is worthwhile noting that this mode cannot decrease the latency which mainly depends on the routing process in the Nostrum NoC. *Minimum Latency* tries to ensure that a packet arrives at its destination without additional delay introduced by error-tolerance measures.

High Reliability. By using codes capable of correcting errors and detecting uncorrectable errors at the same time, the characteristics of the two previously described modes can be combined at the expense of lower bandwidth: Since more redundancy is necessary, less application payload can be transported per packet.

5.4 Error-Control Coding on the Switch-to-Switch Links

We assume a NoC architecture that requires very low area overhead and high performance from the switches. Thus, for error-control between adjacent switches we consider only codes that can process 100-200 bit wide buses in a single clock cycle and that can be implemented in a few gates of hardware for each bit.

In fact, we focus on parity-based, linear block codes, for the necessary encoders/decoders can be implemented by combinatorial logic, namely EXOR-trees. This helps to meet the tight speed constraints discussed above, because neither time-consuming multi-cycle decoding operations are necessary nor requires the implementation the parallelization of a multi-cycle algorithm, which is usually costly in terms of area (e.g., due to loop unrolling etc.).

A code protecting data on the switch-to-switch buses must allow for fast decoding because decoding (at least of the packet header) has to be completed before the switch can make a routing decision. Additionally, area constraints motivate a switch design with as few gates as possible.

Since the requirement of fast decoding can be fulfilled by combinatorial logic circuits of low logic depth, parity-based codes (e.g., Hamming or Hsiao codes [16]) are considered. These can be employed at various coding schemes: A single-error correcting (SEC) code can correct all single errors. All other errors lead to miscorrection. A double-error detecting (DED) code, on the other hand, detects all single and double errors. In fact, a DED code detects even more errors: If there are 2^k valid codewords of length n, $2^n - 2^k$ error patterns are detected (i.e. all patterns that do not represent a valid codeword). An interesting property is that every SEC code can alternatively be used as a DED code. The properties of these two operating modes are combined in SEC-DED codes which require one more bit of redundancy to encode the same amount of information. Due to that, the error detection capability of SEC-DED codes is always slightly less than that of a comparable DED code. Using codes with the ability to detect/correct more than two random errors usually makes decoding slower because arithmetic decoding becomes necessary [16].

Later, it will be shown how these codes can be employed to implement different protection levels corresponding to multiple QoS requirements.

5.5 Fault Model

From the discussion of faults which will effect future DSM circuits it stands to reason that the errors imposed by these faults will appear clustered both in time and in space. While faults may easily affect multiple adjacent wires, or cause errors in subsequent time steps (e.g., due to delay variations), the probability that they influence wires further away or for a longer period of time will be significantly less.

For the simulation results reported later on, we assumed a combination of three statistically independent error sources, each of the errors was assigned a probability of occurrence of 10^{-9} per wire and time-step. The first two error sources generate simple error patterns that account for errors due to multiple random effects. One source causes errors limited to just one wire, the second one almost always affects two (and very seldom three) adjacent wires. While these two models are quite simple, the third one was extracted from a more detailed experiment and hence attempts to be a little more realistic.

Knowing that the inter-wire capacitance C_I between adjacent wires of a bus has a significant influence on the total capacitance which has to be charged during a bus transition, it can be derived that the signal delay depends strongly on the bus transition pattern. Considering three adjacent wires, transition patterns can be classified in 1C…4C sequences depending on how much they slow down the transition of the middle signal [7].

Name	Transitions
4C	$v_1\overline{v_1}v_1 \longrightarrow \overline{v_1}v_1\overline{v_1}$
3C	$v_1\overline{v_1}v_2 \longrightarrow \overline{v_1}v_1v_2$
	$v_2v_1\overline{v_1} \longrightarrow v_2\overline{v_1}v_1$
2C	$v_1v_2v_3 \longrightarrow v_1\overline{v_2}v_3$
1C	$v_1v_2v_2 \longrightarrow v_1\overline{v_2v_2}$
	$v_2v_2v_1 \longrightarrow \overline{v_2v_2}v_1$

Table 6.2. Classification of crosstalk sequences

Out of a large number of random data patterns, we identified all 4C sequences (i.e., the ones causing maximum signal delay) and marked the middle wire as fault location.

The error patterns extracted from this simulation were formulated according to a fault model notation which describes error duration and number of affected adjacent wires [17, 18]. For instance, at most 7 adjacent wires were affected at the same time. These error patterns served as third error source in subsequent simulations. More details on this along with a formal fault model notation can be found in [17, 18]. Obviously, these simulations have to be re-evaluated when more realistic information about fault scenarios in future technology becomes available. While this will change the numbers and thus can have an impact on design decisions, the general observations discussed below will remain valid.

5.6 Tuning Error-Tolerance by Interleaving

As discussed above, linear block codes were chosen for their simple implementation in switches. During the encoding, a certain amount of redundancy is added to the information bits in form of parity bits. Due to this redundancy, the decoder is able to detect errors and/or to recover the sent information even if the codeword has a limited number of erroneous bits.

Since an encoding that can tolerate only a very low number of errors might not provide the necessary robustness, one possibility is to divide the network packet into several smaller blocks encoded separately. The so-called *parallel coding* increases the redundancy, but enhances the overall error detection and correction capabilities and reduces the length of the critical path in the decoder. This approach also enables using different codes for different parts (blocks) of the packet.

Also knowing that errors are likely to appear locally clustered motivates the idea of using interleaving: If the data is split into multiple blocks encoded separately, the n bits of a data block can be assigned to n adjacent wires of a bus or they can be interleaved, so that they will be assigned to the wires $x, x + \delta, x + 2\delta, \ldots, x + (n - 1)\delta$. The distance δ between two wires that belong to the same block is called *interleaving degree*. Based on this definition, assigning a block of n bits to adjacent wires is interleaving with degree 1.

To compare the capabilities of different coding schemes, we use the probability of uncorrected and undetected errors in a packet, $P_{err,UC}$ and $P_{err,UD}$ respectively. It can be seen from Figure 6.1 that in the presence of multi-bit errors, interleaving can lead to the reduction of those probabilities by several orders of magnitude. At small interleaving degrees, multi-wire faults causing errors on adjacent bit-lines may easily affect multiple bits of one block. If this block is protected by a SEC-DED code, more than one erroneous bit will not be correctable and more than two erroneous bits may cause error detection to fail. With increasing interleaving degree, the probability that a fault affects multiple wires of one block decreases. Instead, multiple blocks will have an erroneous bit each. This can be tolerated if each block is protected by an appropriate code. In the fault scenario used, faults can affect up to 7 adjacent wires. This is why in Figure 6.1, $P_{err,UC}$ and $P_{err,UD}$ constantly decrease with increasing interleaving degree. For interleaving degrees greater than 7, no further improvement can be achieved for the fault hypothesis used.

Encoding multiple small blocks of data can considerably increase the amount of parity signals, i.e. the redundancy. It has, however, several advantages: First, the decoder logic will have a low logic depth. Second, the smaller the

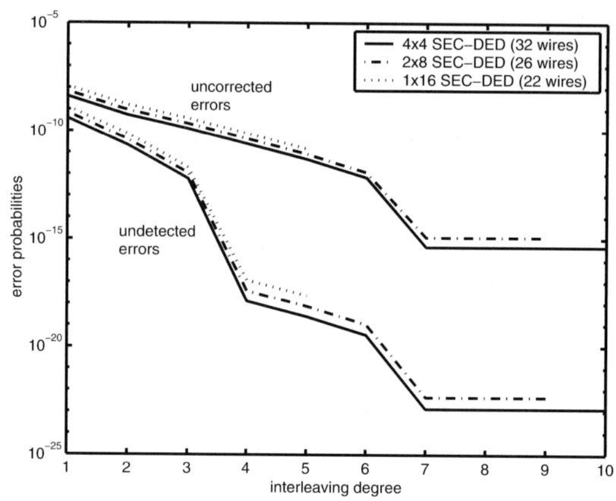

Figure 6.1. Reduction of error probabilities by interleaving

block-size, the higher interleaving degrees are possible. When locally clustered errors are assumed, this usually leads to a significant decrease in the probability of errors that cannot be corrected or detected.

For Nostrum, fast and simple decoder logic was a main requirement. Thus, this method to increase the bus reliability was examined. Given a different set of constraints, a completely different approach might have been chosen. Since the proposed method has the potential of adding a large amount of redundancy to the data transported, the appropriate encoding scheme has to be selected carefully.

5.7 Design Flow

In a packet-switched network, each packet typically carries header and payload. Thus, routing is done on a per-packet basis and each header has to be inspected. The importance of the header's correctness for the network's functionality is obvious. Therefore, the network has to take care of transporting the header information in a way that is as error-tolerant as possible while each application can select a protection level for the data it sends over the network (i.e., the payload part of the packet). Applications cannot, however, influence the header protection scheme.

Arguing that the header is most important naturally leads to the following design flow proposal for selecting protection schemes:

1 Select a header encoding scheme that fulfills a minimum reliability constraint to ensure correct network operation.

2 Determine possible payload encoding schemes and their characteristics.

3 Select appropriate payload encoding schemes to meet the QoS-requirements set by the application class the NoC is designed for.

For the Nostrum architecture which is based on a fixed number of wires connecting adjacent switches, the encoding scheme selected for the header directly influences the possibilities to choose from for payload encoding since transmitting the encoded header needs a certain amount of the fixed number of available wires. This implies at the same time that not necessarily all combinations of header and payload error tolerance requirements can be fulfilled.

5.8 Implementation Example

This example is intended to show how the proposed design flow and encoding scheme can be used in the implementation of a NoC capable of carrying traffic of multiple QoS-classes.

We assume here a 128 bit wide bus between adjacent switches, and 16 bit of header information. The figures given are based on estimation results which were determined using the fault scenario discussed above.

Header Protection. Since the header protection should correct as many errors as possible while at the same time leaving as little errors as possible undetected, the obvious choice is a SEC-DED code.

The packet header (16 bits) can be encoded as one block. In this case, 22 wires are required and the code is referred to as $1 \times (22, 16)$ SEC-DED code. The notation $b \times (n, k)$ is used to denote an encoding scheme in which b blocks are independently encoded by a code which transmits k information bits within codewords of n bits.

Alternatively, better error resilience can be achieved if the 16 bit wide header is encoded as 2 blocks of 8 bits each ($2 \times (13, 8)$ SEC-DED code) or as 4 blocks ($4 \times (8, 4)$ SEC-DED code).

Code	Wires	Max. interl.	Crit. path	$P_{err,UC}$	$P_{err,UD}$
no protection	16	-	0	$5.33 \cdot 10^{-8}$	$5.33 \cdot 10^{-8}$
$1 \times (22, 16)$ SEC-DED	22	5	9	$1.47 \cdot 10^{-11}$	$2.37 \cdot 10^{-18}$
$2 \times (13, 8)$ SEC-DED	26	9	8	$1.21 \cdot 10^{-15}$	$4.63 \cdot 10^{-23}$
$4 \times (8, 4)$ SEC-DED	32	16	6	$4.33 \cdot 10^{-16}$	$1.32 \cdot 10^{-23}$

Table 6.3. Comparison of header protection with different SEC-DED codes

Table 6.3 shows the amount of wires required for these different encoding schemes along with information about the maximum interleaving degree, expected decoder logic depth and the probabilities of uncorrected and undetected errors ($P_{err,UC}$ and $P_{err,UD}$) when maximum interleaving degree is used to allocate bus lines. As a comparison the first line shows the error probability without a protecting code.

A constraint of one undetected error in the header of transported packets per year in a NoC with 100 buses and an operation frequency of 1GHz translates to $P_{err,UD} \leq 3.1710 \cdot 10^{-19}$. Based on the information from Table 6.3 and complementing experiments, the $2 \times (13, 8)$ SEC-DED code was chosen as header encoding scheme for this example. It is a reasonable trade-off between the remaining error probability and the decoder speed (critical path). Since this code requires 26 wires, 102 of the 128 wires remain to transmit the payload.

Payload Protection. Regarding the four Quality-of-Service classes discussed previously, the *Maximum Bandwidth* mode is inherent in every implementation since it does not use any payload encoding. In the current example, this mode's bandwidth is 102 bits/packet and the probability of an erroneous transfer is $3.36 \cdot 10^{-7}$.

Both the *Guaranteed Integrity* and *Minimum Latency* modes can be implemented with one code, using the property that every SEC code can also work as DED code: The same codewords are used for transmission. Hence, the encoder is identical and the DED code just uses the first part of the decoder block (which detects erroneous transmissions) but not the second one which performs correction and thereby alters the received data if necessary. Thus, these two services offer the same bandwidth to the application, but different protection characteristics. It is very advantageous that implementing these two QoS-classes at approximately the cost required for implementing just one of them is possible.

In general, the selection of an appropriate coding scheme is guided by reliability and bandwidth constraints. Tunable parameters are the number of parallel encoded blocks and the interleaving degree. The trade-off for implementing *Guaranteed Integrity* and *Minimum Latency* on 102 wires is shown in Table 6.4. Depending on the code used, the application bandwidth and resilient

Code	Max. interl.	Avail. bandw.	used as SEC $P_{err,UC}$	used as DED $P_{err,UD}$
$1 \times (102, 95)$	1	95	$5.68 \cdot 10^{-8}$	$6.13 \cdot 10^{-9}$
$2 \times (51, 45)$	2	90	$3.77 \cdot 10^{-9}$	$1.71 \cdot 10^{-10}$
$3 \times (34, 28)$	3	84	$5.88 \cdot 10^{-10}$	$3.20 \cdot 10^{-12}$

Table 6.4. Error probabilities for different operation modes of SEC (or DED) codes using 102 physical wires

error probability vary. Out of these coding schemes, the appropriate one can be chosen at design-time to serve as bus encoding for *Guaranteed Integrity* and *Minimum Latency*.

The fourth mode, the *High Reliability* mode is based on a SEC-DED code. Hence, it will require different encoder/decoder logic in parallel to that used for the two previous modes. Note that SEC-DED codes offer one information bit less per block compared to SEC codes occupying the same amount of wires.

For the 128 bit bus of our example, we have finally selected the coding schemes from Table 6.5 to implement the four QoS levels. The selection of services and coding schemes a NoC supports is a design-time decision which depends on the intended application and environment, whereas each packet can be encoded with one of these codes during run-time. Since subsequent packets can be encoded differently, the information about the payload encoding scheme is transmitted as part of the header information.

Mode	Code / Application bandwidth		$P_{err,UC}$	$P_{err,UD}$
maximum bandwidth	none	102	$3.36 \cdot 10^{-7}$	$3.36 \cdot 10^{-7}$
guaranteed integrity	$4 \times (25, 20)$ DED	64	$9.82 \cdot 10^{-8}$	$1.59 \cdot 10^{-17}$
minimum latency	$4 \times (25, 20)$ SEC	64	$8.64 \cdot 10^{-11}$	$8.64 \cdot 10^{-11}$
high reliability	$3 \times (34, 27)$ SEC-DED	65	$5.88 \cdot 10^{-10}$	$3.20 \cdot 10^{-12}$

Table 6.5. Selected coding schemes for payload encoding

Implementation. Figure 6.2 gives an example of how the input stage of a switch could work: the received packet is split into header and payload. After the decoding of the header information, the coding scheme of the payload is known and the appropriate decoder output can be selected. The decoded destination address is used to make the switching decision which finally leads to the selection of the output port the data is forwarded to.

It is important to note that it is not necessary to decode the data and to encode it again in a second block. The decoders can rather work in a way that their output is not the decoded and – if necessary – corrected information, but the associated codeword instead. This means that their input is the received data and their output a valid codeword which can be directly passed to the next bus.

6. Conclusion

We have presented various important aspects in implementing error-tolerant interconnect schemes which are crucial to a NoC since they encapsulate the low-level and implementation details of the NoC-platform. Thereby, they are helpful in offering services with defined characteristics. These services – which form the interface between platform and application – were discussed in detail using the Nostrum NoC as example. It provides a total of eight QoS-classes by combining both link-level and end-to-end error-tolerance. The ser-

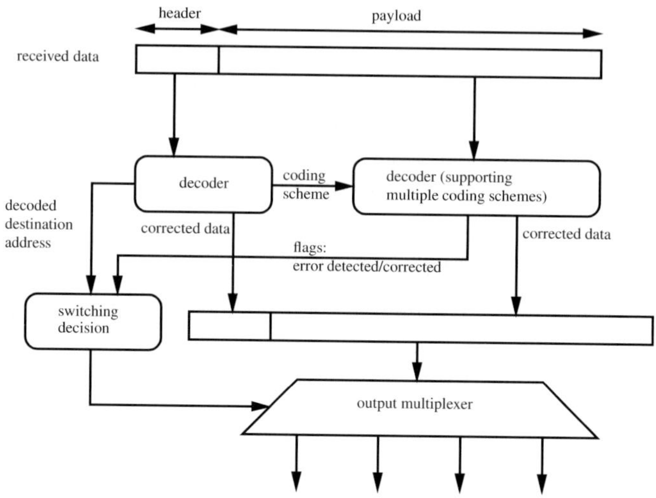

Figure 6.2. Applying error correction to incoming packets in a switch

vices offered by Nostrum were discussed in great detail showing that link-level (i.e., switch-to-switch) error-tolerance within the network in combination with end-to-end error-tolerance allows for efficient adaptive protection schemes.

Since error-tolerance profits from being implemented as low in the protocol stack as possible, we focused on switch-to-switch error protection, presenting a bus encoding scheme that allows for fast and simple decoding while supporting multiple QoS-characteristics. As described in the introductory sections, the development of this error-tolerance scheme was guided by specific constraints and it should be kept in mind that other constraints (especially if pipelining or packet retransmission was possible) could have lead to a different implementation approach.

References

[1] ITRS. International Technology Roadmap for Semiconductors, 2001.

[2] Michael Cuviello, Sujit Dey, Xiaoliang Bai, and Yi Zhao. Fault modeling and simulation for crosstalk in system-on-chip interconnects. *1999 IEEE/ACM International Conference on Computer-Aided Design*, pages 297–303, 1999.

[3] Michele Favalli and Cecilia Metra. Bus crosstalk fault-detection capabilities of error-detecting codes for on-line testing. *IEEE Transactions on Very Large Scale Integration (VLSI) Systems*, 7(3):392–396, September 1999.

[4] Luca Macchiarulo, Enrico Macii, and Massimo Poncino. Wire placement for crosstalk energy minimization in address buses. In *Proceedings of the 2002 Design, Automation and Test in Europe Conference and Exhibition (DATE 2002)*, pages 158 – 162, 2002.

[5] Tiehan Lv, Joerg Henkel, Haris Lekatsas, and Wayne Wolf. An adaptive dictionary encoding scheme for soc data buses. In *Proceedings of the 2002 Design, Automation and Test in Europe Conference and Exhibition (DATE 2002)*, pages 1059 – 1064, 2002.

[6] Davide Bertozzi, Luca Benini, and Giovanni De Micheli. Low power error resilient encoding for on-chip data buses. *Proceedings of the 2002 Design, Automation and Test in Europe Conference and Exhibition (DATE 2002)*, pages 102–109, 2002.

[7] Chinjie Duan, Anup Tirumala, and Sunil P. Khatri. Analysis and avoidance of cross-talk in on-chip buses. *Hot Interconnects 9*, pages 133–138, 2001.

[8] Y. Tosaka, S. Satoh, T. Itakura, K. Suzuki, T. Sugii, H. Ehara, and G. A. Woffinden. Cosmic ray neutron-induced soft errors in sub-half micron cmos circuits. *IEEE Electron Device Letters*, 18(3):99 – 101, March 1997.

[9] Marcello Lajolo, Matteo Sonza Reorda, and Massimo Violante. Early evaluation of bus interconnects dependability for system-on-chip designs. 14th *International Conference on VLSI Design*, pages 371–376, 2001.

[10] Frederic Worm, Patrick Thiran, Paolo Ienne, and Giovanni De Micheli. An adaptive low-power transmission scheme for on-chip networks. In *Proceedings of the* 15th *International Symposium on System Synthesis*, pages 92 – 100, October 2002.

[11] Kees Goossens, John Dielissen, Jef van Meerbergen, Peter Poplavko, Andrei Rădulescu, Edwin Rijpkema, Erwin Waterlander, and Paul Wielage. Guaranteeing the quality of services in networks on chip. In Axel Jantsch and Hannu Tenhunen, editors, *Networks on Chip*. Kluwer Academic Publishers, 2003.

[12] Tudor Dumitras, Sam Kerner, and Radu Mărculescu. Towards on-chip fault-tolerant communication. In *Proceedings of the 2003 Asia and South Pacific Design Automation Conference (ASP-DAC 2003)*, pages 225 – 232, January 2003.

[13] Shashi Kumar, Axel Jantsch, Juha-Pekka Soininen, Martti Forsell, Mikael Millberg, Johnny Öberg, Kari Tiensyrjä, and Ahmed Hemani. A network on chip architecture and design methodology. In *Proceedings of IEEE Computer Society Annual Symposium on VLSI*, pages 105 – 112, April 2002.

[14] Erland Nilsson. Design and implementation of a hot-potato switch in a network on chip. Master of science thesis, Royal Institute of Technology (KTH), Stockholm, Sweden, June 2002.

[15] Erland Nilsson, Mikael Millberg, Johnny Öberg, and Axel Jantsch. Load distribution with the proximity congestion awareness in a network on chip. In *Proceedings of the Design Automation and Test Europe (DATE)*, pages 1126–1127, March 2003.

[16] T.R.N. Rao and E. Fujiwara. *Error-Control Coding for Computer Systems*. Prentice-Hall International, 1989.

[17] Heiko Zimmer. Fault modelling and error-control coding in a network-on-chip. Master of science thesis, Royal Institute of Technology (KTH), Stockholm, Sweden, December 2002.

[18] Heiko Zimmer and Axel Jantsch. A fault model notation and error-control scheme for switch-to-switch buses in a network-on-chip. In *Proceedings of the Frist IEEE/ACM/IFIP International Conference on Hardware/Software Codesign & System Synthesis*, pages 188 – 193, October 2003.

Chapter 7

POWER REDUCTION CODING FOR BUSES

Energy Measures and Performance Limits

Paul P. Sotiriadis
Electrical and Computer Engineering
Johns Hopkins University
pps@jhu.edu

Abstract We know that coding can reduce power consumption in buses. We would like to know how much power reduction is possible. This chapter presents: energy models of deep sub-micron buses, the coupling between energy and transmitted information, the ultimate limits of achievable power reduction using coding, a global theoretical framework of power reduction coding.

Keywords: activity, architecture, bus, coding, deep sub micron, digital circuits, sub micron, energy, entropy, estimation, low power, modeling, power reduction, process, transition, transition activity, transition activity matrix, VLSI

Introduction

Data processing and data transmission through a channel is always associated with power consumption. Power reduction coding is the employment of coding in the reduction of the processing or transmission power consumption that does not alter the total behavior of the processor or the communication device otherwise. The idea is demonstrated in Figure 7.1. Power reduction coding has at least two forms and a long history. One is *data compression*, which has been studied for more than half a century. Consider the transmission of a $1Mbit$ file through a channel, e.g. a piece of wire in the chip or a wireless link, at the cost of $1nJ$ per bit. The total energy required to transmit the file[1] is $1mJ$. If the file is compressed to $100Kbits$ before transmitted then the cost of transmission is $0.1mJ$. In addition to that we have to consider

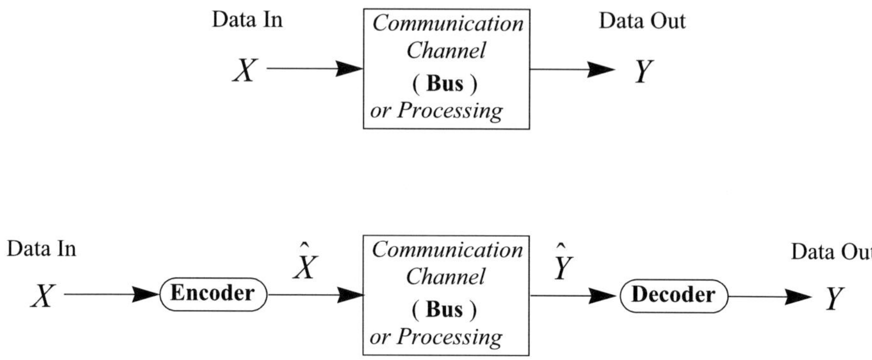

Figure 7.1. Coding for Power Reduction

the amount of energy needed to compress the file before transmitting it, and to decompress it at the receiver side. Let these energy amounts be $0.1mJ$ and $0.1mJ$ respectively. Therefore, in this hypothetical example we can reduce the total required energy for the transmission of the file from $1mJ$ to $0.3mJ$. In a real system, we also need to consider the time it takes to compress and decompress the file (latency and delay), the size of the encoder and decoder, the reliability of the expanded system, and maybe a few other characteristics.

In the case of data compression coding for power reduction, *the size of the data decreases.* Although it is counterintuitive, power/energy consumption can be reduced by *increasing the size of the data* as well. To see this, let's consider the following example[2]. Suppose we have a serial, binary channel consuming energy E for every 1 transmitted and zero energy for every 0 transmitted. Suppose we want to transmit a sequence of 8 bits $b_0, b_1, \ldots, b_7$. If we do this in brute force the cost will be (*Number of* 1*'s*)$\times E$. Alternatively we may encode $b_0, b_1, \ldots, b_7$ into $2^8 = 256$ physically-transmitted bits $f_0, f_1, \ldots, f_{255}$ so that: all bits $f_0, f_1, \ldots, f_{255}$ are zero except the one in the position $i = b_0 + b_1 2 + \ldots + b_7 2^7$.

In this case the cost is E, which, in average, is much less than (*Number of* 1*'s*)$\times E$. You may agree that this simple scheme can reduce power dramatically; but, you may also wonder about the huge "redundancy" in bit transmissions that is required, i.e. sending 256, instead of 8 bits, through the serial link; this reduces the bit rate to 1/32. Indeed, although this is an extreme example, power reduction may not be free of reduction in the speed of transmission, in many cases.

Now consider another example. Suppose the channel we discussed before was formed out of a single wire carrying a stream of bits. Suppose now that instead of one wire, we place 256 of them in parallel. In this case we can encode the data, $b_0, b_1, \ldots, b_7$, *spatially*, i.e. use a parallel transmission of bits through the 256 wires and send a 1 through wire number $i = b_0 + b_1 2 + \ldots + b_7 2^7$, and zeros through all other wires. By doing so we can transmit 8 bits within one clock cycle[3] instead of 8 clock cycles that would be needed if we had only one wire available. In addition, we achieve significant energy reduction as before[4]. How do we pay for the joint bit rate increase and power reduction? We pay in chip area, i.e. number of wires.

The conclusion of the previous examples is that there are three major quantities: *Power*, *Speed* and *Area*[5]. Coding trades off one for the others.

This chapter presents the fundamental relation between *Power*, *Speed* and *Area* in the case of modern *Deep Sub-Micron* technology (DSM) buses. To this end, energy models of DSM buses and their corresponding statistical energy measures are discussed.

It is not possible to cover all the developments in bus coding for power reduction within one chapter. Research on this topic dates back to the early $80's$, possibly even earlier depending upon one's view of the subject. We could consider the works in [1]-[5] as the first efforts in the field. There, the bus lines were modeled as decoupled, grounded capacitances; and, energy was consumed every time there was a bit change. This bus model was reasonable for older (large scale) technologies (see Section 7.1). A significant amount of work followed using the same bus model [6]-[21][6] including theoretical results in [22] and more recently in [23].

Power reduction coding in the deep sub-micron (DSM) era was initiated in [24] where a DSM bus energy model was introduced to support coding schemes appropriate for DSM technologies. Changing the bus energy model was unavoidable. The strong interline coupling due to nearness and relatively higher aspect ratio of the lines could not be ignored. The DSM bus energy model is available in detail in [25]-[26]. Power reduction coding techniques using the new energy measures followed immediately in [27], [28], [29] and later in [30]-[35].

The derivation of the fundamental relation between *Power*, *Speed* and *Area* was done in [36] and [26]. The class of Finite State Machine (FSM) coding schemes has been analyzed in detail in [37] where closed form expressions of the power savings are available.

It is important to mention that work in power reduction coding has been supported by work on the modeling and characterization of buses [38]-[50] as well as general power modeling and CAD tools [51]-[56].

1. Bus Energy Model

A tool to estimate the energy consumption in the bus is the first thing we need in order to study the efficiency of coding schemes for power/energy reduction. Simulation is the obvious tool but not a useful one when a large number of energy estimates is needed. For example, in a 16-line bus there are $2^{16} \times 2^{16} = 2^{32}$ possible pairs of consecutive vectors; although it is almost impossible to make 2^{32} simulation runs, it is relatively easy to estimate all these energies using a simple analytical model. Moreover, an analytical model can be the guideline for the design of power reduction schemes [37].

Before we proceed with the discussion of the general DSM bus energy model, we consider first the simple case of the 1-line bus shown in Figure 7.2. For simplicity let's assume that the line can be considered as a

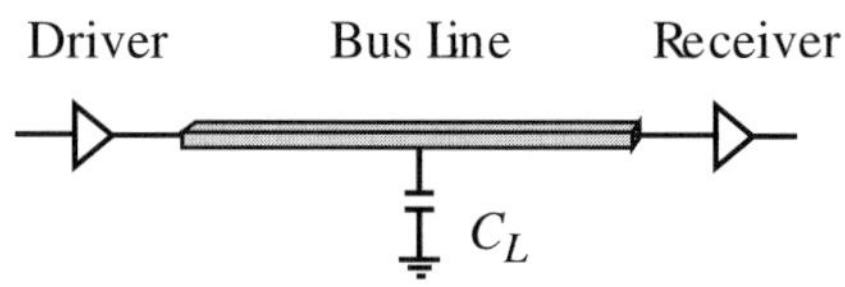

Figure 7.2. One-line bus

lumped circuit node, i.e. no distributed phenomena take place, with a total parasitic capacitance to ground (environment) equal to C_L.

Let's examine what happens during the clock cycle $[0, T]$, where, T is the clock period of the bus. Suppose that at $t = 0$, the voltage, V, on the line, is zero, $V(0) = 0$, and that during the cycle $[0, T]$, the line voltage changes to V_{dd}. The current $i(t)$ charging C_L flows from the power source, V_{dd}, to the line through the (final) CMOS inverter of the driver. Therefore, the instant power drawn from V_{dd} to charge C_L is $V_{dd}\, i(t)$. Note that $i(t) = C_L \frac{dV(t)}{dt}$. The total energy drawn from V_{dd} during the cycle[7] is

$$E^{Vdd} = \int_0^T V_{dd}\, C_L \frac{dV(t)}{dt} dt \tag{7.1}$$

which implies $E^{V_{dd}} = V_{dd}\, C_L\, (V(T) - V(0))$. In most digital systems we can assume that the time period, T, is sufficiently long for $V(T)$ to reach its final value, here, V_{dd}. Therefore, during the transition we consider, from $V(0) = 0$ to $V(T) = V_{dd}$, it is $E^{V_{dd}} = V_{dd}^2\, C_L$.

It is emphasized that $E^{V_{dd}}$ is the *energy drawn from the power source* required to change the voltage of the capacitor, C_L, from 0 to V_{dd} during clock cycle $[0, T]$. Note that there is zero energy stored in the capacitor

C_L at $t = 0$, while at the end of the period, $t = T$, energy $\frac{1}{2}C_L V_{dd}^2$ is stored in it. Energy balance implies that the *energy consumed* (transformed to heat) during the transition is $E = \frac{1}{2}C_L V_{dd}^2$.

Now consider the case where the voltage of the line changes from V_{dd} to 0 during the cycle $[0, T]$. During this transition, there is no current flowing from the power source to the line. Instead, there is current flow from the line to ground, discharging the capacitor C_L. This means that no energy is drawn from the power source, i.e. $E^{V_{dd}} = 0$, while the energy consumption (energy transformed to heat) is equal to the energy initially stored in the capacitor, so $E = \frac{1}{2}V_{dd}^2 C_L$.

Finally, if the voltage of the line, V, remains unchanged during the clock cycle, then there is no current flow and both the energy drawn from the power source, $E^{V_{dd}}$, and, the energy consumed, E, are zero. The results of the discussion are summarized in the following tables.

Table 7.1a. Energy drawn from the power source during the transitions of the 1-line bus

↱	0	V_{dd}
0	0	$V_{dd}^2 C_L$
V_{dd}	0	0

Table 7.1b. Energy consumed (transformed to heat) during the transitions of the 1-line bus

↱	0	V_{dd}
0	0	$\frac{1}{2}V_{dd}^2 C_L$
V_{dd}	$\frac{1}{2}V_{dd}^2 C_L$	0

Let's name the initial voltage of the line V^i and the final voltage V^f, $V^i, V^f \in \{0, V_{dd}\}$. The two energy functions presented in Table 7.1a and Table 7.1b can be expressed algebraically as

$$E^{Vdd}(V^i, V^f) = V^f \, C_L \, (V^f - V^i) \tag{7.2}$$

and

$$E(V^i, V^f) = \frac{1}{2} \, (V^f - V^i) \, C_L \, (V^f - V^i) \tag{7.3}$$

respectively. The capacitive term, C_L, was placed in the middle of the expressions (7.2) and (7.3) on purpose. The reason will become obvious, shortly.

Energy expressions (7.2) and (7.3) are the starting point of rigorous bus energy modeling and coding-scheme performance evaluation. In Section 7.2, it is shown how they lead to *transition activity*, a well known statistical power measure.

Now let's generalize the energy results we have found. Consider a bus with n lines that are electrically decoupled and can be treated as lumped circuit nodes. The bus is shown in Figure 7.3 and is a juxtaposition of replicas of the 1-line bus shown in Figure 7.2.

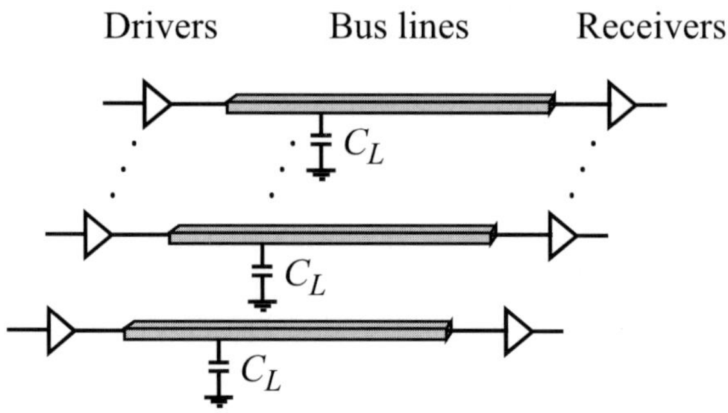

Figure 7.3. Simple bus with n decoupled and lumped lines

Again, we examine the energy behavior of the bus during the clock cycle $[0, T]$. Let V_r^i and V_r^f be the initial and final voltages of the rth line respectively, $r = 0, 1, \ldots, n$. It is $V_r^i, V_r^f \in \{0, V_{dd}\}$. The (total) energy drawn from the power source and the (total) energy consumed are the sums of these of the individual lines; so,

$$E^{Vdd}(V^i, V^f) = \sum_{r=1}^{n} V_r^f \, C_L \, (V_r^f - V_r^i) \tag{7.4}$$

and

$$E(V^i, V^f) = \sum_{r=1}^{n} \frac{1}{2} \, (V_r^f - V_r^i) \, C_L \, (V_r^f - V_r^i) \tag{7.5}$$

Expressions (7.4) and (7.5) can be written more compactly if we define, the initial voltage vector, $V^i = (V_1^i, V_2^i, \cdots, V_n^i)^T$, the final voltage vector, $V^f = (V_1^f, V_2^f, \cdots, V_n^f)^T$, and the $n \times n$ diagonal matrix $\mathcal{A}$,

$$\mathcal{A} = \begin{bmatrix} 1 & 0 & 0 & \cdots & 0 \\ 0 & 1 & 0 & \cdots & 0 \\ 0 & 0 & 1 & \cdots & 0 \\ \vdots & \vdots & \vdots & \ddots & \vdots \\ 0 & 0 & 0 & \cdots & 1 \end{bmatrix} \cdot C_L \tag{7.6}$$

Then we have

$$E^{Vdd}(V^i, V^f) = (V^f)^T \mathcal{A} \, (V^f - V^i) \tag{7.7}$$

and

$$E(V^i, V^f) = \frac{1}{2} \, (V^f - V^i)^T \mathcal{A} \, (V^f - V^i) \tag{7.8}$$

Matrix $\mathcal{A}$ is the *admittance* matrix of the total-capacitance network of the bus; i.e. the admittance matrix of the set of n nodes connected to ground through n capacitors of size C_L. This is a trivial extension of the admittance C_L in expression (7.3).

In the above analysis, we have ignored two other causes of energy consumption. One is the parasitic capacitances of the drivers of the lines and the parasitic capacitances of the registers at the other end of the bus. Although, in most cases, these are small compared to C_L, they can also be considered as part of C_L; and, the energy expressions remain valid. Another source of energy consumption is the short circuit current of the drivers of the lines. The energy loss due to short circuit currents is usually very small compared to the energy consumed by the bus. Estimates of the short-circuit current can be found, e.g., in [55].

1.1 DSM Bus Circuit and Energy Models

In DSM technologies, the lines have: smaller width, larger aspect ratio (height/width); and, in most cases, they are placed closer to each other as compared to older, larger-scale technologies. Moreover, the chip area has increased and so has the ratio of the expected bus length over the cross sectional area of the lines. All of these give rise to more parasitic elements than the simple grounded capacitor making a distributed model of the bus lines with capacitive and inductive inter-line coupling more appropriate than the basic one we considered before. The circuit model of Figure 7.4 has been used extensively for delay and signal integrity evaluation [47–50] as a (simple) electrical equivalent to modern DSM technology buses. The lines are distributed[8], laid in parallel along the x axis, and have physical length L. They have serial resistance, $r_i(x)$, $i = 1, 2, \ldots, n$. The capacitance density between the ith line and ground is $c_{i,i}(x)$, and that between lines i and j is $c_{i,j}(x)$. Moreover, $\mu_{i,i}(x)$ is the density of the self inductance of the ith line and $\mu_{i,j}(x)$ is the density of the mutual inductance between lines i and j. The densities may depend upon x. Lumped parasitics, if they exist, can be considered as limiting cases of distributed ones. Details of the electrical characterization of the bus lines and their modeling can be found in [38]-[45].

As before, we examine the bus during the clock cycle $[0, T]$. Let V_r^i and V_r^f be the voltages along line r at the beginning and at the end of the cycle respectively, $r = 1, 2, \ldots, n$. Again we make the realistic assumption that T is sufficient for the voltages along the lines to settle to their final values. The vector of the initial voltages is $V^i = (V_1^i, V_2^i, \cdots, V_n^i)^T$; and, the vector of the final voltages is $V^f = (V_1^f, V_2^f, \cdots, V_n^f)^T$.

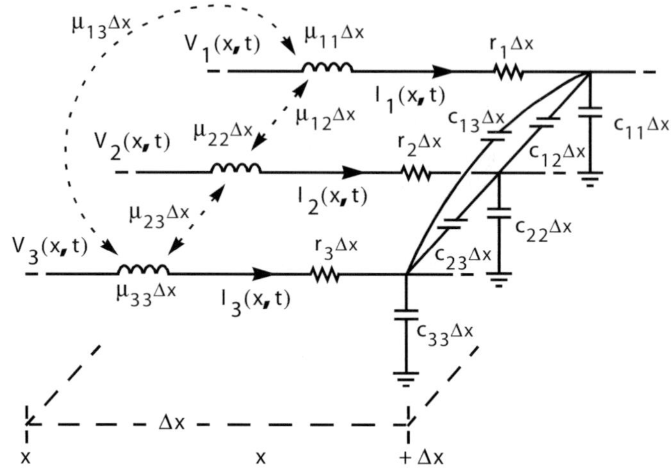

Figure 7.4. DSM distributed model of the bus lines (coupled transmission lines) [from Ref. [25] @ 2002 IEEE]

The energy drawn from the power source, V_{dd}, during the clock cycle is

$$E^{Vdd}(V^i, V^f) = (V^f)^T \mathcal{A} (V^f - V^i) \tag{7.9}$$

A detailed derivation of the formula can be found in [25] or [26]. Matrix $\mathcal{A} = [\mathcal{A}]_{i,j=1}^n$ is given by

$$[\mathcal{A}]_{i,j} = \begin{cases} \sum_{k=1}^n C_{i,k} & \text{if} \quad i = j \\ -C_{i,j} & \text{if} \quad i \neq j \end{cases}, \tag{7.10}$$

where $C_{i,j}$, $i \neq j$ is the total capacitance between the ith and j^{th} bus lines; and, $C_{i,i}$ is the total capacitance between the ith bus line and ground [9]. The energy consumed during the clock cycle is

$$E(V^i, V^f) = \frac{1}{2} (V^f - V^i)^T \mathcal{A} (V^f - V^i) \tag{7.11}$$

The details of the derivation are available in [25] and [26]. Again, matrix $\mathcal{A}$ is given by (7.10). The similarity between (7.2), (7.7) and (7.9) becomes clear after observing that $\mathcal{A}$ is the admittance matrix of the total-capacitance part of the network in Figure 7.4. For example, if $n = 4$, the total-capacitance part of the network in Figure 7.4 is shown in Figure 7.5.

Finally, note the similarity between the expressions of the consumed energy, (7.3), (7.8) and (7.10).

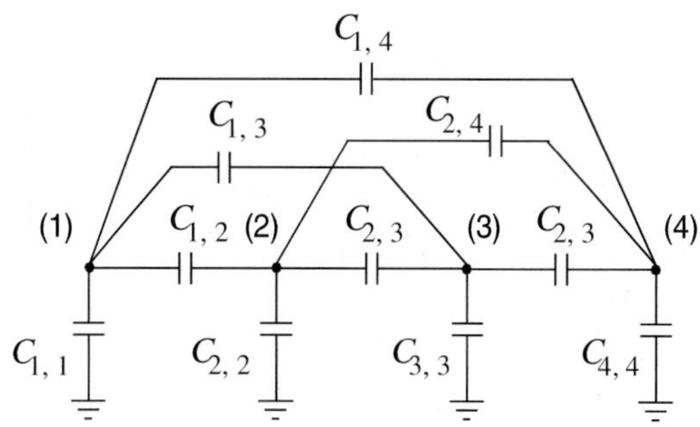

Figure 7.5. Equivalent total-capacitance network for $n = 4$ [from Ref. [25] @ 2002 IEEE]

EXAMPLE 1 *Let's calculate the energy consumed on a four-line bus during the transition from initial voltages $V^i = V_{dd} \cdot (0, 1, 0, 1)$ to final voltages $V^f = V_{dd} \cdot (1, 0, 0, 0)$. Suppose the bus has total parasitic capacitances like those of the network in Figure 7.5. We have*

$$E = \tfrac{1}{2}\,(V^f - V^i)^T\,\mathcal{A}\,(V^f - V^i)$$

$$= \frac{V_{dd}^2}{2}\,(1,-1,0,-1)\begin{bmatrix} \sum_{k=1}^4 C_{1,k} & -C_{1,2} & -C_{1,3} & -C_{1,4} \\ -C_{2,1} & \sum_{k=1}^4 C_{2,k} & -C_{2,3} & -C_{2,4} \\ -C_{3,1} & -C_{3,2} & \sum_{k=1}^4 C_{3,k} & -C_{3,4} \\ -C_{4,1} & -C_{4,2} & -C_{4,3} & \sum_{k=1}^4 C_{4,k} \end{bmatrix}\begin{pmatrix} 1 \\ -1 \\ 0 \\ -1 \end{pmatrix}$$

EXAMPLE 2 *In a bus with a structure as in Figure 7.4, the capacitive coupling between non-consecutive lines is usually very weak relatively to that between consecutive lines. An approximate model of the total-capacitance network is shown in Figure 7.6.*

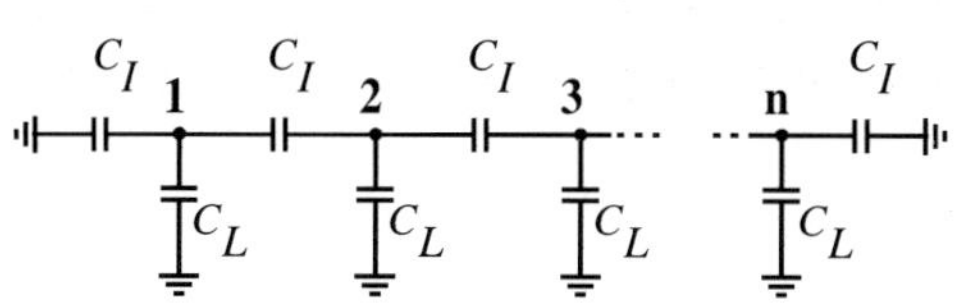

Figure 7.6. Equivalent total-capacitance network ignoring coupling between non adjacent lines [from Ref. [26] @ 2002 WSPC]

The exact value of the boundary capacitances depends on what is placed next to the bus and whether the bus is laterally shielded or not. Here, we assume that the boundary capacitances are equal to C_I. In this case, the admittance matrix, $\mathcal{A}$, is simplified to (7.12) with $\lambda = C_I/C_L$.

$$\mathcal{A} = \begin{bmatrix} 1+2\lambda & -\lambda & 0 & \cdots & 0 & 0 \\ -\lambda & 1+2\lambda & -\lambda & \cdots & 0 & 0 \\ 0 & -\lambda & 1+2\lambda & \cdots & 0 & 0 \\ \vdots & \vdots & \vdots & \ddots & \vdots & \vdots \\ 0 & 0 & 0 & \cdots & 1+2\lambda & -\lambda \\ 0 & 0 & 0 & \cdots & -\lambda & 1+2\lambda \end{bmatrix} \cdot C_L \tag{7.12}$$

In modern bus designs, wire arrangements, like that of Figure 7.7, are used to reduce coupling between the lines (mostly inductive coupling). The lines carrying data are grouped into blocks of one, two, three, or more (usually less than eight) that are separated by dummy lines connected to ground or V_{dd}. The energy formulas (7.9) and (7.11) are valid

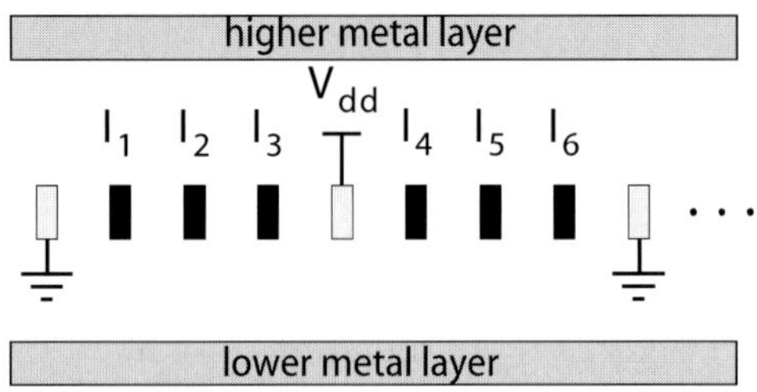

Figure 7.7. Bus-lines fabric [from Ref. [26] @ 2002 WSPC]

in this case, as well. To apply them, we only need to fix the voltages of the dummy lines in the vectors V^i and V^f to 0 or V_{dd} appropriately. This is illustrated in the following example.

EXAMPLE 3 *In the case of a bus with a capacitive structure, as in Figure 7.7, (and with negligible capacitive coupling between non-consecutive lines) the admittance matrix, $\mathcal{A}$, is given by (7.12). The energies are given by (7.9) and (7.11). The initial and final voltage vectors are*

$$V^i = (0, V_1^i, V_2^i, V_3^i, V_{dd}, V_4^i, V_5^i, V_6^i, 0, \ldots)^T$$

and

$$V^f = (0, V_1^f, V_2^f, V_3^f, V_{dd}, V_4^f, V_5^f, V_6^f, 0, \ldots)^T$$

respectively.

2. Statistical Energy Estimation

Now, we couple the energy measures we derived in the previous section with the statistics of the signals in the bus. We consider first the simple case of the 1-line bus of Figure 7.2. It is convenient to denote the (final) voltage of the line at the end of the kth cycle as $V(k)$. In other words, during the kth cycle, the voltage of the line changes from $V(k-1)$ to $V(k)$. Also, we denote the consumed energy during the kth cycle as $E(k)$. Then, expression (7.3) gives $E(k) = \frac{1}{2}(V(k) - V(k-1))^2 C_L$. If, in addition, we write $V(k) = V_{dd}\, l(k)$, where $l(k) \in \{0, 1\}$ is the binary value of the signal at the end of the cycle, then[10]

$$E(k) = \frac{1}{2}\,(l(k) - l(k-1))^2\, V_{dd}^2\, C_L \tag{7.13}$$

The bits $l(1), l(2), \ldots, l(k), \ldots$ of the sequence transmitted though the line are random variables taking the values 0 or 1. They may or may not be statistically dependent. Since $E(k)$ depends on random bits, it is a random variable itself. It terms of power consumption, we would like to estimate the expected value $\overline{E(k)}$ of $E(k)$ where the *overbar* means expectation with respect to all associated random variables. We have

$$\begin{aligned} \overline{E(k)} &= \frac{1}{2}\,\overline{(l(k) - l(k-1))^2}\, V_{dd}^2\, C_L \\ &= \frac{1}{2}\left(\overline{l^2(k)} + \overline{l^2(k-1)} - 2\overline{l(k)l(k-1)}\right) V_{dd}^2\, C_L \end{aligned} \tag{7.14}$$

In many practical situations, the random bit sequence is stationary, at least in the wide sense. Assuming this, let $R(r) = \overline{l(k+r)\, l(k)}$ be its autocorrelation function[11]. Then (7.14) gives $\overline{E} = (R(0) - R(1))\, V_{dd}^2\, C_L$. Note that $R(0) = \overline{l(k)^2} = \overline{l(k)}$. The quantity,

$$T^a = R(0) - R(1) \tag{7.15}$$

is called the *transition activity* of the line[12]. Finally, if f is the clock frequency of the bus, the *expected energy consumed per cycle* and *expected power consumption* can be written, using expression (7.13) and definition (7.15), respectively as

$$\overline{E} = T^a\, V_{dd}^2\, C_L\,, \quad P = T^a\, f\, V_{dd}^2\, C_L \tag{7.16}$$

Remark: It is important to mention that formulas (7.16) and the notion of transition activity as an energy/power measure in general, *cannot* be used[13] when the line is capacitively coupled to other active lines or nodes of the circuit. Where, by "active", we mean that the line carries a signal, i.e. it is not fixed to ground or V_{dd}. Some authors have attempted to treat such cases by replacing the value of C_L by another "effective" capacitance. This can easily lead to inaccurate results. The way to deal with coupled active lines is to use the *transition activity matrix* that is discussed later.

If a bus has more than one lines *and they are electrically decoupled (to each other)*, then, expressions (7.16) can be used for each of the lines. The total energy/power of the bus is the sum of the energy/power of the lines. This approach has been used extensively in the past.

Finally, the transition activity can be expressed in another interesting way if we note that for every k, $\overline{(l(k)-l(k-1))^2}$ is equal to the probability of observing a transition during the kth clock cycle, i.e.,

$$\begin{aligned}\overline{(l(k)-l(k-1))^2} &= \Pr(l(k-1)=0 \textit{ and } l(k)=1)\\ &+ \Pr(l(k-1)=1 \textit{ and } l(k)=0).\end{aligned} \tag{7.17}$$

If we make the assumption that the random sequence is stationary and Markov with the stationary distribution, $(1/2, 1/2)$, *a very strong assumption compared to wide sense stationarity, we assumed before*[14], then we have

$$\Pr(l(k-1)=0 \textit{ and } l(k)=1) = \Pr(l(k-1)=1 \textit{ and } l(k)=0),$$

and $\Pr(l(k-1)=0 \textit{ and } l(k)=1)$ is independent of k. These two conditions, along with (7.17), validate the alternative expression, (7.18), for the transition activity.

$$T^a = \Pr(l(k-1)=0 \textit{ and } l(k)=1). \tag{7.18}$$

Now, we extend the definition of transition activity to a statistical measure that is applicable to buses with coupled lines. Consider a bus with n lines as in Figure 7.4. Let $l_r(k)$ be the binary value of the rth line at the end of the kth cycle, $r = 1, 2, \ldots, n$. We define the the sequence of vectors

$$L(k) = (l_1(k), l_2(k), \cdots, l_n(k))^T, \quad k = 1, 2, \ldots$$

which are random because their entries are random variables. Also, let $V(k) = (V_1(k), V_2(k), \ldots, V_n(k))^T$ be the vector of the voltages of the lines at the end of the kth cycle. As before we have $V(k) = V_{dd}\, L(k)$. Moreover, the energy consumed during the kth cycle is given by expression (7.11), i.e.

$$E(k) = \frac{1}{2}\,(V(k)-V(k-1))^T\,\mathcal{A}\,(V(k)-V(k-1)),$$

where $\mathcal{A}$ is given by (7.10). If we assume that the bit sequences $l_r(1)$, $l_r(2)$, ..., $l_r(k)$..., $r = 1, 2, \ldots, n$ are jointly stationary in the wide sense, then we can define the autocorrelation matrix, $R(r)$, of the vector sequence $L(1), L(2), L(3), \ldots$:

$$R(r) = [R_{i,j}(r)]_{i,j=1}^n = \overline{L(k+r) \cdot L^T(k)} \tag{7.19}$$

and for $i, j = 1, 2 \ldots, n$

$$R_{i,j}(r) = \overline{l_i(k+r)\, l_j(k)}. \tag{7.20}$$

It can be shown [25] that the expected energy consumed, in a bus with coupled lines, during a clock cycle is

$$\overline{E} = V_{dd}^2 \cdot trace\,(\mathcal{A} \cdot \mathcal{T}^a) \tag{7.21}$$

where $\mathcal{T}^a = \left[\mathcal{T}_{i,j}^a\right]_{i,j=1}^n$ is called the *transition activity matrix* of the bus [25], [26] and is defined as

$$\mathcal{T}^a = R(0) - \frac{1}{2}\left(R(1) + R^T(1)\right) \tag{7.22}$$

The function *trace* is such that $trace\,(X) = \sum_{j=1}^n x_{j,j}$ for every $n \times n$ matrix X. Note that the transition matrix is a generalization of the transition activity and that expression (7.21) is a generalization of the left expression in (7.16). The elements of the transition matrix are:

$$\mathcal{T}_{i,j}^a = \overline{l_i(k) \cdot l_j(k)} - \frac{1}{2}\left(\overline{l_i(k) \cdot l_j(k-1)} + \overline{l_i(k-1) \cdot l_j(k)}\right)$$

EXAMPLE 4 *Neglecting the coupling between non-adjacent lines in the bus, we get a total-capacitance network like that of Figure 7.6 and an admittance matrix, $\mathcal{A}$, given by expression (7.12). In this case, because of the symmetry of $\mathcal{A}$, formula (7.21) of the expected energy becomes*

$$\overline{E} = \left\{(1+2\lambda)\sum_{i=1}^{n} \mathcal{T}_{i,i}^a - 2\lambda \sum_{i=1}^{n-1} \mathcal{T}_{i,i+1}^a\right\} \cdot V_{dd}^2 \cdot C_L$$

EXAMPLE 5 *Let's assume the scenario of the previous example. If the transmitted bits are independent and uniformly distributed in $\{0, 1\}$ then*

$$\overline{l_i(k) \cdot l_j(k+r)} = \begin{cases} 1/2 & \text{if } i = j \text{ and } r = 0 \\ 1/4 & \text{otherwise} \end{cases}$$

and the transition activity matrix is $\mathcal{T}^a = \frac{1}{4}\mathcal{I}$, where $\mathcal{I}$ is the identity matrix. In this case the expected energy is

$$\overline{E} = \frac{n(1+2\lambda)}{4} \cdot V_{dd}^2 \cdot C_L \tag{7.23}$$

EXAMPLE 6 *In recent work, fixed permutation of the data bits was proposed as an approach to reduce the expected power consumption [32]. Instead of transmitting the sequence of vectors,* $L(k) = (l_1(k), l_2(k), \cdots, l_n(k))^T$, $k = 1, 2, \ldots$ *we can transmit the sequence of vectors with permuted entries*

$$L_\pi(k) = (l_{\pi(1)}(k), l_{\pi(2)}(k), \cdots, l_{\pi(n)}(k))^T, \quad k = 1, 2, \ldots$$

where π *is a permutation of the indices* $1, 2, \ldots, n$. *The goal in this approach is to minimize opposite and maximize concurrent transitions in adjacent lines by choosing the appropriate permutation. A heuristic approach was presented in [32]. Using the formulation introduced in the previous sections, the problem can be written formally as follows. Let* Π *be the* $n \times n$ *permutation matrix corresponding to* π, *then* $L_\pi(k) = \Pi \cdot L(k)$. *If* $\mathcal{T}^a$ *is the transition activity matrix of the original bus, then the transition activity matrix of the bus with the permuted data bits is*

$$\mathcal{T}_\pi^a = \Pi \cdot \mathcal{T}^a \cdot \Pi^T.$$

Using (7.21), the expected energy consumption is given by the expression:

$$\overline{E}_\pi = V_{dd}^2 \cdot trace\left(\mathcal{A} \cdot \Pi \cdot \mathcal{T}^a \cdot \Pi^T\right) \tag{7.24}$$

It is desirable to minimize expression (7.24) with respect to the permutation matrix Π. *An analytic lower bound of the minimum energy,* $\min_\pi \overline{E}_\pi$, *can be derived if we allow* Π *to be a doubly-stochastic matrix.*

3. The Limit of Power Reduction using Coding

We know that coding can be used to reduce power consumption. We would also like to know *how much power reduction is possible.* The question can be posed differently: *what is the minimum amount of energy we have to spend per bit of information sent through the bus?*

Here we are looking for the theoretical limit of what is achievable and *not* for particular implementations. Knowing what the limit is, we can evaluate the possible benefits of using coding for power reduction. Moreover, we can avoid seeking coding schemes that result in impossible power reduction!

Since the topic involves some notions from information and communication theory, we should note that there is a slight incompatibility between the communication/information theory terminology and digital architecture terminology. From a communications perspective, *addresses* and *data* are both *data,* i.e. information. Similarly, *address buses* are in some sense, *data buses* because they carry *address* vectors, which are one form of "data". The essential difference between *address sequences* and *data sequences* is in their statistical properties.

Consider a bus with n lines. For $r = 1, 2, \ldots, n$ let $l_r(k)$ be the binary value of the rth line at the end of the kth cycle. Also, let $L(k) = (l_1(k), l_2(k), \cdots, l_n(k))^T$, $k = 1, 2, \ldots$ be the bit vector and $V(k) = (V_1(k), V_2(k), \ldots, V_n(k))^T$ be the voltage vector of the lines at the end of the kth cycle. For every k, it is $V(k) = V_{dd}\, L(k)$. We know that the energy consumed during the kth cycle is given by expression (7.11), i.e. $E(k) = \frac{1}{2}\,(V(k) - V(k-1))^T\, \mathcal{A}\,(V(k) - V(k-1))$, where $\mathcal{A}$ is given by (7.10).

We have the necessary energy measures. We also need to introduce a measure of the amount of information, that is transmitted through the bus every cycle. Note first that, although, the n bits transmitted during the kth cycle are random, they may not be independent of each other.

EXAMPLE 7 *Consider the trivial case where the n-bit vector $(0, 0, \ldots, 0)$ is transmitted with probability $1/4$ and the n-bit vector $(1, 1, \ldots, 1)$ is transmitted with probability $3/4$. If we know what the first bit of the vector is, we automatically know the values of the other bits as well. Therefore, in this example, the amount of pure information that is carried by each vector is definitely not n bits, but rather one bit or less.*

The redundancy observed in the vector sequence of the example above can be removed using coding. For this reason, we must use a measure of information that is insensitive to redundancy. The best candidate is the *entropy*, $\mathcal{H}$, of random variables. If X is a binary random variable, then its entropy is defined as

$$\mathcal{H}(X) = -\sum_{v=0,1} \Pr(X = v) \cdot \log_2\Big(\Pr(X = v)\Big) \tag{7.25}$$

The definition is generalized directly to the case that X is a random vector, $X = (x_1, x_2, \ldots, x_n)$.

$$\mathcal{H}(X) = -\sum_{\substack{v_1, v_2, \ldots, v_n = 0,1 \\ V = (v_1, v_2, \ldots, v_n)}} \Pr(X = V) \cdot \log_2\Big(\Pr(X = V)\Big) \tag{7.26}$$

For simplicity we write $\mathcal{H}(X) = -\sum_X \Pr(X) \cdot \log_2(\Pr(X))$. In Example 7 above, the entropy of the random vector X, where $X = (0, 0, \ldots, 0)$ with probability $1/4$ and $X = (1, 1, \ldots, 1)$ with probability $3/4$, is $\mathcal{H}(X) = -1/4 \cdot \log_2(1/4) - 3/4 \cdot \log_2(3/4) \cong 0.81$ bits.

Until now, we have ignored the possible temporal dependance between vectors transmitted through the bus. For example, it is very common that the probability distribution of vector $L(k)$ depends upon the value of the previous vector, $L(k-1)$.

EXAMPLE 8 *Sequences of address vectors are highly predictable. In most cycles* k, $L(k) = L(k-1)+1$ [15]. *We can approximately model the sequence* $L(1), L(2), L(3)\ldots$ *as a (first order) Markov chain [58]. Then, for* $x, y = 0, 1, \ldots, 2^n - 1$

$$\Pr(L(k) = y|L(k-1) = x) = p_{x,y}$$

with $p_{x,x+1}$ *very close to 1. In some cases, we can simplify the model of the process even further by assuming that, for some small* δ, $p_{x,x+1} = 1-(2^n-1)\delta$ *and* $p_{x,y} = \delta$ *if* $y \neq x+1$.

In Example 8, we expect that the amount of information transmitted during the kth cycle is not as large as $\mathcal{H}(L(k))$ because $L(k)$ depends strongly on the previously transmitted vector[16], $L(k-1)$, and, therefore, it is not "as random as it looks when considered alone". To deal with this temporary redundancy we use the long-term-average information rate measure: the *entropy rate* $\mathcal{H}(L)$ of the random process $L : L(1), L(2), L(3), \ldots$. The entropy rate is insensitive to temporary, as well as, spatial redundancy. It is defined as [60]

$$\mathcal{H}(L) = \lim_{k\to\infty} \frac{H\left(L(1), L(2), \ldots, L(k)\right)}{k} \tag{7.27}$$

where $H\left(L(1), L(2), \ldots, L(k)\right)$ is the entropy of the partial sequence $L(1)$,$L(2)$,..., $L(k)$, i.e.

$$H\left(L(1), \ldots, L(k)\right) = \\ - \sum_{L(1),\ldots,L(k)} \Pr\left(L(1), \ldots, L(k)\right) \cdot \log_2\left(\Pr\left(L(1), \ldots, L(k)\right)\right) \tag{7.28}$$

In the sense of the Shannon-McMillan-Breiman theorem [60], $\mathcal{H}(L)$ equals the *expected number of bits needed to express the information content of vector* $L(k)$ *when the previous vectors* $L(1), L(2), \ldots, L(k-1)$ *are given and* $k \to \infty$[17]. So, it is possible to encode a long segment $L(1), L(2), \ldots,$ $L(K)$, containing $n \times K$ physical bits, into only $n \times \mathcal{H}(L)$ bits, i.e. $\mathcal{H}(L)$ bits/cycle, on average.

EXAMPLE 9 *For the Markov address process of Example 8, we have*

$$\begin{aligned} H\left(L(1), L(2), \ldots, L(k)\right) &= H(L(1)) + H(L(2)|L(1)) + \ldots \\ &\quad + H(L(k)|L(k-1)) \\ &= H(L(1)) + (k-1)H(L(2)|L(1)) \end{aligned}$$

where $H(\cdot|\cdot)$ is the conditional entropy function [60]. Let's assume that, $L(k)$ is uniformly distributed[18] *in* $0, 1, \ldots, 2^n - 1$. *Then,*

$$
\begin{aligned}
H\left(L(2)|L(1)\right) &= -\frac{1}{2^n}\sum_{x,y} p_{x,y}\cdot\log_2\left(p_{x,y}\right) \\
&= -(2^n-1)\delta\cdot\log_2(\delta) - (1-(2^n-1)\delta)\cdot\log_2(1-(2^n-1)\delta)
\end{aligned}
$$

So, the entropy rate of the address process is given by

$$
\mathcal{H}(L) = -(2^n-1)\delta\log_2(\delta) - (1-(2^n-1)\delta)\log_2(1-(2^n-1)\delta) \quad (7.29)
$$

Now we use the energy formula derived in Section 7.1.1 to study the relationship between energy consumption and information transmission in buses, particularly in DSM buses. The complete energy-information theory, that applies to more general problems, is available in [36], [46] and [26]. The problem is treated using information theoretic tools and a general mathematical framework is established in [36]. In [26], a more application oriented presentation is given. The special case of a 1-line bus has been studied in [22] as well.

Finally, data and address sequences are random sequences. Their statistics and information rates can be approximately estimated or measured. See for example [56].

Example 10 *Consider a bus with the capacitive structure of Figure 7.6, $n = 8$, $\lambda = 5$, $V_{dd} = 1V$ and $C_L = 100fF$. The transition energy is given by equation (7.11), where the admittance matrix, $\mathcal{A}$, is given by expression (7.10). Suppose the bus carries a sequence of address vectors that can be modeled as a Markov process, like that of Example 8, with $\delta = 1/2^{10}$. From expression (7.29), we get that the process L carries $\mathcal{H}(L) = 2.8$ bits per cycle (or per transmission, or per bus transition) on average. In addition, the expected energy cost per cycle is*

$$
\overline{E} = \sum_{X,Y}\frac{V_{dd}^2}{2}\left(Y-X\right)^T\mathcal{A}\left(Y-X\right)\cdot\Pr(L(k-1)=X, L(k)=Y)
$$

where X, Y take all $n-$bit vector values. Evaluating the above expression ($\Pr(X,Y) = \delta/2^8$ if $Y \neq X+1$, $\Pr(X, X+1) = [1-(2^8-1)\delta]/2^8$ and $\Pr(X) = 1/2^8$), we get $\overline{E} = 1.37pJ$. Therefore,

$$
\frac{\overline{E}}{\mathcal{H}(L)} = \frac{1.37pJ}{2.8\ bits} = 0.49\frac{pJ}{bit}.
$$

On average, $0.49pJ$ must be dissipated per bit of information that is transmitted through the bus.

Following Example 10, if $L : L(1), L(2), L(3) \ldots$ is a sequence of n-bit random vectors transmitted through the bus, we define the *expected energy per information bit* of L, $E_b(L)$, as the ratio of the expected energy consumption per cycle, $\overline{E(L)}$, over the expected number of information bits transmitted per cycle, $\mathcal{H}(L)$ [19]:

$$E_b(L) = \frac{\overline{E(L)}}{\mathcal{H}(L)}. \tag{7.30}$$

We write $\overline{E(L)}$ to indicate that the energy per cycle is associated with the transition of sequence L through the given bus. Using (7.11), we have that $\overline{E(L)} = \overline{E(V_{dd}\,L(k-1), V_{dd}\,L(k))}$.

There is a misconception in the literature about the "efficiency" of coding schemes. Saying only that a particular coding scheme can save 20% in power does *not* mean much. More parameters of the system must be given, e.g.: the number of additional lines and the structure of the expanded bus, if the original bus has been expanded to accommodate coding; the rate, $\mathcal{H}(L)$, at which information is transmitted through the bus, if the bus has not changed[20]. The point is illustrated by the following examples that make clear the *dependance between the expected energy and entropy rate, as well as, the difference between the expected energy, and, the expected energy per information bit.*

EXAMPLE 11 *Consider a 1-line bus carrying a sequence of bits, $b(1)$, $b(2)$, $b(3), \ldots$. Furthermore, suppose that, for some (large) number m and for every $r = 1, 2, 3, \ldots$, exactly one of the bits $b(r2^m), b(r2^m + 1), \ldots, b(r2^m + (2^m - 1))$ is 1. By choosing the position of "1" among the 2^m possible ones, it is clear that we can encode m information bits (data) into 2^m bits that are physically transmitted. In this case, the entropy rate, $\mathcal{H}(b)$, of the sequence b, is $m/2^m$ bits (per cycle). The expected energy per cycle is $\overline{E} \cong \frac{1}{2^m} \cdot C \cdot V_{dd}^2$, approximately, since there are about two transitions ($0 \to 1$ and $1 \to 0$) in every 2^m bits. Letting m become large, $\overline{E}$ becomes arbitrarily small! and the energy per information bit, $\overline{E}/\mathcal{H}(b) = \frac{1}{m} \cdot C \cdot V_{dd}^2$, becomes arbitrarily small as well! Unfortunately, the entropy rate $\mathcal{H}(b)$ tends to zero, too.*

EXAMPLE 12 *Let's alter the encoding in the previous example. Let's assume that, for every r, m, we allow exactly one $k \in \{0, 1, \ldots, 2^m - 1\}$ such that $b(r2^m + k) = b(r2^m + k + 1)$. All other consecutive (physical) bits have complementary binary values. Again, we can encode m information bits into 2^m physical bits. Letting m become large, we have that $\overline{E} \to \frac{1}{2} C V_{dd}^2$, $\mathcal{H}(b) \to 0$ and $\overline{E}/\mathcal{H}(b) \to \infty$.*

The size of the bus, that is "the bandwidth of the communication channel", is an important parameter that must be mentioned explicitly. The following example shows that the information rate cannot be considered independently of the size of the bus.

EXAMPLE 13 *Let's consider the one-hot encoding scheme [59] where m bits are encoded into 2^m bus lines simply by having exactly one line carrying a* 1 *and the rest of them carrying zeros. The energy loss per cycle is bounded above by a constant e, i.e. $\overline{E} < e$. Letting m become large we have that the energy per bit, which is less than e/m, can become arbitrarily small. At the same time, the information rate, which is equal to m, becomes arbitrarily large. It appears to be a win-win situation but unfortunately, it is not. We need to consider the size of the bus (2^m lines) which grows exponentially with m. Therefore,* **the energy per information bit must be considered with respect to the information rate and the size of the bus.**

The above example motivates one more definition. Let L be a sequence of $n - bit$ random vectors $L(k)$, $k = 1, 2, \ldots$, that is transmitted through a bus with n lines. The *utilization* α, $0 \leq \alpha \leq 1$, of the bus by the random sequence is:

$$\alpha = \frac{\mathcal{H}(L)}{n}. \tag{7.31}$$

The utilization, α, is the percentage of the "bandwidth" of the bus that is occupied by the transmission of the information contained in sequence L. In Example 13, the utilization approaches zero as the parameter m tends to infinity.

It is desirable to transmit information at the lowest energy cost possible. This poses an important question: *What is the minimum energy, $E_b^*(a)$, required to transmit a bit of information through the bus when $a \cdot n$ information bits are transmitted per cycle on average?* Written in mathematical terms:

$$E_b^*(a) = \min_{L:\, \mathcal{H}(L)/n = a} E_b(L) \tag{7.32}$$

The following theorem provides the answer [36],[26]. To apply the result of the theorem, the knowledge of the energy cost, $E(x, y)$, of the transition from any vector x to any other vector y, is required. To this end, expression (7.11) may be used. The result of the theorem is not limited to buses or a particular energy function.

THEOREM 1 *When the bus is utilized by a factor α, $0 \leq \alpha \leq 1$, the minimum energy per information bit is given by*[21]

$$E_b^*(\alpha) = \ln(2) \cdot \left(\gamma - \frac{1}{\frac{\partial}{\partial \gamma} \ln\left(\ln\left(\mu(\gamma)\right)\right)}\right)^{-1} \tag{7.33}$$

where γ is the positive solution of the equation:

$$\alpha = -\frac{1}{n\,\ln(2)} \gamma^2 \frac{\partial}{\partial \gamma} \left(\frac{\ln\left(\mu(\gamma)\right)}{\gamma}\right) \tag{7.34}$$

and $\mu(\gamma)$ is the maximal eigenvalue of the matrix

$$W(\gamma) = \left[e^{-\gamma\, E(x,y)}\right]_{x,y=0}^{2^n-1} \tag{7.35}$$

Although the above expressions for α and E_b^* are useful for analytical use, the alternative ones below are more appropriate for numerical calculations.

$$\alpha = \frac{1}{n\ln(2)} \left(\ln(\mu(\gamma)) + \frac{\gamma}{\mu(\gamma)} \sum_{i,j} \frac{g_i\, g_j}{\|g\|^2}\, e^{-\gamma E(i,j)}\, E(i,j)\right) \tag{7.36}$$

$$E_b^*(\alpha) = \frac{\ln(2)}{\gamma} - \frac{\ln(\mu(\gamma))}{\alpha n \gamma} \tag{7.37}$$

Where $g = (g_0, g_1, \ldots, g_{2^n-1})^T$ is an (the) eigenvector of $W(\gamma)$ corresponding to the maximal eigenvalue, μ. The proof of the theorem is very lengthy and technical and can be found in [36] and [26]. A set of MATLAB routines related to the theorem is available in [57]. The following example illustrates the use of the theorem.

EXAMPLE 14 *Suppose we want to transmit a sequence, L, of 2-bit vectors through a bus. For the bus, we have two options: 1) use a 2-line bus whose equivalent total-capacitance network is shown in Figure 7.6 with $V_{dd} = 1$, $C_L = 1$, and $\lambda = 5$; 2) use a similar 3-line bus with $V_{dd} = 1$, $C_L = 1$, $\lambda = 5$ and coding*[22]*. For the random vectors of the sequence, L, we assume that they are independent and uniformly distributed in $\{0,1\}^2$, i.e. all of the bits are temporally and spatially independent, and take the values 0 and 1 with probability 1/2. In option (1) the admittance matrix, $\mathcal{A}_2$, is given by (7.12),*

$$\mathcal{A}_2 = \begin{bmatrix} 11 & -5 \\ -5 & 11 \end{bmatrix} \cdot 1 = \begin{bmatrix} 11 & -5 \\ -5 & 11 \end{bmatrix} \tag{7.38}$$

All quantities are chosen dimensionless for simplicity. The expected energy per cycle is derived using (7.11).

$$\begin{aligned}\overline{E} &= \frac{1}{2} \sum_{X,Y \in \{0,1\}^2} (Y-X)^T \mathcal{A}_2 (Y-X) \cdot \Pr(X,Y) \\ &= \frac{1}{2} \sum_{x_i, y_j = 0,1} \begin{pmatrix} y_1 - x_1 \\ y_2 - x_2 \end{pmatrix}^T \begin{bmatrix} 11 & -5 \\ -5 & 11 \end{bmatrix} \begin{pmatrix} y_1 - x_1 \\ y_2 - x_2 \end{pmatrix} \cdot \frac{1}{16} \\ &= 5.5\end{aligned}$$

Since there is no statistical dependence between the bits of the sequence L, $\mathcal{H}(L) = 2$ *(bits/cycle), and, therefore, for option (1), the energy per (information) bit is* $\overline{E}/\mathcal{H}(L) = 2.75$*. For the same reason, the utilization of the bus is* $\alpha_2 = \mathcal{H}(L)/2 = 1$*.*

Now consider option (2). In this case, the utilization is $\alpha_3 = \mathcal{H}(L)/3 = 2/3$*; because, the information rate is the same as before, while, the size of the bus has been increased. What we would like to know is the maximum possible energy reduction we can achieve by using this larger-than-needed bus along with coding. The answer is given by expression (7.37) above. First, we need to find the positive solution* γ *of (7.36) for* $\alpha = \alpha_3$*. This can be done using the MATLAB routines available in [57]. Also, it can be shown [36] that* α *is a strictly decreasing function of* γ*, for* $\gamma > 0$*; and so, (7.36) (and (7.34)) has a unique positive solution. Moreover, the solution is easy to find. It turns out that* $\gamma = 0.3201$*,* $\mu = 1.7549$*, and from (7.37) we have:*

$$E_b^*(\alpha) = \frac{\ln(2)}{\gamma} - \frac{\ln(\mu)}{\alpha n \gamma} = 1.2866. \tag{7.39}$$

Therefore, in theory we can reduce the power consumption using coding by

$$100 \left(\frac{2.75 - 1.2866}{2.75} \right) \% = 53.2\% \tag{7.40}$$

The result is impressive but may not be easy to approach in practice. The theory tells us nothing about the complexity of the encoder/decoder that achieve the optimum.

EXAMPLE 15 *Consider the setup of Example 14; but, this time, we use a 4-line bus and coding to transmit the 2-bit vector sequence. In this case, the utilization of the bus is* $\alpha_4 = 2/4 = 0.5$*, and the theoretically minimum achievable limit of the energy per information bit is:*

$$E_b^*(\alpha) = \frac{\ln(2)}{\gamma} - \frac{\ln(\mu)}{\alpha n \gamma} = 1.0748. \tag{7.41}$$

Therefore, the maximum achievable power <u>*reduction*</u> *is*

$$100\left(\frac{2.75-1.0748}{2.75}\right)\% = 61\%. \qquad (7.42)$$

By increasing the number of additional lines further, **we can achieve <u>any</u> power reduction**[23].

EXAMPLE 16 *Consider, again, a bus with the capacitive structure of Figure 7.6. The energy function of the bus is given by expression (7.11) with the admittance matrix, $\mathcal{A}$, given by (7.12). In Figure 7.8 we see the normalized minimum energy per information bit, $E_b^*(\alpha)/E_b^*(1)$, as a function of the bus utilization, α. The three graphs correspond to the following cases: $n = 2, 4$ and 8, with $\lambda = 5$. Energy increases rapidly around $\alpha = 0$ and $\alpha = 1$, is zero at $\alpha = 0$ and maximal at $\alpha = 1$. This confirms Example 15; that is, the energy per information bit can become arbitrarily low for sufficiently small rate. Note that this would not be true if the energy expression (7.11) included leakage or other terms that would contribute to energy consumption even when there is no transition in the bus. If leakage is present, then there is an optimal utilization.*

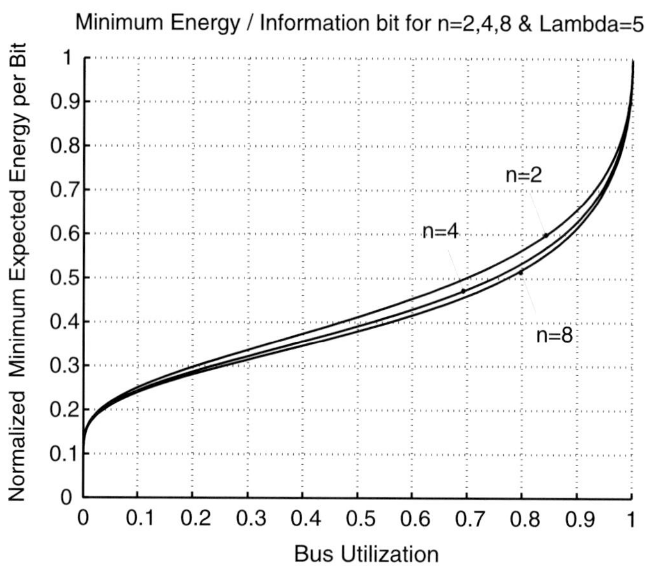

Figure 7.8. Minimum (normalized) Energy per Bit for a family of buses [from Ref. [36] @ 2003 IEEE]

4. Conclusions

Three topics have been discussed in this chapter: Energy models appropriate for Deep Sub-Micron buses; the corresponding Statistical Energy models and the extension of transition activity to the Transition Activity Matrix that can be used in buses with interline coupling; the fundamental relationship between the minimum possible power consumption, the utilization of the bus (speed), and the number of bus lines (area). These theoretical tools can be used to analyze practical coding schemes, estimate their energy savings and compare them with that of the optimal coding scheme for a given bit rate and bus size.

5. Acknowledgement

The author would like to thank Drs. Timo D.Hämäläinen, Jari Nurmi, Lars Wanhammar, Timo Yli-Pietilä, and Heiko Zimmer; and Mr. Mounir Khatib for their useful comments on this chapter.

Notes

1. Throughout the chapter we consider noiseless systems.
2. Pulse position modulation
3. Suppose each transmission takes a clock cycle.
4. In the examples, we assumed that the energy required for the encoding and decoding is insignificant compared to the transmission energy we save.
5. The area is determined by the number of wires and their geometries
6. The author put significant effort to complete the list. He apologizes if he missed reporting some significant work.
7. Drawn from V_{dd}, *not* consumed by the circuit.
8. We treat them a coupled transmission lines
9. The parasitic capacitance at the output of the inverter driving the ith line, say C_i^d, and the parasitic capacitance at the input of the register at the other end of the bus, say C_i^r, can be integrated into $C_{i,i}$. Then, according to the model in Figure 7.4, we have $C_{i,j} = \int_0^L c_{i,j}(x)\,dx$, if $i \neq j$, and $C_{i,i} = \int_0^L c_{i,i}(x)\,dx + C_i^d + C_i^r$.
10. Throughout the chapter, when binary values are involved in numerical calculations, they are treated as the real numbers $0, 1$, i.e. $0 - 1 = -1$ etc.
11. It is stationary, in the wide sense; so, $\overline{l(k+r)\,l(k)}$ is independent of k.
12. The name is misleading since T^a is associated with the sequence $\{l(k)\}$ rather than the line itself. Some authors define it as $2(R(0) - R(1))$. Symbol "a" is used some times to denote the transition activity. In this chapter we reserve "a" for another quantity.
13. except in some very special cases
14. Some researchers consider this given without providing any supporting explanation.
15. For notational purposes, the numbers $0, 1, \ldots, 2^n - 1$ are identified with their binary expansions. This is done throughout the paper.
16. In general we should consider the dependance of $L(k)$ on $L(k-1)$, $L(k-2)$, ...
17. When, for example, $L(k)$ is stationary and ergodic.
18. Actually, for $0 < \delta < 1$, the uniform distribution is the steady state distribution of the process.

19. To avoid technicalities, we assume that the random sequence L is stationary. The definitions and results can be directly expanded to more general cases.

20. In many circumstances, especially in address buses, a significant amount of power can be saved by exploiting the strong temporal correlation of the address vectors. Since the information rate is very low compared to the size of the bus (see Example 9) an efficient coding scheme could reduce power dramatically.

21. Although expression (7.33) holds for the case of energy function, (7.11), for general cost functions there is one exception: if there are some c, θ_x, $x = 0, 1, \ldots, 2^n - 1$ such that $E(x, y) = c + \theta_x - \theta_y$, for every x, y, then, $E_b^*(\alpha) = c/(\alpha\, n)$.

22. No matter what the statistics of the data are, the additional line introduces communication bandwidth redundancy.

23. Note that the energy model we used does not account for leakage. The statement is true only when leakage is not present. If leakage is present, then there is an optimal number of lines. The extension of the model to include leakage is trivial

References

[1] A. Brunin, G. D'Hervilly, *Method and network for improving transmission of data signals between integrated circuit chips*, US Patent 4,495,626 January 22, 1985

[2] R. Fletcher, *Integrated circuit having outputs configured for reduced state changes*, US Patent 4,667,337, May 19, 1987.

[3] J. Tabor, *Noise reduction using low weight and constant weight coding techniques*, Master Thesis, MIT, May 1990.

[4] M. Stan, W. Burleson, "Coding a terminated bus for low power", *Proceedings Fifth Great Lakes Symposium on VLSI, Page(s): 70 -73, March, 1995.*

[5] M. Stan, W. Burleson,"Bus-invert coding for low-power I/O", *IEEE Transactions on VLSI, Volume: 3 Issue: 1 , March 1995 Page(s): 49 -58*

[6] Youngsoo Shin; Soo-Ik Chae; Kiyoung Choi, "Partial bus-invert coding for power optimization of system level bus", *Low Power Electronics and Design, 1998. Proceedings. 1998 International Symposium on , 10-12 Aug. 1998 Page(s): 127 -129*

[7] Lang, T.; Musoll, E.; Cortadella, J., "Extension of the working-zone-encoding method to reduce the energy on the microprocessor data bus", *VLSI in Computers and Processors, 1998. ICCD '98. Proceedings., International Conference on , 5-7 Oct. 1998 Page(s): 414 -419.*

[8] Benini, L.; De Micheli, G.; Macii, E.; Sciuto, D.; Silvano, C., "Address bus encoding techniques for system-level power optimization", *Design, Automation and Test in Europe, 1998., Proceedings , 23-26 Feb. 1998 Page(s): 861 -866.*

[9] Naehyuck Chang; Kwan-Ho Kim; Heonshik Shin, "Dual-mode low-power bus encoding for high-performance bus drivers", *TENCON 99.*

Proceedings of the IEEE Region 10 Conference , Volume: 2 , 15-17 Sept. 1999 Page(s): 876 -879 vol.2.

[10] Sungjoo Yoo; Kiyoung Choi, "Interleaving partial bus-invert coding for low power reconfiguration of FPGAs", *ICVC '99. 6th International Conference on , 26-27 Oct. 1999 Page(s): 549 -552*

[11] Sunpack Hong; Narayanan, U.; Ki-Seok Chung; Taewhan Kim, "Bus-invert coding for low-power I/O - a decomposition approach", *Proceedings of the 43rd IEEE Midwest Symposium on , Volume: 2 , 8-11 Aug. 2000 Page(s): 750 -753 vol.2.*

[12] Komatsu, S.; Ikeda, M.; Asada, K., "Low power chip interface based on bus data encoding with adaptive code-book method", *Proceedings. Ninth Great Lakes Symposium on , 4-6 March 1999 Page(s): 368 - 371*

[13] Yeshik Shin; Deog-Kyoon Jeong, "Precoded two-dimensional coding method for low-power serial bus", *Electronics Letters , Volume: 35 Issue: 18 , 2 Sept. 1999 Page(s): 1524 -1526.*

[14] Wei-Chung Cheng; Pedram, M., "Power-optimal encoding for DRAM address bus", *ISLPED '00. Proceedings of the 2000 International Symposium on , 26-27 July 2000 Page(s): 250 -252.*

[15] Youngsoo Shin; Kiyoung Choi, "Narrow bus encoding for low power systems", *Design Automation Conference, 2000. Proceedings of the ASP-DAC 2000. Asia and South Pacific , 25-28 Jan. 2000 Page(s): 217 -220.*

[16] Wei-Chung Cheng; Pedram, M., "Memory bus encoding for low power: a tutorial", *Quality Electronic Design, 2001 International Symposium on , 26-28 March 2001 Page(s): 199 -204*

[17] Bishop, B.; Bahuman, A., "A low-energy adaptive bus coding scheme", *VLSI, 2001. Proceedings. IEEE Computer Society Workshop on , 19-20 April 2001 Page(s): 118 -122.*

[18] Rossi, D.; van Dijk, V.E.S.; Kleihorst, R.P.; Nieuwland, A.H.; Metra, C., "Coding scheme for low energy consumption fault-tolerant bus", *On-Line Testing Workshop, 2002. Proceedings of the Eighth IEEE International , 8-10 July 2002 Page(s): 8 -12.*

[19] Rung-Bin Lin; Chi-Ming Tsai, "Weight-based bus-invert coding for low-power applications", *Design Automation Conference, 2002. Proceedings of ASP-DAC 2002. 7th Asia and South Pacific and the 15th International Conference on VLSI Design. Proceedings. , 7-11 Jan. 2002 Page(s): 121 -125.*

[20] Osborne, S.; Erdogan, A.T.; Arslan, T.; Robinson, D., "Bus encoding architecture for low-power implementation of an AMBA-based

SoC platform", *Computers and Digital Techniques, IEE Proceedings-, Volume: 149 Issue: 4 , July 2002 Page(s): 152 -156.*

[21] Aghaghiri, Y.; Fallah, F.; Pedram, M., "BEAM: bus encoding based on instruction-sef-aware memories", *Design Automation Conference, 2003. Proceedings of the ASP-DAC 2003. Asia and South Pacific , Jan. 21-24, 2003 Page(s): 3 -8.*

[22] S. Ramprasad, N. Shanbhag, I. Hajj, "Information-Theoretic Bounds on Average Signal Transition Activity", *IEEE Trans. on VLSI, Vol. 7, No. 3, Sept. 1999.*

[23] Rung-Bin Lin; Chi-Ming Tsai, "Theoretical analysis of bus-invert coding", *Circuits and Systems, 2000. Proceedings of the 43rd IEEE Midwest Symposium on , Volume: 2 , 8-11 Aug. 2000 Page(s): 742 -745 vol.2.*

[24] P. Sotiriadis and A. Chandrakasan, "Low Power Bus Coding Techniques Considering Inter-wire Capacitances", *Proceedings of the Custom Integrated Circuits Conference, May 2000.*

[25] P. Sotiriadis and A. Chandrakasan, "A bus energy model for deep submicron technology", *IEEE Transactions on VLSI Systems, pp. 341-350, Vol. 10, No. 3, June 2002.*

[26] P. Sotiriadis and A. Chandrakasan, "Power Estimation and Power Optimal Communication in Deep Sub-Micron Buses: Analytical Models and Statistical Measures", *Journal of Circuits, Systems, and Computers, Vol. 11, No. 6, 2002, 637-658.*

[27] P. Sotiriadis , A. Wang, A. Chandrakasan, "Transition Pattern Coding: An Approach to Reduce Energy in Interconnect", *26th European Solid-State Circuit Conference, (ESSCIRC 00), Stockholm, 2000, pp. 320-323.*

[28] P. Sotiriadis , A. Chandrakasan, "Bus Energy Minimization by Transition Pattern Coding (TPC) in Deep Sub-Micron Technologies", *IEEE/ACM International Conference on CAD, San Jose, 2000, pp. 322-327.*

[29] K.W. Kim, K.H. Baek, N. Shanbhag, C.L. Liu, S.M. Kang, "Coupling-Driven Signal Encoding Scheme for Low-Power Interface Design", *IEEE/ACM International Conference on CAD, San Jose, 2000.*

[30] Bishop, B.; Bahuman, A.; VLSI, "A low-energy adaptive bus coding scheme", *Proceedings. IEEE Computer Society Workshop on , 19-20 April 2001 Page(s): 118 -122.*

[31] Henkel, J.; Lekatsas, H.,"A2BC: adaptive address bus coding for low power deep sub-micron designs", *Design Automation Conference, 2001. Proceedings , 18-22 June 2001 Page(s): 744 -749.*

[32] L. Macchiarulo, E. Macii, M. Poncino, "Low-Energy Encoding for Deep-Submicron Address Buses", *IEEE/ACM International Symposium on Low Power Electronics and Design 2001, pp. 176-181.*

[33] Taylor, C.N.; Dey, S.; Yi Zhao, "Modeling and minimization of interconnect energy dissipation in nanometer technologies", *Design Automation Conference, 2001. Proceedings , 18-22 June 2001 Page(s): 754 -757.*

[34] Komatsu, S.; Fujita, M, "Irredundant address bus encoding techniques based on adaptive codebooks for low power", *Design Automation Conference, 2003. Proceedings of the ASP-DAC 2003. Asia and South Pacific , Jan. 21-24, 2003 Page(s): 9 -14.*

[35] Madhu, M.; Murty, V.S.; Kamakoti, V., "Dynamic coding technique for low-power data bus", *VLSI, 2003. Proceedings. IEEE Computer Society Annual Symposium on , 20-21 Feb. 2003 Page(s): 252 -253.*

[36] P. Sotiriadis , V. Tarokh, A. Chandrakasan, "Energy Reduction in VLSI Computation Modules: An Information-Theoretic Approach", *IEEE Trans. on Information Theory, April 2003, Vol. 49, No. 4, pp. 790-808.*

[37] P. Sotiriadis, A. Chandrakasan, "Coding for Energy Reduction in Deep Sub-Micron Technologies The Transition Pattern Coding Approach", *IEEE Transactions on Circuits and Systems I, Vol. 50, No. 10, Oct. 2003, pp. 1280-1295.*

[38] S. Das, W. Smith, C. Paul, "Modeling of data bus structures using numerical methods", *International Symposium on Electromagnetic Compatibility, 1993, pp. 409-414.*

[39] L. Pileggi, "Coping with RC(L) interconnect design headaches", *IEEE/ACM International Conference on CAD 1995, pp. 246-253.*

[40] Li-Rong Zheng; Bingxin Li; Tenhunen, H., "Global interconnect design for high speed ULSI and system-on-package", *ASIC/SOC Conference, 1999. Proceedings. Twelfth Annual IEEE International , 15-18 Sept. 1999 Page(s): 251 -256.*

[41] J. Davis, J. Meindl, "Compact Distributed RLC Interconnect Models Part II: Coupled Line Transient Expressions and Peak Crosstalk in Multilevel Networks", *IEEE Transactions on Electron Devices, Vol. 47, No. 11, Nov. 2000.*

[42] T. Sakurai, "Design Challenges for 0.1μm and Beyond", *Asia and South Pacific Design Automation Conference 2000, pp. 553-558.*

[43] C. Cheng, J. Lillis, S. Lin, N. Chang, *Interconnect Analysis and Synthesis*, John Wiley and Sons, 2000.

[44] K. Yamashita, S. Odanaka, "Interconnect Scaling Scenario Using a Chip Level Interconnect Model", *IEEE transactions on electron devices Vol. 47, No. 1, Jan. 2000.*

[45] N. Menezes, L. Pillegi, "*Analyzing on-chips Interconnect effects*", Design of High Performance Microprocessor Circuits, A. Chandrakasan, W.J. Bowhill, F. Fox, Eds., IEEE Editions 2001, Chapter 16.

[46] P. Sotiriadis, *Interconnect Modeling and Optimization in Deep Sub-Micron Technologies*, Ph.D. Thesis, Massachusetts Institute of Technology, May 2002.

[47] T. Sakurai, "Closed-form expressions for interconnection delay, coupling, and crosstalk in VLSIs", *IEEE Transactions on Electron Devices, Vol. 40, Issue 1, Jan. 1993, pp. 118-124.*

[48] K. Tang, E. Friedman, "Peak noise prediction in loosely coupled interconnect [VLSI circuits]", *International Symposium on Circuit and Systems 1999, Vol.1, pp. 541-544.*

[49] T. Sakurai, S. Kobayashi, M. Noda, "Simple expressions for interconnection delay, coupling and crosstalk in VLSI", *International Symposium on Circuit and Systems 1991, Vol. 4, pp. 2375-2378.*

[50] M. Becer, I. Hajj, "An analytical model for delay and crosstalk estimation in interconnects under general switching conditions", *IEEE International Conference on Electronics, Circuits and Systems, 2000, vol.2, pp. 831-834.*

[51] A. Ghosh, S. Devadas, K. Keutzer, J. White, "Estimation of average switching activity in combinatorial and sequential circuits", *Proceedings of ACM/IEEE Design Automation Conference, pp. 253-259, June 1992.*

[52] F. Najm, "A survey of power estimation techniques in VLSI circuits", *IEEE Transactions on VLSI Systems, Vol. 2, pp. 446-455, Dec. 1994.*

[53] R. Marculescu, D. Marculescu, M. Pedram, "Efficient power estimation for highly correlated input streams", *Proceedings of ACM/IEEE Design Automation Conference, pp. 628-634, June 1995.*

[54] M. Pedram, "Power minimization in IC design: Principles and applications", *ACM Transactions on Design Automation Electronic Systems, Vol. 1, No. 1, pp. 1-54, Jan. 1996.*

[55] Y.I. Ismail, E.G. Friedman, J.L. Neves, "Transient power in CMOS gates driving LC transmission lines", *IEEE International Conference on Electronics, Circuits and Systems, 1998, pp. 337-340.*

[56] R. Marculescu, D. Marculescu, M. Pedram, " Probabilistic Modeling of Dependencies During Switching Activity Analysis", *IEEE Transactions Computer-Aided Design of Integrated Circuits and Systems*, vol. 17, pp. 73-83, June 1998.

[57] *http://www.ece.jhu.edu/~pps/*

[58] R. Gallager, *Discrete Stochastic Processes*, Kluwer Academic Publishers, 1998.

[59] A. Chandrakasan (Ed.) and R. Brodersen (Ed.), *Low power CMOS Design*, IEEE Press, 1998.

[60] T. Cover, J. Thomas, *Elements of Information Theory*, John Wiley and Sons, 1991.

[61] R. Horn, C.Johnson, *Matrix Analysis*, Cambridge University Press 1994. J. Rabaey, *Digital Integrated circuits*, Prentice Hall 1996.

Chapter 8

BUS STRUCTURES IN NETWORK-ON-CHIPS

Vesa Lahtinen, Erno Salminen, Kimmo Kuusilinna and Timo D. Hämäläinen
Institute of Digital and Computer Systems, Tampere University of Technology

1. INTRODUCTION

Buses have been the preferred interconnection in most of the processor-based systems in the past thirty years. They have also been the basic building blocks of almost all implemented system-on-chips (SoC). Buses offer a simple and efficient way to transfer data between the components of a system. Their simple signaling and structure facilitate both manual and automatic generation of bus-based architectures. Moreover, the simple programming model of bus-based systems has been favored by the application designers and has led to the development of advanced programming tools.

A single bus is a good choice for many systems when the number of connected components is small. However, in the future SoCs, the complexity of a single component is not likely to significantly increase because only relatively simple components can be scaled with technology [1, 2]. Therefore, the number of components in a SoC is increasing. On the other hand, there has been only a modest increase in the ability of physical wires to transfer signals, although the performance of logic gates has been rapidly improving. The delay of a fixed length wire is even estimated to increase with decreasing feature sizes [1, 3]. Based on this analysis, it seems that a simple bus is no longer a preferable solution for SoC interconnection requirements. Sometimes network-on-chips (NoC), such as the one depicted in Figure 8-1, are proposed to solve this problem. In this Chapter, a SoC architecture that is based on more complex structures than a single bus or fully connected point-to-point links is defined as a network-on-chip.

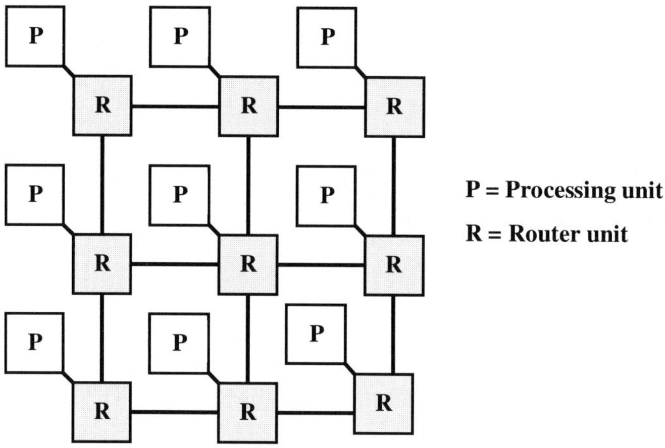

Figure 8-1. An example NoC structure based on a mesh topology

The homogeneous architectures, such as the mesh of Figure 8-1, are one alternative NoC architecture. They have been used to run fine-grained parallel scientific algorithms at very high speeds. However, the logical limits of parallelism inherent in most practical applications result in redundancies and inefficiencies that cannot always be tolerated in on-chip systems because of increased area and power consumption. In addition, the structures utilized in scientific computing are usually optimized for fine-grain granularity computation. Sometimes NoC architectures are benchmarked with a fairly simple fine-grained algorithm with good results although the distribution of computation for such a large system may not be meaningful. The NoC scheme implies that the components are part of a fairly large coarse-grain granularity system where the components are smaller systems of their own.

A computer local area network (LAN) is another possible analogy for the design of NoCs. Although the on-chip architectures have a lot in common with these systems, they also have considerable differences. The relatively complex network architectures and protocols of LANs as well as their large network buffer memories are designed to make the systems reliable with particular emphasis on connectivity and performance requirements. In contrast, large buffers and arbitrary connectivity are not tolerable in SoCs that prefer limited complexity and real-time operation.

This Chapter advocates a NoC approach that utilizes a hierarchy of bus structures and augments this strategy with more specialized interconnection topologies on case to case basis. In this type of architecture, the local interconnections are buses making use of their described good properties. On the other hand, the global interconnections use a more complex network structure containing buffering elements resembling the router units of Figure 8-1. This type of a heterogeneous architecture can be modified according to the requirements of the application. The following Sections illustrate that buses and NoCs do not exclude each other and how the single bus architecture can be modified and extended to fit the NoC scheme.

2. SOME BASIC PROPERTIES OF SHARED INTERCONNECTIONS

Buses [4] are shared interconnection paths between the components of an electronic system. They are defined by their electrical signaling, the required interface logic, and the utilized transaction protocols. The basic properties of bus interconnections have remained almost unchanged during their evolvement from computer backplanes to on-chip interconnections.

2.1 Structure

Interconnections can be classified as static or dynamic. Dynamic interconnections use switching elements whereas their static counterparts, for example buses, have a fixed path between the sender and the receiver. By definition, a bus uses shared interconnection lines as opposed to dedicated point-to-point signals used in other static interconnection types. Traditionally, buses have been implemented with three-state buffers which drive bidirectional signal lines as illustrated in Figure 8-2 a). Three-state buffers are a good approach for backplane buses but their high power consumption and the difficulty of debugging make them less suitable for SoCs. The basic alternatives for three-state buses are multiplexer-based buses and AND-OR structures depicted in Figure 8-2 b) and Figure 8-2 c), respectively. They both use unidirectional signal lines.

The components connected to a bus are called agents. An agent that can initiate and control a transfer is called a master or an initiator. Agents that respond to transfer requests initiated by masters are called slaves or targets. An agent can also be a master and a slave at the same time. An agent can contain, for example, a processor, its local bus, memory units, and an interface wrapper to connect to the global bus architecture.

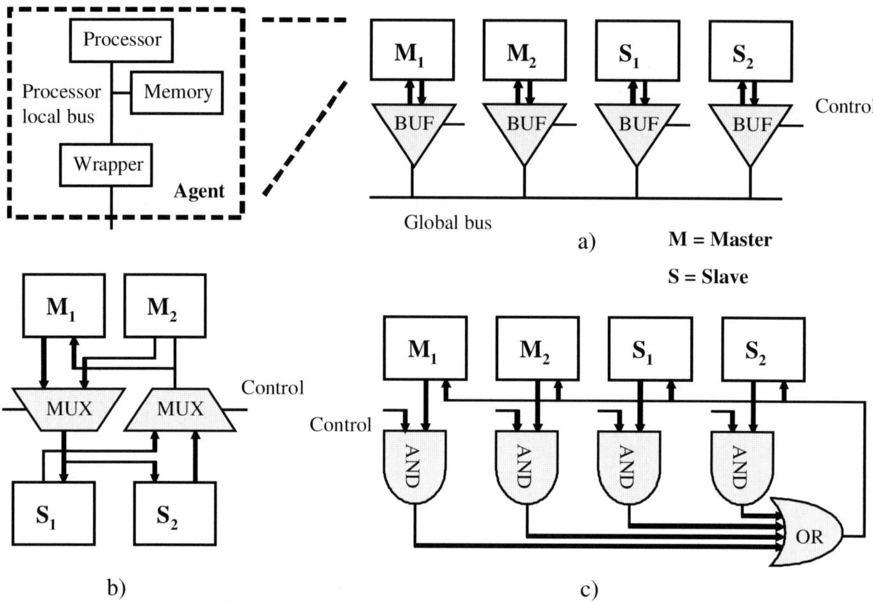

Figure 8-2. a) Three-state, b) multiplexer-based, and c) AND-OR bus implementations

2.2 Transfers

On the basis of supported transfer types, buses can be either synchronous or asynchronous. Synchronous transfers are based on a global clock signal while the asynchronous transfers are self-timing. Transfers in a bus normally require some sort of handshaking between the sender and the receiver. A generic, two-phase synchronous read operation consists of a read request (master) and a data response phase (slave). In a synchronous bus, the handshaking does not need to be as extensive but the problem is that the slowest agent defines the clock frequency of the whole system. Furthermore, distributing a global clock signal in a large SoC can be problematic.

Asynchronous transfers do not need a global clock signal. Instead, each phase of the transmission process takes only as long as is required, which makes it easier for devices with different response times to interact. An additional benefit is that signal propagation delays do not affect the correct operation of asynchronous systems. However, because of the prolonged handshaking, asynchronous transfers are usually more complex than their synchronous counterparts. The asynchronous transfers frequently use a four-phase handshaking scheme. The four phases of a generic read operation are: read request (master), acknowledge (slave), data response (slave), and end of transfer process (master).

Traditionally, a transfer between two agents reserves the bus for the duration of the transfer. If, for example, a processor reads from a memory it first sends a read request, stalls its own processing, and waits for the memory to respond. The memory is usually slow when compared to the processor and the shared medium is reserved during the whole process. In a split transfer, the processor first sends a read request, after which it frees the bus for other transfers. When the memory is ready, it uses a normal write operation to respond. The processor does not get data any faster but the advantage is that the shared medium is available for other agents in the middle of the transmission (e.g. memory read) process.

In a single bus, only one agent can transmit data at a time. On the other hand, all connected agents can read the transferred data making broadcast a simple operation. Buses can have separated data and address lines, but they can also be multiplexed onto the same lines. In a multiplexed bus, typically the master first sends the address and then the data. One way to speed up the operation with separate bus signals for data and address is to put the address of the next transfer to the address lines before the previous transfer has ended. This technique is called transfer pipelining and it is particularly good with block or burst transfers in which a single address cycle is followed by multiple data cycles.

2.3 Arbitration

Arbitration is the mechanism for resolving the bus owner if more than one agents request the ownership at the same time. Arbitration can be centralized in which case request and grant signals are needed between every agent and the arbiter unit. In a distributed version, all the information needed for the arbitration process is stored in each bus agent and can be updated by monitoring the bus signals. In addition to an arbiter, a decoder is required. The decoder does not need to be a separate unit, it can be integrated as a part of a slave. The task of the decoder is to direct the correct slave to respond to a transaction. In the centralized scheme of Figure 8-2 b), the multiplexer on the left is controlled by the arbiter and the multiplexer on the right by the decoder.

The most popular arbitration algorithms are based on priority or utilize the round-robin method. In priority-based systems, the requesting agent with the highest priority gets the bus ownership. This process can, therefore, lead to starvation meaning that lower priority agents are not necessarily able to access the bus. Priority arbitration can be pre-emptive, that is, a lower priority transfer is interrupted when a higher priority agent requests the communication channel. Non pre-emptive transfers are allowed to complete before starting new transfers. Pre-emption can also occur when the maximum allowed transfer length has been exceeded decreasing the risk of starvation. In round-robin systems, the bus owner is changed in a circular manner providing fair arbitration because every agent eventually gets the bus ownership. Each agent passes the ownership to the next agent in line when it no longer has data to send or it has held the bus for the maximum allowed time. One interesting approach to arbitration is to utilize static methods where bus owners are decided before the system runtime. An example of these methods is TDMA (time division multiple access) in which timeframes are divided into time slots that are assigned to bus agents.

3. OVERVIEW OF BUS-BASED SOC INTERCONNECTIONS

In this Section, we study the basic properties of the following nine on-chip buses: AMBA [5], CoreConnect [6], CoreFrame [7], HIBI [8, 9], Marble [10], PI-bus [11], SiliconBackplane [12], STBus [13], and Wishbone [14]. Some of their properties are summarized in Tables 8-1, 8-2, and 8-3. A number of other SoC buses exist also but not enough details of their properties are publicly available to allow a fair summarization. This analysis is partially based on [15].

The advanced microcontroller bus architecture (AMBA) by Advanced RISC machines Limited is one of the most frequently used on-chip buses. Reasons for this popularity include its easily available specification and its close connection to the popular ARM RISC processors. AMBA is a typical multiplexer-based bus and it utilizes a central arbiter and decoder architecture. It has three versions for different transmission needs, namely the advanced high-performance (AHB), advanced system (ASB), and the advanced peripheral (APB) buses. AHB is a system backbone bus whereas ASB is a system bus that does not have all the features of AHB. APB is targeted for low-power and low-complexity peripheral devices and can be used together with the other two AMBA bus types.

CoreConnect is another widely used on-chip bus developed by IBM Corporation. It is an open standard with publicly available specifications and it is used, for example, in conjunction with the PowerPC processors. CoreConnect specification contains two buses for different purposes, namely the processor local bus (PLB) and the on-chip peripheral bus (OCP). In addition, it has a separate device control register (DCR) bus and an arbiter unit supporting address pipelining and bridges between PLB and OCP.

CoreFrame, by Palmchip Corporation, is also a two-level on-chip bus architecture. It consists of PalmBus which is used for connections between processors and peripherals and MBus for high-speed memory and peripheral accesses. The two buses are actually independent parallel buses allowing concurrent activity and increasing the throughput of the system. CoreFrame uses point-to-point signals and multiplexing to avoid three-state buffering. It does not use shared signal lines, instead the communication is carried out through shared variables in memory.

Heterogeneous IP block interconnection (HIBI) is a versatile interconnection scheme from Tampere University of Technology. It is primarily targeted towards continuous-media systems having large regular data transfers for which it utilizes static arbitration methods. In HIBI-based systems, the arbitration is distributed to the connected agents. In addition, HIBI supports the basic high-performance features of on-chip buses. At the moment the HIBI development is focused on complex bus and network structures in addition to multi-priority transmissions.

Developed at the University of Manchester, Marble (Manchester asynchronous bus for low energy) is an original on-chip bus design lacking a global clock signal. The bus supports split transfers, test structures, and bridges, and demonstrates that the requirements of a SoC bus can be met by a fully asynchronous design style. For arbitration, Marble uses a centralized, tree-based structure that can be modified to adjust the latency and bandwidth allocated to each agent of the system.

The PI-bus is an open standard published by the Open microprocessor systems initiative (OMI). VHDL codes for master, slave, master/slave, and bus control units are freely distributed. Also, synthesis scripts for different ASIC and FPGA technologies and system examples are available. The bus control is a centralized unit taking care of master arbitration and slave selection (decoding). It also logs all the error messages coming from the slave units. The PI-bus is a synchronous bus utilizing non-multiplexed address and data signals and it supports multiple masters. In addition, bridges have been used in systems based on the PI-bus [16].

The μNetworks designed by Sonics Incorporated contains a set of architectures and tools for SoC design. The defined architectures are the SiliconBackplane for on-chip and MultiChip for off-chip interconnections. SiliconBackplane has a two-level arbitration scheme based on TDMA and round-robin. Furthermore, dynamic reconfiguration of a set of system parameters is supported in SiliconBackplane. The system parameters are implemented in the agents as registers that are visible to the application software.

The Split transaction bus (STBus), used in the scalable Daytona DSP architecture, has many interesting properties that will be discussed later. It was jointly designed by Lucent technologies and Massachusetts Institute of Technology. The STBus was designed to minimize average latencies in systems where large amounts of data are transferred. This is achieved through a memory control unit that supports multiple outstanding transactions and the use of large cache memories. A programmable round-robin scheme is used for arbitration.

Wishbone, by Silicore Corporation, is another on-chip bus standard with many typical features included in its specification. Wishbone supports the basic transfer types and many of the possible bus implementation techniques, including multiplexer and three-state-based structures. The Wishbone interconnection can also implement other topologies than a bus such as a crossbar. The arbitration mechanism of Wishbone can be chosen by the system designer. The Wishbone VHDL source codes are freely distributed. The specification also includes a list of rules with which the Wishbone compatible components and systems have to comply, for example documentation and naming conventions rules.

Table 8-1 tabulates the basic structural properties of the aforementioned buses, whereas Tables 8-2 and 8-3 tabulate the transfer and arbitration properties, respectively. The notation is 'x' for supported features and 'n/a' if the feature is not supported or the information was not available or clearly stated in the specification. In addition, the tables include the Lotterybus [17] specification that mainly defines a partly random arbitration scheme, the Virtual component interface (VCI) [18], and the Open core protocol (OCP) [19] interface standards that are discussed in Section 4. The VCI and the OCP only define an interface to an interconnection but not the implementation. Sometimes the determination between supported vs. unsupported features is very difficult and a specific functionality can be achieved by using the interconnection a little differently.

There are many structural properties that are typical to almost all the SoC buses. Hierarchical structures are supported by many of the buses, and the rest of the specifications did not make a clear statement about the issue. The signal line implementations include both uni- and bidirectional versions, as well as combinations of these two. Most of the presented SoC buses utilize shared signal lines. Only one of the buses uses asynchronous techniques, the rest are fully synchronous. Multiple clock domains and test structures are also supported by several of the bus specifications.

Table 8-1. On-chip bus structures

Interconnection	1. Hierar.	2. Sig.	3. Sha.	4. Sync.	5. Mul.	6. Test
AMBA AHB	x (APB)	U	S	S	x	x
CoreConnect PLB	x (OCP)	U	*	S	x	x
CoreFrame MBus	x (PalmBus)	U	P	S	x	n/a
HIBI	x	B/U	S	S	x	n/a
Marble	x	B	S	A	x	x
PI-bus	x	B	S	S	x	n/a
SiliconBackplane	n/a	U	**	S	x	x
STBus	n/a	n/a	S	S	n/a	n/a
Wishbone	n/a	B/U	S	S	n/a	n/a
Lotterybus	x	n/a	S	S	n/a	n/a
OCP	n/a	U	P	S	x	x
VCI	n/a	U	P	S	x	n/a

1. Hierarchical structures
2. Unidirectional ('U') or bidirectional ('B') links
3. Shared ('S') or point-to-point signals ('P')
   Exceptions: * In CoreConnect, data lines are shared, control lines form a ring
   ** In SiliconBackplane, data lines are shared, control flags are point-to-point
4. Synchronous ('S') or asynchronous ('A') transfers
5. Support for multiple clock domains
6. Test structures

Almost all the presented SoC buses use low-level handshaking based on dedicated handshaking signals. The typical SoC transfer types include split transactions and pipelined transfers. Although broadcast is almost a trivial application of the basic physical bus properties, its use is not explicitly stated in most of the specifications. This does not necessarily mean that the feature is impossible to implement with these schemes.

Table 8-2. On-chip bus transfers

Interconnection	1. Handsh.	2. Split tra.	3. Pipelined	4. Broadcast
AMBA AHB	x	x	x	n/a
CoreConnect PLB	x	x	x	n/a
CoreFrame MBus	n/a	n/a	n/a	n/a
HIBI	*	x	n/a	x
Marble	x	x	x	x
PI-bus	x	n/a	n/a	n/a
SiliconBackplane	x	n/a	x	x
STBus	x	x	x	n/a
Wishbone	x	n/a	n/a	n/a
Lotterybus	n/a	n/a	n/a	n/a
OCP	x	x	x	x
VCI	x	x	x	n/a

1. Dedicated bus control signals used for handshaking
   Exceptions: * Depends on the HIBI version
2. Split transfers
3. Pipelined transfers
4. Broadcast support

The arbitration mechanism is a major differentiator between the bus interconnections. The bus specifications included one- and two-level techniques based on priorities, round-robin techniques, and TDMA. Application specific arbitration means that the protocol for requesting the bus ownership is specified, but the algorithm for granting it is left unspecified. The arbitration process was pipelined in most of the buses for performance reasons. Most of the SoC buses use centralized arbitration but some opted for distributed versions. Dynamic reconfiguration of at least some of the arbitration (e.g. priorities and TDMA parameters) and system parameters (e.g. address ranges and interrupt sensitivity information) was specified for five of the bus specifications.

Table 8-3. On-chip bus arbitration and reconfiguration

Interconnection	1. Scheme	2. Pipelined	3. Cent./Dist.	4. Reconf.
AMBA AHB	as	x	C	n/a
CoreConnect PLB	1	x	C	x
CoreFrame MBus	as	x	C	x
HIBI	2	x	D	x
Marble	1	x	C	n/a
PI-bus	as	x	C	n/a
SiliconBackplane	2	x	D	x
STBus	1	x	C	n/a
Wishbone	as	n/a	C	n/a
Lotterybus	1	x	C	x
OCP	as	n/a	C/D	n/a
VCI	as	n/a	C/D	n/a

1. Application specific ('as'), one-level ('1') or two-level ('2') arbitration scheme
2. Arbitration done during previous transfer (pipelined arbitration)
3. Centralized arbitration ('C') or distributed arbitration ('D')
4. Dynamic reconfiguration

4. ENHANCING BUS PERFORMANCE TO FIT INTO SOC AND NOC SCHEMES

Many publications speculate that when the complexity of systems reaches a certain limit, buses are no longer a viable option [2, 20, 21]. In the following Subsections, some of the perceived and published bus limitations are listed together with some solution proposals.

4.1 Application space

By definition, buses are static networks since they do not have dynamic switching elements in them. This can lead to the conclusion that buses cannot adapt to fit multiple applications. Although the structure cannot be modified, there are several interconnection related parameters that can be tuned during system operation. This increases the performance of the interconnection and makes it more flexible.

In HIBI [8, 9], system parameters are initialized during system synthesis but they can also be modified at runtime with dynamic reconfiguration. Reconfiguration uses specific write commands that modify the configuration memories located in each interconnection wrapper and bridge of a system. The parameters include arbitration parameters (priorities, TDMA parameters, and longest allowable bus reservation times), power mode, and address ranges of the agents. This way, the latency and the allocated bandwidth of the components can be modified at runtime, following the variable communication requirements placed on the connected components.

Single bus architectures can also be extended to hierarchical and multiple buses; in effect forming a network-on-chip. Hierarchical buses have multiple bus segments that are connected to each other through bridges. Multiple bus systems, on the other hand, have several separate buses between components. These two methods help extending a bus system to handle multiple applications at the same time. A single bus, a hierarchical bus, and a multiple bus architecture are depicted in Figure 8-3. The fourth basic structure is the split-bus shown in Figure 8-3 d). In a split-bus architecture, the bus is divided into segments with three-state buffers. It can be utilized to achieve multiple simultaneous transfers, but has additional advantageous properties that are discussed in subsequent Subsections.

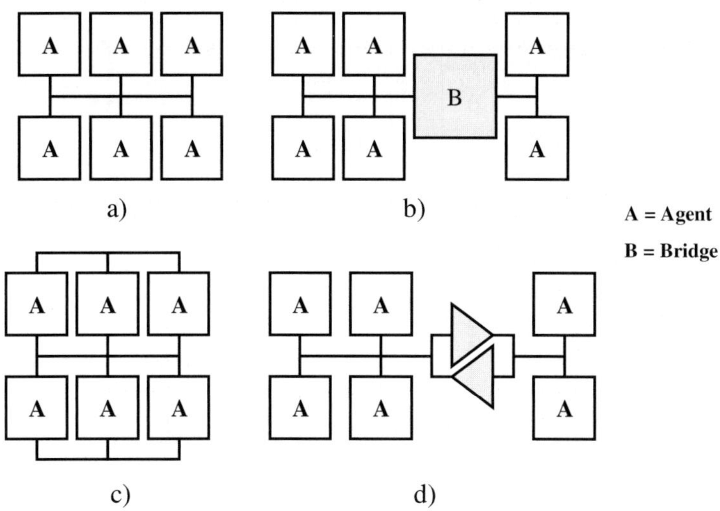

Figure 8-3. a) Single bus, b) hierarchical bus, c) multiple bus, and d) split-bus structures

4.2 Concurrent transfers

One limitation of buses is that since they use a shared transmission medium, they only allow the execution of one transfer at a time. This limitation applies also to the typical hierarchical structures where a transmission from one bus segment to another reserves both buses. This lack of parallel transfers makes also the scaling of bus-based systems difficult, because each added agent decreases the bandwidth available to the other agents. It should be noted that the parallelism found in many practical applications is quite limited making even a single bus sufficient for these applications. However, the number of parallel transfers in a bus-based system can be increased by utilizing multiple bus structures as depicted in Figure 8-3 c) where three parallel transfers are possible. In practice, the amount of agent I/O and interconnection area limit the use of this technique.

One solution to the problem is the hierarchical structure used in HIBI [8, 9]. In this scheme, the bus segments are separated by bridges that have buffer memory. The bridges can be seen as agents that have to arbitrate for a bus segment since they are implemented with two modified wrappers as depicted in Figure 8-4. A circuit-switched approach reserves all the segments between the source and the destination. Because the HIBI interconnection has bridging or routing components with buffer memory, a hierarchical HIBI bus could also be defined as a packet-switched network. It is possible to have parallel transfers in every segment simultaneously. An example of a HIBI hierarchical structure is illustrated in Figure 8-4.

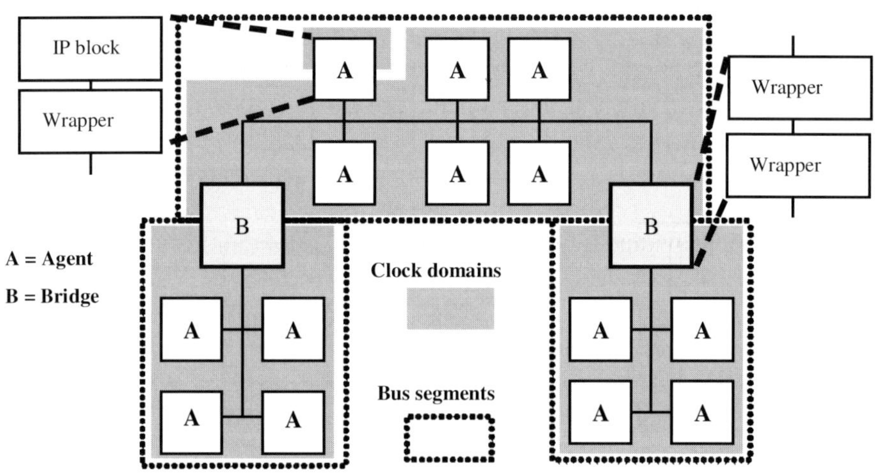

Figure 8-4. A hierarchical HIBI structure with four clock domains and three bus segments

4.3 Signaling

The speed of electromagnetic waves has a finite upper limit. The size of the chip designs scale up very rapidly, as does the clock frequency, and therefore, it will not be possible to time the signal transfers of a future NoC within a single clock cycle solely based on a global clock. It has been estimated that the propagation delay from edge to edge of 50 nm technology chips will be between six to ten clock cycles [3]. There are at least four ways to tackle this problem: using asynchronous techniques [10], multiple clock domains [8, 9], latency insensitive protocols [22], and split-bus techniques [23]. The long interconnection wires of buses also cause other problems, such as significant power consumption and increased noise sensitivity. One method proposed for addressing the power problem is the bus-invert scheme [24]. In addition, buses that transmit data serially are sometimes used to help the signal routing and minimize the bus signal line area. Problems with interconnect signaling are discussed further in Chapters 2 and 4 and the difficulties of clock distribution in Chapter 5 of this book.

An example of an asynchronous transfer technique is the Marble bus [10]. It uses a four-phase handshaking scheme similar to the one presented in Section 2.2. instead of a global clock and still provides all the required functions of a high-performance on-chip bus. All this can be achieved with reduced power consumption because of the asynchronous transfers lacking the clock distribution needed in its synchronous counterparts. Different subsystems can also run at different rates, completely independent of each other. This feature also helps the utilization of modular design techniques. Although Marble uses fully asynchronous techniques, another option would be to use globally asynchronous, locally synchronous designs (GALS), where transmissions within components are synchronous and the data traffic between components asynchronous.

Locally synchronous operations are a basic practical requirement for efficiently designing correct functionality in large digital systems. The difficulty in providing a global clock can be overcome with the use of multiple clock domains. For example, in the HIBI bus, the clock domains are isolated from each other by bridges or single wrappers that have FIFO buffers inside them. The wrappers take care of the handshaking between the bus segments operating on different clock frequencies. An example HIBI structure is depicted in Figure 8-4. A noteworthy issue is that all the agents of a bus segment are not required to use the same internal clock frequency. This is achieved by letting the wrapper unit of the agent operate at the same frequency as the rest of the bus segment but using another clock frequency for the rest of the components in the agent. By exploiting FIFO buffers and handshaking, the attached IP block can operate at any suitable frequency.

The latency insensitive design lets the system components use a single clock, that is, to apply the globally synchronous model. The global signals that cannot travel the whole distance in one clock cycle are pipelined to be on flight several clock cycles. One example are the methods proposed in [22], where single components, or pearls, are encapsulated in shells that contain buffers and control logic for making the transfers latency insensitive. Next, the physical layout is designed after which pipeline elements (relay stations) are inserted to the lines having a latency of over one clock cycle. This method has been developed for point-to-point connections but could also be applied to many bus schemes.

The split-bus technique [23] depicted in Figure 8-3 d) can also be used for reducing the problems with long bus lines. If the delay and energy consumption of the dual three-state buffer structure is smaller than the part of the bus that is being disconnected, then split-bus architecture is preferable. With split-bus technique, the parasitic loads and therefore energy consumption and noise problems can be decreased considerably. For example [23] reports energy savings of 16 to 50% with the split-bus technique by using heuristic techniques to optimize the bus splitting. Split-bus technique is particularly suitable for asynchronous buses which can also utilize the shorter propagation times. Synchronous buses, on the other hand, usually set the clock frequency according to the worst case propagation time.

The power consumption of long interconnection wires can be lowered by different signal encoding techniques because the activity of the bus signals considerably affects the power consumption. One proposed technique for bus encoding is the bus-invert scheme presented in [24]. In this scheme, the Hamming distance between the data previously on the bus and the data being transmitted is first calculated. If the Hamming distance shows that more than half of the data has changed, the inverse of the data is sent. The receiver knows the encoding of the data from a dedicated invert bit. If the Hamming distance of the data shows that less than half of the data has changed, the data is sent as it is. This simple technique has been used to achieve a 25% average power consumption reduction in bus interconnections [24]. Some power reduction coding methods are presented in this book's Chapter 7.

4.4 Reliable signals

Long, parallel lines needed in buses increase the rate of faulty behavior. This is particularly due to crosstalk and dynamic delay which is caused by the changing capacitance seen by a gate. The fact that long wires need to be narrow (for high wiring densities) and thick (for lower resistance) only makes these problems worse by increasing the coupling capacitance [1].

One solution is to use bus guardians as in [25]. In this method, dedicated modules constantly monitor the bus looking for errors. The bus guardians also provide error correction mechanisms. Long lines are frequently also broken into smaller ones by utilizing hierarchy or repeaters. In hierarchical systems, bridges split buses into segments. In this process, the grouping of the components forms the basis after which layout level techniques can be used to optimize the transmission line lengths. In single bus solutions, one has the option of utilizing repeaters and other low-level optimization techniques as presented, for example, in [26].

4.5 Efficient arbitration

Centrally arbitrating a bus with multiple agents can be a very time consuming task. This is also one of the features of buses that make them hard to scale, because the more agents there are, the longer the arbitration tends to take. The agents of a bus can also have completely diverging bandwidth requirements, and therefore the arbitration scheme can be very complex to implement.

In the STBus [13], this problem has been solved with data and address buses that are arbitrated separately. Each transmission has to first request an access to the transmission medium on the address bus. Upon being granted an access to the bus, the agent gets a transaction ID and can begin the actual transaction on the data bus. In this scheme, the data bus is used for data transmissions and at the same time the next transfer is already being arbitrated on the address bus. During the arbitration process, the agents can also negotiate different transmission parameters, such as the maximum length of the forthcoming transfer. This scheme works best in systems having long transfers.

Some arbitration algorithms, like priority, round-robin, and TDMA, can also be implemented in a distributed manner. Distributed arbitration does not require the existence of a high-level, centralized, and possibly very complex, arbiter. In addition, distributed arbitration lacks the point-to-point signals which improves scalability. A distributed TDMA and round-robin/priority-based arbitration is implemented for example in [12] and [8, 9]. The times during which the agents are guaranteed a bus access and the mechanism for acquiring the bus during other time periods is coded into a dedicated memory block located in the agents.

4.6 Quality of service

Many modern SoC applications would benefit from an interconnection that could give some guarantees for the transmissions. A frequently used term is quality of service (QoS), which implies that different agents need different guarantees for data transmissions. Here, two examples are shown. One demonstrates the applicability of TDMA for guaranteeing bandwidth and the other for providing smaller latency.

Real-time requirements demand that a system will respond to certain events during a specified time interval. In many systems, the occurrences of these events can be predicted in advance with some accuracy. SiliconBackplane [12] uses TDMA as an arbitration method to guarantee real-time requirements. In this method, time is divided into a repeating time frame that is again divided into time slots that are allocated to each agent according to their bandwidth needs. This way, all agents are guaranteed the use of the interconnection medium when they need it. The unallocated time slots and the time slots that are not used (at run-time) by their owner are arbitrated with a second-level arbitration scheme that is based on the round-robin scheme. The second tier method can be utilized by components requiring unpredictable, irregular, or low-priority transfers.

Almost all on-chip buses incorporate a type of a priority scheme. Usually, this concerns only the transmission order of the agents. An individual agent might have data transmissions that demand higher priority. The HIBI bus [8, 9] has two different priorities for two types of transfers: data transfers and messages. All the interconnection components (wrappers and bridges) have two sets of FIFO buffers, one for each type. The idea is that the higher priority messages can bypass the data transfers by using their own dedicated buffers. This concept can be extended by adding more priority specifiers and FIFO buffers to implement multiple virtual channels with different priority.

4.7 Standardization

Many system buses exist for SoCs and the processors inside the SoCs incorporate their own local buses. It is usually possible to fit them together, but having too many interfaces and adapters usually leads to problems. From this point of view, it would be better if all buses would comply to a single interface standard. Two proposed standards are the VCI [18] and the OCP [19]. VCI defines a protocol which can be used by components in communicating with each other regardless of the physical bus implementation. OCP is a superset of the VCI specification. It has the same data flow aspects as VCI, but includes additional control and test signaling.

Another way of dealing with these complexities is to generate the interconnection architecture automatically. One example is the component-based design automation approach presented in [27]. It is meant to be used with multiprocessor SoC platforms. In this approach, the architecture is first specified in an abstract way, after which the system is analyzed to determine a suitable hardware/software partition. Following this, the interconnection components are generated automatically with the help of hardware and software wrapper generators that utilize pre-defined library components.

5. EXAMPLES OF UTILIZING BUS STRUCTURES IN SOCS AND NOCS

The previous Sections described a number of ways in which a bus system can be extended to a relatively large NoC. The communication and connectivity requirements of NoCs are usually quite versatile; typically there are only a few components that require large bandwidths and not all of the components need to communicate with each other. The hierarchical bus architecture is a good fit for this type of communication profile. The following Subsections present bus-based design examples emphasizing their innovative design solutions. The first example is a quite traditional single bus, but the following systems start to have more complex network features such as hierarchical structures and buffering bridges.

5.1 Amulet3i

The Amulet3i telecommunication controller SoC [10, 28], depicted in Figure 8-5, incorporates an Amulet3 core and a Marble bus. It was presented by the University of Manchester, Cogency technology Inc., and ASIC alliance corporation in 1999. The Amulet3 core is capable of executing the code of ARM v4T instruction set architecture. In addition, the system has RAM and ROM memory, DMA controller, off-chip interface, analog-to-digital converter, instruction and data bridges, synchronous peripheral bridge, and control and test registers. Furthermore, the bus control unit takes care of bus arbitration and decoding.

Amulet3i was first used in a wireless communications device (DRACO) that has relatively large synchronous parts to which Amulet3i has access through the synchronous peripheral bridge. The DRACO chip was implemented with 0.35 μm technology and its area was about 7.0 x 3.5 mm^2. About half of this area was taken by the Amulet3i system, the rest of the area was consumed by the telecommunications peripherals. Simulations showed that the utilized Marble bus could transmit data at a speed relative to an operating frequency of 85 MHz giving the Amulet3i the processing power of about 120 MIPS (million instructions per second) with average power consumption of 215 mW [28].

One of the major advantages of Amulet3i design is the absence of a global clock signal. In addition, the handshaking scheme of Marble helps in building systems using a number of different IP blocks operating at different speeds. Although Amulet3i is completely asynchronous, possible synchronous components can be accessed through the synchronous peripheral bridge.

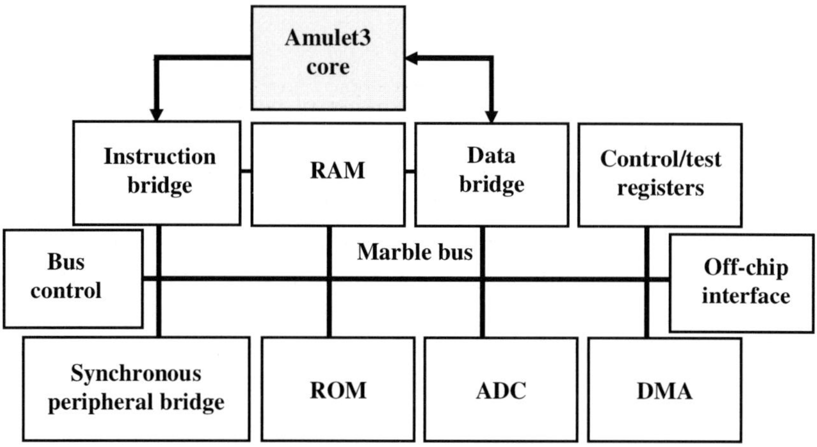

Figure 8-5. Amulet3i

5.2 MoVa

MoVa [29] is an MPEG-4 video codec designed at the ETRI design center. It is based on Hyundai ARM7TDMI processor cell, memory elements and soft cores (synthesizable VHDL) connected together with AMBA ASB and APB buses as depicted in Figure 8-6. The video coding IP blocks include motion estimation, discrete cosine transform, variable length decoding, and reconstruction functions. In addition, the design has buffers for storing bus data. The utilized buses were somewhat modified versions of the ones described in the AMBA specification. The whole design was based on pre-designed components, so the implementation was done using a standard ASIC flow (HDL description and synthesis) and C/assembler coding. A key factor in the design was the scheduling that was used to perform hardware and software operations concurrently.

MoVa was implemented with a 0.35 μm CMOS technology. It contains 220k equivalent gates (NAND2), 412 Kbits of static RAM memory, and the ARM7 core. The area of the chip is estimated to be 110.25 mm^2. All in all, it contains 1.7 million gates and has a power consumption of 0.5 W at the operating voltage of 3.3 V. The performance of MoVa is 30 frames/s for

QCIF and 15 frames/s for CIF pictures at the operating frequency of 27 MHz and, therefore, it meets the requirements of the MPEG-4 SP@L2 standard.

A standardized interconnection enables the use of pre-designed IP blocks. In this case there were six video codec IP clocks that had been previously designed and verified. The use of IP blocks having a simple standardized bus interface considerably shortened the design time.

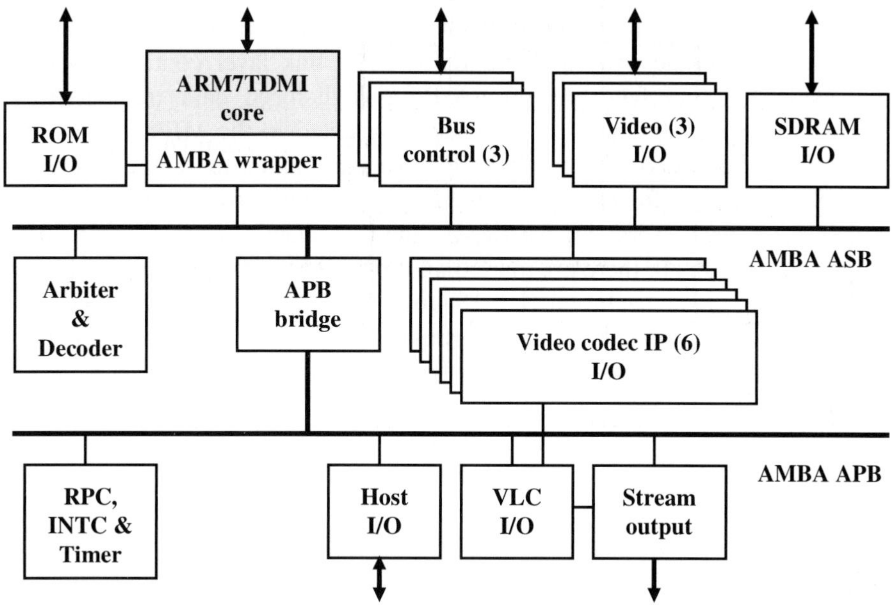

Figure 8-6. The MoVa MPEG-4 codec

5.3 Viper

The Philips Nexperia PNX8500, also called Viper [16], is an example of a SoC targeted for set-top boxes and digital television systems. It is one of the first Philips Nexperia-DVP (digital video platform) SoC solutions that are optimized for concurrent processing of audio, video, graphics, and communication data and is depicted in Figure 8-7. Viper contains a MIPS and a TriMedia processor. The MIPS 3940 is a typical 32-bit RISC processor that was chosen because of its high performance, ability to run popular embedded operating systems, and efficient peripheral control. The TriMedia 32, on the other hand, is a high performance, multimedia enhanced VLIW (very long instruction word) DSP processor optimized for audio and video processing. In addition to these two processors, Viper has a large amount of very complex hardware accelerators. The internal data transfers of the SoC are done through four bus segments. Three of these segments are PI-buses. In addition, the chip has also a very high speed 64-bit memory bus (DVP) to access off-chip SDRAM memory. Five bridges connect the bus segments together. Some functional units are connected to two bus segments. However, all units do not share a common bus. Therefore, the structure is a hybrid of hierarchical and multiple bus architectures.

The general peripherals of Viper include a USB controller, three UART interfaces, two I2C interfaces, a synchronous serial interface, and a general-purpose I/O module. An enhanced IEEE 1394 link layer controller and an expansion bus interface are provided for high-speed data transmissions. There are also other interface IP blocks connected to the MIPS PI-bus. All the other audio, video, and graphics IP blocks are connected to the TM32 PI-bus. They contain, for example, three MPEG system processors, two image composition processors, an MPEG-2 video decoder, and two video input processors. All in all, there are more than thirty IP blocks in the Viper chip.

For Viper implementation, a 0.18 μm CMOS technology was used. The whole chip had about 35 million transistors (8 million gates). The total memory implemented on the chip was 750 Kbits. The chip had 82 clock domains, for example the MIPS processor, TM32 processor, and the SDRAM operated on 200 MHz, 150 MHz, and 143 MHz clock frequencies, respectively. With an operating voltage of 1.8 V, the average power consumption of the chip was 4.5 W.

The Viper chip utilizes a versatile, hierarchical structure which has enabled the interconnection of a large amount of IP blocks. A hierarchical structure is a good fit to heterogeneous systems having different blocks with different demands from the architecture. In this case, the existing legacy IP blocks dictated the used interconnection but made also the design of the system much faster. Multiple clock domains were utilized to enable the connection of components operating at different frequencies. The Viper chips are further described in Chapter 15 of this book.

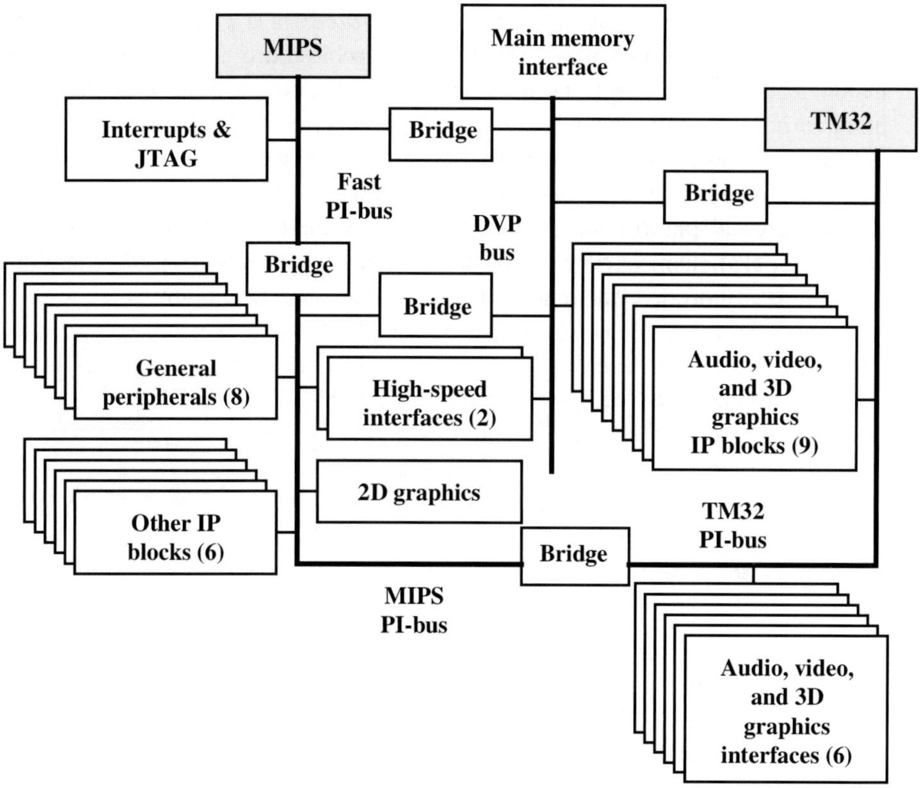

Figure 8-7. Viper

5.4 HIBI architecture

Three of the previous bus systems, Marble, AMBA, and PI-bus, supported hierarchical structures. However, the practical architectures based on circuit-switched bridges have limitations. To achieve a heterogeneous architecture, bridges or routers with buffer memory are required. These components receive data from one segment, possibly storing it before accessing the next communication channel, and then transmitting the data onwards. Therefore, they do not reserve the whole communication link from the transmitter to the receiver for the whole duration of the transfer.

HIBI interconnection is designed based on the principle that the application should dictate the architecture of the system. This approach is different from the topologies acquired from scientific multiprocessor systems and fixed bus architectures where an application is mapped to an existing architecture. The idea with heterogeneous systems is to allow a number of possible topologies in a hierarchical manner enabling the construction of systems with topologies optimized for a given application. In Figure 8-8, a heterogeneous HIBI-based architecture is presented which includes a single bus segment, a multiple bus structure, a tree structure, some point-to-point signals, and a buffering bridge between two segments.

The heterogeneous architecture of Figure 8-8 is a trade-off between a circuit-switched and a packet-switched network. The transmitted data is stored only in the bridges and not in every element of the network. In addition, parallelism can be utilized only where it is required and therefore the complexity of the system does not become unnecessary high.

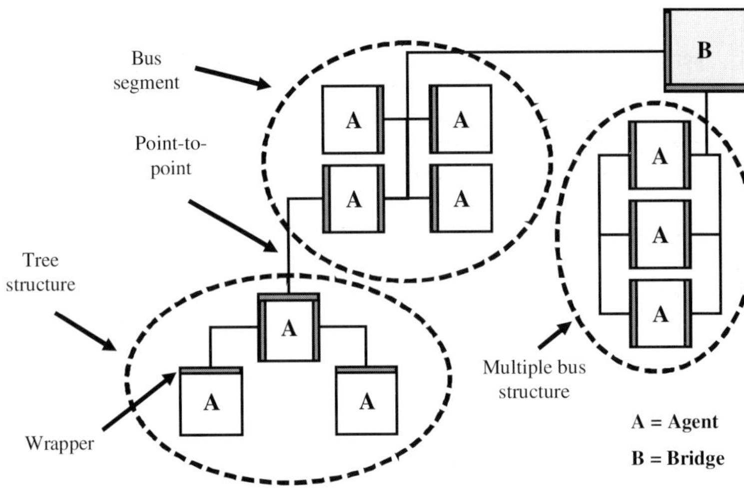

Figure 8-8. A heterogeneous HIBI architecture

6. CONCLUSIONS

It seems obvious that there is no general interconnection that perfectly fits every arbitrary application. Particularly, the proposed homogeneous network topologies and fixed bus architectures have many limitations. Because application is rarely an exact fit to the architecture, the ratio of average throughput to maximum available throughput in these systems is relatively small. A heterogeneous architecture, which makes a further distinction between local and global communication, addresses some of these problems. Locally, in segments having only a few agents, the communication can be accomplished via a bus. On the other hand, the global communication topology between these segments should be based on application specific bus and network structures.

ACKNOWLEDGEMENTS

All trademarks or registered trademarks are the property of their respective owner companies. This research work was supported by the Tampere Graduate School in Information Science and Engineering (TISE), the Nokia foundation, the Foundation of advancement of technology (TES), and the Ulla Tuominen foundation.

REFERENCES

1. D. Sylvester and K. Keutzer, "Impact of small process geometries on microarchitectures in systems on a chip," *Proceedings of the IEEE*, Vol. 89, No. 4, Apr. 2001, pp. 467-489.
2. P. Wielage and K. Goossens "Networks on silicon: blessing or nightmare?," *Symp. Digital system design*, Dortmund, Germany, 4-6 Sep. 2002, pp. 196-200.
3. R. Ho, K.W. Mai, and M.A. Horowitz, "The future of wires," *Proceedings of the IEEE*, Vol. 89, No. 4, Apr. 2001, pp. 490-504.
4. D.B. Gustavson, "Computer buses — a tutorial," in *Advanced multiprocessor bus architectures*, Janusz Zalewski (ed.), IEEE Computer society press, 1995. pp. 10-25.
5. ARM, *AMBA Specification Rev 2.0*, ARM Limited, 1999.
6. IBM, *32-bit Processor local bus architecture specification, Version 2.9*, IBM Corporation, 2001.
7. B. Cordan, "An efficient bus architecture for system-on-chip design," *IEEE Custom integrated circuits conference*, San Diego, California, 16-19 May 1999, pp. 623-626.
8. K. Kuusilinna *et. al.*, "Low latency interconnection for IP-block based multimedia chips," *IASTED Int'l conf. Parallel and distributed computing and networks*, Brisbane, Australia, 14-16 Dec. 1998, pp. 411-416.
9. V. Lahtinen *et. al.*, "Interconnection scheme for continuous-media systems-on-a-chip," *Microprocessors and microsystems*, Vol. 26, No. 3, April 2002, pp. 123-138.
10. W.J. Bainbridge and S.B. Furber, "MARBLE: an asynchronous on-chip macrocell bus," *Microprocessors and microsystems*, Vol. 24, No. 4, Aug. 2000, pp. 213-222.
11. OMI, *PI-bus VHDL toolkit, Version 3.1*, Open microprocessor systems initiative, 1997.
12. Sonics, *Sonics μNetworks technical overview*, Sonics inc., June 2000.
13. B. Ackland *et. al.*, "A single-chip, 1.6-billion, 16-b MAC/s multiprocessor DSP," *IEEE Journal of solid state circuits*, Vol. 35, No. 3, Mar. 2000, pp. 412-424.
14. Silicore, *Wishbone system-on-chip (SoC) interconnection architecture for portable IP cores, Revision: B.1*, Silicore corporation, 2001.
15. E. Salminen *et. al.*, "Overview of Bus-based System-on-Chip Interconnections," *Int'l symp. Circuits and systems*, Scottsdale, Arizona, 26-29 May 2002, pp. II-372-II-375.
16. S. Dutta, R. Jensen, and A. Rieckmann, "Viper: a multiprocessor SoC for advanced set-top box and digital TV systems," *IEEE Design and test of computers*, Vol. 8, No. 5, Sep./Oct. 2001, pp. 21-31.
17. K. Lahiri, A. Raghunathan, and G. Lakshminarayana, "Lotterybus: a new high-performance communication architecture for system-on-chip designs," *Design automation conference*, Las Vegas, Nevada, 18-22 June 2001, pp. 15-20.
18. VSIA, *Virtual component interface specification (OCB 2 1.0)*, VSI alliance, 1999.
19. OCP international partnership, *Open core protocol specification, release 1.0*, OCP-IP association, 2001.

20. L. Benini and G. De Micheli, "Networks on chips: a new SoC paradigm," *Computer*, Vol. 35, No. 1, Jan. 2002, pp. 70-78.
21. A. Boxer, "Where buses cannot go," *IEEE Spectrum*, Vol. 32, No. 2, Feb. 1995, pp. 41-45.
22. L.P. Carloni and A.L. Sangiovanni-Vincentelli, "Coping with latency in SoC design," *IEEE Micro*, Vol. 22, No. 5, Sep./Oct. 2002, pp. 24-35.
23. Cheng-Ta Hsieh and M. Pedram, "Architectural energy optimization by bus splitting," *IEEE Transactions on computer-aided design of integrated circuits and systems*, Vol. 21, No. 4, Apr. 2002, pp. 408-414.
24. M.R. Stan and W.P. Burleson, "Bus-invert coding for low-power I/O," *IEEE Transactions on very large scale integration (VLSI) systems*, Vol. 3, No. 1, Mar. 1995, pp. 49-58.
25. M. Lajolo, "Bus guardians: an effective solution for online detection and correction of faults affecting system-on-chip buses," *IEEE Transactions on very large scale integration (VLSI) systems*, Vol. 9, No. 6, Dec. 2001, pp. 974-982.
26. A.B. Kahng, S. Muddu, and E. Sarto, "Interconnect optimization strategies for high-performance VLSI designs," *Int'l conf. VLSI design*, Goa, India, 7-10 Jan. 1999, pp. 464-469.
27. W.O. Cesario *et. al.*, "Multiprocessor SoC platforms: a component-based design approach," *IEEE Design and test of computers*, Vol. 19, No. 6, Nov./Dec. 2002, pp. 52-63.
28. J.D. Garside *et. al.*, "AMULET3i - an asyncronous system-on-chip," *Int'l symp. Advanced research in asynchronous circuits and systems*, Eilat, Israel, 2-6 Apr. 2000, pp. 162-175.
29. J.H. Park *et. al.*, "MPEG-4 Video codec on an ARM core and AMBA," *Works. MPEG-4*, San Jose, California, 18-20 June 2001, pp. 95-98.

Chapter 9

FROM BUSES TO NETWORKS

David Sigüenza-Tortosa and Jari Nurmi

Abstract This chapter tries to make the case for the adoption of packet-switching networks as the main type of interconnect for future System-on-Chip. A set of requirements for Network-on-Chip is presented. A brief description of the Proteo Network-on-Chip being developed at Tampere University of Technology is included.

Keywords: Network-on-Chip, bus, circuit-switching, packet-switching, Proteo.

1. Introduction

The topic of Network-on-Chip (NoC) is being actively researched around the world by many groups and individuals. There exists the conviction that NoC will be one of the cornerstones of future highly-integrated System-on-Chip (SoC) devices, along with block-based design methodologies and hardware-software co-design.

The goal of this chapter is to give an understanding of the principles of networked communication from the perspective of NoC design. Since there are many publications dealing with networks and protocols in general, we will focus on summarizing the most important concepts and discussing the required qualities for a practical NoC architecture.

1.1 Physical and Technological Issues

As seen in Part I of this book, when electronic designs are scaled down to Deep Sub-Micron (DSM) regime, several non-ideal effects that were considered quantitatively small in past technologies appear now as serious obstacles for straightforward downscaling. Other problems arise because old techniques and materials are pushed to the limit, e.g. the power dissipation capabilities of packaging [1].

In this environment, long interconnects must be thought of as inherently noisy and slow. Part of the solution to the interconnection problems is the adoption of the Globally-Asynchronous Locally-Synchronous (GALS) paradigm, in which the system is partitioned into different subsystems that use unrelated clocks and communicate asynchronously. Removing the global clock brings several advantages: suppression of clock skew and synchronization problems and getting rid of one of the most important power consuming elements – the clock distribution network. Since most available Intellectual Property blocks (IPs) are currently designed as synchronous blocks, they will require wrappers to interface the asynchronous environment. Figure 9.1 represents a generic functional IP and its wrapper, following in part the work of [2].

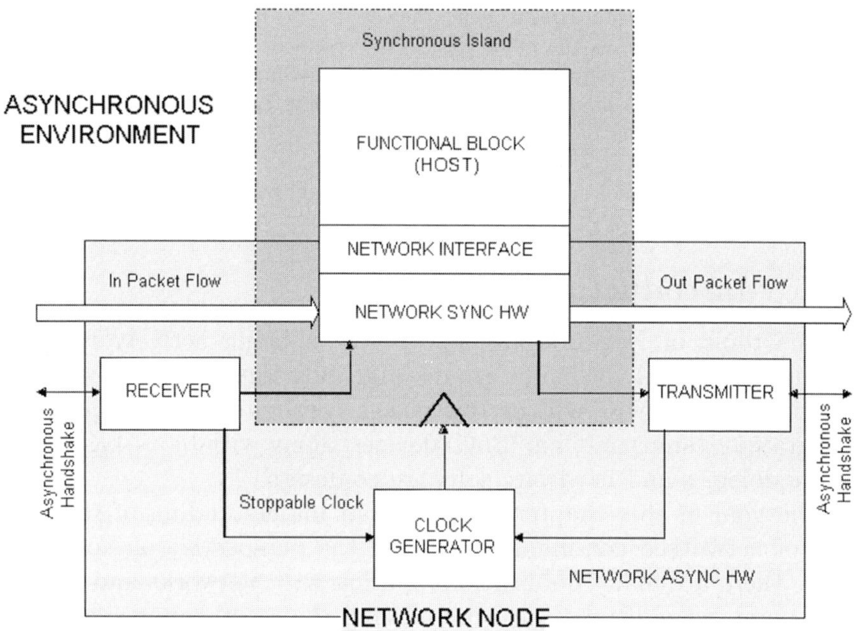

Figure 9.1. Network node general architecture.

NoC conceives the interconnection infrastructure as a pre-made element in the design process, with some degree of independence from the specific functionality of the system. This independence allows optimization of the interconnect to keep disturbances under control, and enforces modularity [3]. There have been studies about separated design of the computation and the communication parts of a system [4]. More information is available in Part III of this book.

Another major concern is the increasing costs of testing an IC. It is being argued that current design goals of performance and small size may not be economical in the near future and a stronger emphasis in manufacturability is needed [5]. The introduction of fault tolerance and test infrastructures in a standardized and possibly automated way is very appealing in this context, as a means to enhance production yield. The introduction of IP-based methodologies enables the development of these aspects as part of the services offered by the network. Using pre-made computational blocks and a pre-designed NoC in a new design will not be the optimum choice from the point of view of performance, but it may be the most economical, saving most of the work of implementing system-level mechanisms.

1.2 IP-Based and Platform-Based Design

The success of DSM technology depends on the availability of new design methodologies and tools. One key concept would be the reuse of previously designed blocks from in-house repositories or purchased from an external source. IP blocks are delivered in different formats, most of which allow customization[6]. One author suggests that future SoCs could contain a number in the order of hundreds of IP blocks, with an individual size between 50K and 100K gates [7].

With the availability of pre-made blocks, the design effort should shift to the interaction problem. Interconnection and system verification are the main issues. Relaying on a pre-designed network with known characteristics, can help in the process. Obviously, such an interconnect must be extremely flexible for it to be of moderately wide application and must use a predefined standard interface for its integration.

The Virtual Socket Interface Alliance (VSIA)[8] is an organization whose goal is to promote IP block reuse and integration solutions [9]. One of their first efforts has been the definition of a standard interface for IP blocks. The goal of this interface recommendation is to ease the interconnection of IP blocks from different suppliers, using a basic set of signals, protocols and communication semantics. Although VSIA's interface is not very common in commercial products, it is a good, neutral selection in design cases where choosing a proprietary solution is not desirable. The Open Core Protocol International Partnership [10] is another association that has defined an interface standard. The Open Core Protocol (OCP) specification is in fact a superset of the VSIA recommendation, with some useful features added.

Open standards like VSIA and OCP introduce some concepts that may be of great use for the on-chip interconnect designer, like split transactions, assignment of identifiers to individual packets and threading. Tasks that are usually left for the network hardware are: packet buffering and reordering, error detection and correction, and flow control. All these imply the use of hardware resources, like storage buffers, control logic, etc.

2. Characteristics of on-Chip Interconnects

First, let us fix some of the terminology used in this chapter. We will consider an on-chip interconnect as an entity consisting of a *medium*, which provides one or multiple *channels*, and *nodes* which interface to those channels. More specifically, we will call *network node* or simply *node* to a component exclusively dedicated to communication duties. A system component that does not interface the interconnect medium directly but through a network node is called a *host*.

Obviously, the most inmediate advantage of using a pre-defined interconnection scheme is that we save the effort of designing a new one. If the on-chip interconnect is delivered as a set IP-blocks, the job is reduced to verify the instantiation. This is one of the roles of on-chip interconnects. Also, their introduction has to allow the economical interconnection of a number of devices, which may possibly be quite high. This may imply the inclusion of standard test, debug and/or fault-tolerance mechanisms. Another important role is to structure communication inside the chip to obtain a certain degree of control on the interconnection performance. Finally, an on-chip interconnect must carry out the communication tasks in a efficient way, power- and resource-wise.

In order to fulfill the above goals, the requirements for a reusable on-chip interconnect can be summarized as follows (classified in five groups):

Performance Requirements. A tentative expression of generic performance goals (somewhat obvious choices) could be:

- *Latency* must be low, both in the sense of time to access the network and in delay of end-to-end communication. A latency of less than $1\mu s$ may be acceptable.
- *Bandwidth* (capacity) must be high. A peak bandwidth in the order of a few gigabytes per second may be good for most applications.
- *Power consumption* must be low. At least, the network should not increase the power demands of the system excessively.
- It must be *performance-wise scalable*, i.e. efficient for a range of performance levels. This requirement could have been placed also

under the next heading, but we wanted to stress the idea that the network should be a reasonable choice for both high- and low-performance designs.

Architecture Requirements. They include all the topology and protocol requirements. If the interconnect must be reusable, it must be:

- Compatible with the GALS paradigm, in the sense that it is possible to implement it using asynchronous links for global communications.
- *Architecturally scalable*, i.e. independent of the number of nodes present in the system.
- *General purpose*, preferably not confined in a specific domain of application, and supporting different types of transactions.
- *Programmable*, i.e. dynamically adaptable to varying traffic patterns, in case the system software is updated or the system topology is changed during operation, due to the activity of a power-saving or fault tolerance mechanism.

Quality-of-Service (QoS) Requirements. In order to design a reliable network, the following conditions should be met:

- The delivered performance is always within some predetermined boundaries. This is a basic feature needed for early performance estimation, real time system design and other time related issues. Working with asynchronous hardware makes this requirement non-trivial.
- The interconnect must be *resistant to noise*, because it is expected to work in a very noisy environment. In order to decrease sensitivity to noise, redundancy can be introduced at several levels in the protocol stack, involving duplicated hardware, retransmissions, etc.
- It must be *fault tolerant*, that is, offering a gracefully degrading service in the presence of faults that may appear in the system after production (transient or static).

Technology and Methodology Requirements. The interconnect must meet the following requirements to be suitable to future design environments:

- It should be possible to estimate the performance of the network at an early time during the design cycle or tune the network components to a specific performance level.

- Use of as few resources as possible.

- Compatible with an IP-based design flow – this suggest that the interconnect itself should be an IP-block.

Manufacturability requirements. This group of requirements may render the interconnect design totally useless if not met:

- The interconnect must be highly testable, providing an error coverage of over 95% .

- It must be fault tolerant, which in this context means that it must include enough redundancy to provide static fault recovery after production or at least the possibility of using the interconnect in a downgraded form in the presence of faults.

Fulfilling all the above requirements is not trivial. They are interrelated and careful consideration of trade-offs and left-outs is required.

3. Buses

A straightforward means for interconnecting a set of nodes is to tend a wire between each pair of components. This is not economical nor feasible in most situations. It is not efficient because the nodes will not communicate with each other all the time, and thus the medium becomes seriously underutilized. We must reduce the number of wires and rationalize the sharing of the medium without degrading performance excessively. Arbitration mechanisms and guarantees that each block is able to connect to other blocks must be provided. The Open Systems Interconnection (OSI) reference model[11] shows how to structure the communication tasks.

The bus is the simplest and also the most extensively used interconnection scheme. It is based on the sharing of one single channel. The advantages of the bus are clear:

- It is a very cheap structure and everybody understands how it works.

- Bus protocols have easy semantics. The resources dedicated to communication tasks within the boundaries of the computational blocks connected to a bus are few.

- Priorities can be asssigned to some components and reduce latency to a minimum for the highest priority ones (at the expense of the other components, of course).
- The communication model the bus favours is the memory-mapping of components, where all components in the system share a common address space and answer to a range of addresses, as opposed to the message passing model. Most IP-blocks today are memory-mapped and standard interfaces are defined following this model. This means that nowadays any on-chip interconnect must provide support for this model if it is intended to be used together with readily available components.

The bus has very important disadvantages too:

- It is not *scalable*. As nodes are added, performance degrades due to capacitances and parasitics. There is a practical limit for the number of components in a system using a bus as its communication scheme, and this limit is not high.
- It is not very *testable*. In a bus there are many possible states for the interfaces between nodes and medium and a full test of the structure is almost impossible for a high number of components and high coverage.
- There could be an important degradation of access time due to *contention* and arbitration. The low priority components may suffer severe delays. The whole of the bus must be designed for the worst case which may lead to inefficiencies.
- Another problem is its *lack of modularity*. This causes the poor fault tolerance performance of buses.

Salminen[12] has published a review of bus interconnects that are used in SoC designs. Buses are useful in systems with up to a dozen or so nodes. By checking against the list of requirements in the previous section, we realize that the bus is expected to fail to meet the scalability and testability constraints of future environments. Hierarchical buses may be able to patch up the scalability limit, but, in general, they are in disadvantage when compared to switched networks. Still, buses will be used as local interconnects in SoCs implementing hierarchical networks, in which components are grouped in clusters.

4. Circuit/Packet-Switching Networks

To solve the scalability and testability problems of buses, interaction between nodes in the system must be limited. The method is to introduce point-to-point connections instead of shared channels. In practice, the medium is divided in segments called *links.* The information flows from node to node passing through one or several of these links. Each time an information unit crosses from one link to the next, it is said that it has performed a *hop.* The network architecture defines how the different links are interconnected and a means of guiding or *routing* the data through the network from source to destination. If the routing decisions are taken only once at the beginning of the transaction, and all data in that transaction is forced to follow the thus establlished path or *circuit*, we talk about a *circuit-switching* network. If each data unit or *packet* is routed independently, we talk about a *packet-switching* network. Each type of network provides a more dynamic interconnection scheme than the bus. In this section we will introduce their main characteristics and their value in the NoC context.

4.1 Circuit-Switching

In figure 9.2 a basic example of the operation of a circuit-switching network is shown. In this kind of network, a dedicated circuit must be established before the transmission of data begins. Then, data flows from source to destination using the selected path. When the transaction is complete, the circuit is released for another use, possibly involving other nodes. Normally several circuits can be established simultaneously, as long as they use different links. Note that in figure 9.2, node **B** can not communicate with node **C** until node **A** frees the circuit, because switch *b* is busy. In this case, it is said that the connection **B**-**C** is being *blocked.* Node **B** can, however, establish a circuit to node **D** and carry out a transaction in parallel with the transaction **A**-**C**. The use of the network resources is decoupled. This was not the case in a bus.

The accumulated bandwidth of the circuit-switching network is high, although the actual usage may be quite low, due to the exclussiveness of the circuits. Any idle period is wasted bandwidth since, until the circuit is terminated, no other nodes apart from the ones that set up the connection can use it. Also, the time spent setting up and liberating the circuit penalizes performance. But once the transmission starts, the paremeters of the connection are very stable. This is a very interesting characteristic. In consequence, circuit-switching is a good candidate for infrequent but big transactions, or constant streaming of data between two components.

An important problem is the poor fault tolerance of this kind of connection. A faulty link affects all connections that use it and may inter-

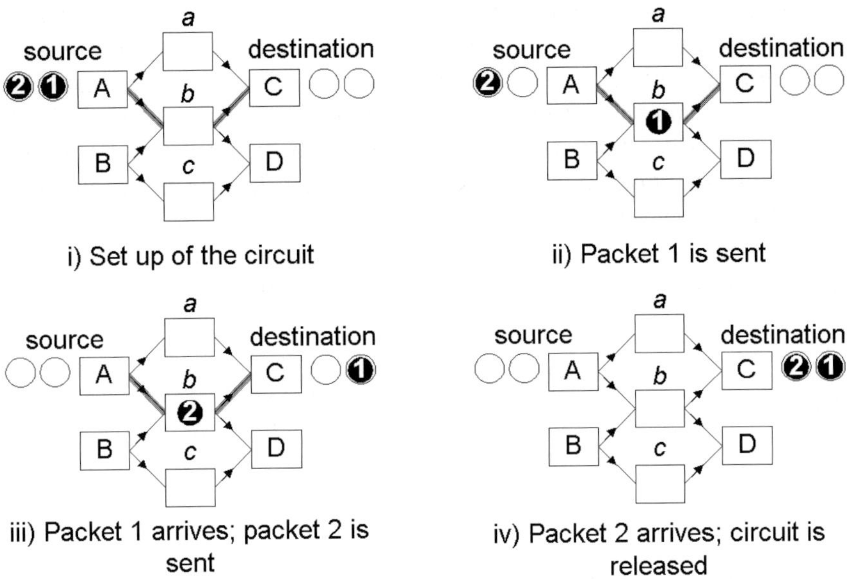

Figure 9.2. Operation of a circuit-switching network.

rupt completely communication between some of the components in the system.

4.2 Packet-Switching

In a packet-switching network there are no dedicated circuits: each link may participate in many different transmissions running simultaneously. Data sent over the network is organized in finite size units called *packets*. A packet includes a *header* section containing routing information and a *payload* section. The nodes laying between the source node and the destination node determine how to route the packets by examining the headers. This implies that a percentage of the total transmitted information is dedicated to protocol operation and should be considered an overhead. On the other hand, there is no time penalty for creating any circuit. As a consequence, a packet-switched network seems better suited for communication in which transactions occur sparsely in time.

In figure 9.3 we have represented the basic operation of a packet-switched network. It shows that packets transmitted between two given nodes do not have to use always the same links. The possibility to use alternate routing is useful in congestion avoidance, by selecting intelligently a different path each time a potential blocking situation is detected. Another application is fault tolerance: faulty links or nodes can be disconnected without affecting connectivity by providing a substitute for the eliminated path. On the other hand, this redundancy makes the routing decisions more complex and introduces the possibility that packets arive to their destinations in a different order than they were originally sent. A re-ordering mechanism has to be implemented in this case.

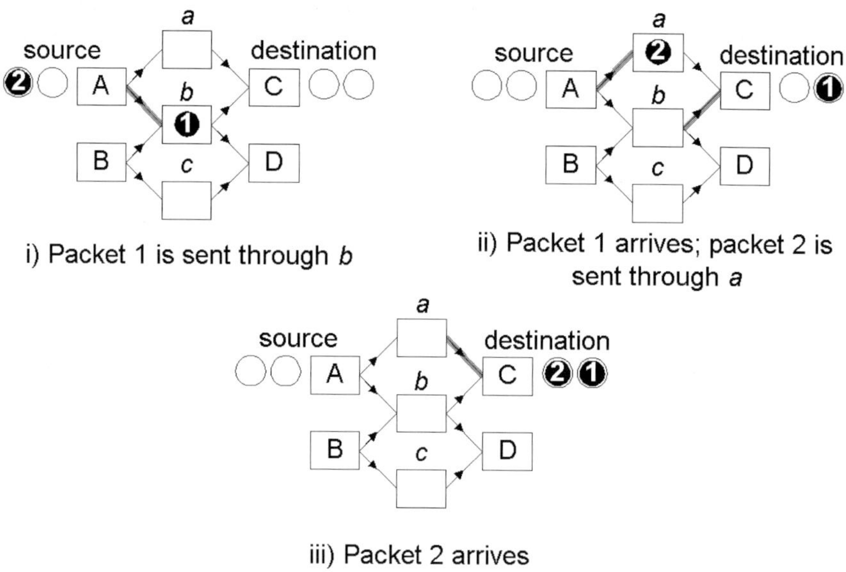

Figure 9.3. Operation of a packet-switching network.

As a consequence of the packetization of the data, blocking of connections is less likely to occur. For example, in figure 9.3, node **B** can already start a transaction at the same time node **A** is sending packet **2**, before **A** has finished its transaction.

A packet-switching network can be designed in a very modular way, and this is the foundation of its most important characteristics. Its main disadvantage is its high complexity. Automated tools are esential in its design and deployment. Nodes in a packet-switching network require an important amount of hardware resources. Careful tuning and optimization of this type of network is mandatory.

4.3 Comparison

The bus is very economic and simple, and is suitable for most applications, but it has some insurmontable flaws: it is not scalable and it does not meet the demands of manufacturability for the future.

Circuit-switching and packet-switching networks require more resources, but software tools can help optimizing the implementation. A packet-switching network has the very important advantages of being both more flexible and scalable than a circuit-switching network, and although for a specific application it would be possible to find a better candidate, it has a general purpose character. With the availability of adequate software tools, packet-switching would be the better option for the design of a generic NoC.

5. Proteo

Several researchers have already proposed packet switching networks in [13] and [3]. In [14], one author presents a NoC architecture in which guaranteeing QoS requirements is the essential drive. A brief description of some NoC experiments can be found in [15]. An overview of other proposed NoCs can be found in [16].

At Tampere University of Technology (TUT), we are developing a NoC architecture that we call *Proteo*, the name of an ancient Greek god who could change his form at will. He also had the gift of prophecy, although he was not very inclined to give oracles to mortals... With this name we want to express the idea of flexibility: based on a small library of predefined, parameterized components, we are able to implement a range of different topologies, protocols and configurations.

The problem is to design an interconnection mechanism for a relatively wide range of systems built from heterogeneus blocks, providing an adequate level of performance for a given application. The solution will consist of a network of some type, a set of protocols and a standard interface for accessing, along with the implementation of software tools for the automation of the integration process. In our project, the focus is on researching new protocols, architectures and the implementation of synthesizable blocks, and initiate the development of a set of prototype software tools.

The network will be built from parameterized IP blocks, providing a very flexible structure. Nodes will be customized by sizing their internal buffers, enabling/disabling protocol features, etc. One of our most important goals is to design a highly scalable network, both in terms of number of nodes and in performance. We do not want to focus (initially) on any specific application and our desire is to create a general purpose NoC generator.

We have defined a basic architecture and implemented the library blocks and obtained some initial synthesis figures. We are currently exploring an automation environment. A presentation of our project can be found in [17] and [18], and an overview of the simulation environment in [19]. A comprehensive description was published in [20].

5.1 Proteo Overview

Our vision is that SoCs will be built from heterogeneous clusters of blocks. In this scenario, at least a large part of total communication involve only neighbouring nodes. These clusters carry out related functions and are in fact subsystems composed of a handful of blocks. A bus is an adequate communication scheme for intra-cluster activities. Only certain blocks inside each cluster would need to communicate with blocks in other clusters. Inter-cluster communication is handled by dedicated *hubs* which implement high level functionality which is not needed for intra-cluster buses, like packet reordering and assignement of individual packet identifiers. For inter-cluster connectivity, a full-fledged network is implemented using packet-switching in a ring, mesh or tree topology. In figure 9.4 an example topology is represented. This is the kind of system we are using as a template in our development.

5.2 Proteo Hardware Elements

The basic hardware elements in our network are:

Hosts Every host corresponds to a functional IP or a multicomponent subsystem that will be connected to the network using a dedicated node as a wrapper. Any host compliant with the core VSIA-AVCI recommendation is supported at this moment.

Nodes The nodes regulate the interaction of the different packet flows in the network and interface the synchronous and asynchronous domains.

Links The links are the elements that actually move the information from node to node in the network in a clock-independent manner.

For development purposes, we have been only considering the VSIA standards, as a representative of a valid proposal for IP blocks' interface standardization. Currently we are moving to the OCP standard, because it is more practical. Depending on the type of interface, we will connect the IP blocks using different topologies. The reason is that, in

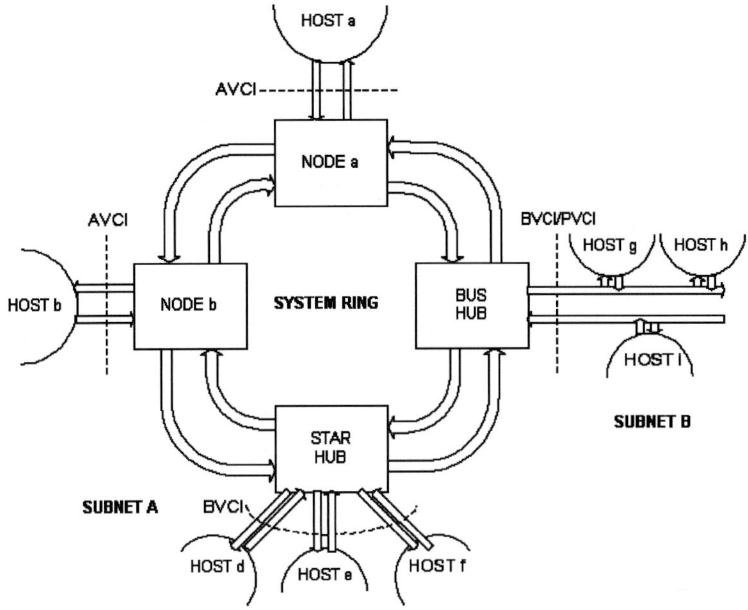

Figure 9.4. Example topology.

the basic version of the interface recommendations, there is not enough explicit information available at the interface to support a complex protocol stack. In order to keep network nodes as simple as possible, it is better to group similar blocks in clusters, and leave the network protocol operations to specialized hubs.

5.2.1 Nodes. The architecture of a typical node is inspired in Scalable Coherent Interface (SCI) standard[21], which is a standard interface developed for multiprocessor systems. SCI implements a rich set of mechanisms covering most of the needs of high performance systems. Our node architecture extends the basic SCI architecture to allow a configurable number of dimensions (figure 9.5). We have chosen a highly modular structure that makes easy its configuration and tuning. This node architecture has the advantage of being scalable because complex functionality like routing and flow control is confined to specific blocks.

The *Output-*, *Input-* and *Bypass-FIFOs*, plus the *link mux* and *link demux* blocks, form what we call a *port-module* block. The node architecture is built using one port-module and an *interface block*. This interface block translates the functional IP's (host) interface signals, packetizes the bursts coming from the host, and supply the output buffers with the correct packet format. Conversely, it translates the format of the

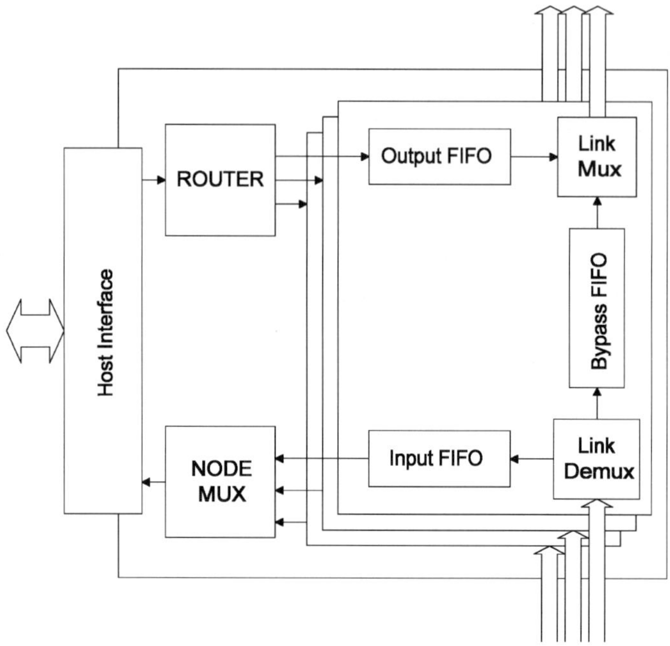

Figure 9.5. Extended node architecture with three I/O links.

packets placed in the input buffers to the correct interface signals. All components are implemented as independent IP blocks. For the moment we have only implemented a VCI-compliant interface block, but other modules are also possible.

The link-mux block inside the port-module takes care of the flow-control. The tasks of a flow-control algorithm in the context of NoC are:

- Ensuring the host is allowed to use its share of bandwidth.

- In case there is the possibility of taking up unused bandwidth, restraining the host from using too much of it, because it could starve other nodes.

- Divide the available bandwidth among nodes, according to the designer's intentions.

- Minimize buffer size.

Currently we are using a variation of the Persistence With Break (PWB) algorithm mentioned in [22].

In order to provide support for multidimensional networks (figure 9.5) we connect together several port-module blocks, together with a *node-router* and a *node-mux*. The node-router block contains a configurable routing table used to decide to which port-module to send each packet presented by the interface block. The node-mux block selects the next packet from the ones offered by the pool of port-modules, based on some local priority scheme. Currently it uses a round-robin policy. Multidimensional nodes grow bigger and slower, but in general lower dimensional networks seem to work more efficiently than high-dimensional networks[23], so we expect to find more applications for the smaller nodes. The configuration parameters include number of port-modules for each network node, FIFO geometry, supported protocols and packet formats.

The architecture just presented limits the number of possible topologies. For example, tree networks are not directly implementable using this kind of nodes. We are developing a couple of new types of nodes that allow the use of new topologies. These new types are not attached to any host and their functionality is equivalent to a 2-to-1 multiplexer or a 1-to-2 demultiplexer.

In reference to figure 9.1, the description above only covers the blocks labelled as "network synchronous hardware" and "network interface". We will not describe in this article the asynchronous block. It might be beneficial to reduce the amount of synchronous logic in the node design, up to the point in which only the network interface is synchronous. Advantages include: easier inter-clock domain communication, better noise inmunity and low power. Some research is being done in this direction.

Hub nodes are built basically in the same way, providing the adequate interface block for the kind of subnet it serves.

5.2.2 Links. In our terminology, a link block includes also the asynchronous transmitter and receiver shown in figure 9.1. Links interface the synchronous and the asynchronous parts of the network, using a special technique based on clock-stopping, similar to the one described in [2], to avoid metastability.

A link contain a group of signals that transmit the packet bits in parallel. There are additional signals for the asynchronous handshake and sideband signaling. Sideband signaling is used to mark the start and end of a Proteo packet.

Links provide a high level interface for the design tools, so they are effectively treated as modular elements and independently tuned.

5.3 Proteo Architecture

5.3.1 Topology. Currently, the topology being explored is a hierarchical network built from a system-wide bidirectional ring and several subnets with bus, ring or star topology (figure 9.4). The use of regular topologies allows easy routing and direct replication of blocks throughout the system.

5.3.2 Protocol. In the ring, transactions are split and out-of-order responses are allowed. The AVCI-like interface presented by the blocks attached to the ring provides more information about the transaction than the bus interfaces do. The nodes can be made quite simple and still form complex networks, because their functionality is restricted to the lower levels of the protocol stack.

Inside the Proteo network, the packets are forwarded in virtual cut-through fashion. In one router the start of the packet is routed forward immediately, before the whole packet has arrived, if the output link is free. If the output is not available, packets are stored in the FIFO buffers. The amount of buffering is one generic design parameter.

If the communication needs are characterized correctly, we can enable/disable protocol features at each node. Just a basic packet format has to be defined and kept throughout the network.

5.3.3 Packet format. Packets create a communication overhead. Therefore Proteo packets are kept as simple as possible. A packet consist of two areas: the header and the payload. The header fields are needed to deliver data from the source node to the target node. The payload section carries the data and the rest of the address as generated by the initiator host. Other VCI signals, like *Byte Enable* (BE), can be included in the payload section to allow the reconstruction of the original interface signals at the target host interface, although they are not relevant in the network context. The structure of the packet and the length of each field are configurable at design time.

6. Conclusion

Although buses will play a very important role in any electronic system as far in time as we can foresee, it is unavoidable the adoption of Networks-on-Chip in a non-distant future. In this chapter we have tried to summarize the basic relevant concepts in networking and identify the possible requirements of future NoCs. We tried also to present a reasoned argument for the adoption of packet-switching networks in future designs.

NoC are not quite the same as previous multiprocessor interconnection schemes or common local area networks. It presents some new characteristics, like its vast possibilities for customization, and several very important methodology questions. Also, the kind of systems we will have to deal with will be different: they will present different contraints and a much more diverse range of applications.

We have presented our own NoC proposal, called Proteo. Still not fully developed, it is an exercise on network design that allows us to try our concepts and ideas in a realistic set up.

References

[1] D. Sylvester and Himanshu Kaul. Future performance challenges in nanometer design. In *Proceedings of the 38th conference on Design automation*, Las Vegas, Nevada, United States, 2001.

[2] J. Muttersbach, T. Villiger, H. Kaeslin, N. Felber, and W. Fichtner. Globally-asynchronous locally-synchronous architectures to simplify the design of on-chip systems. In *Proceedings of the 12th Annual IEEE International ASIC/SOC Conference*, Washington DC, USA, September 1999.

[3] W. J. Dally and B. Towles. Route packets, not wires: On-chip interconnection networks. In *Proceedings of DAC*, Las Vegas, USA, June 2001.

[4] J. A. Rowson and A. Sangiovanni-Vincentelli. Interface-based design. In *Proceedings of DAC*, Anaheim, California, USA, June 1997.

[5] W. Maly, H. Heineken, J. Khare, and P. K. Nag. Design for manufacturability in submicron domain. In *Proceedings of ICCAD*, San Jose, USA, November 1996.

[6] A. M. rincon, C. Cherichetti, J. A. Monzel, D. R. Stauffer, and M. T. Trick. Core design and system-on-a-chip integration. *IEEE Design and Test of Computers*, 14(4):26–35, October-December 1997.

[7] D. Sylvester and K. Keutzer. A global wiring paradigm for deep submicron design. *IEEE Transactions on Computer Aided Design of Integrated Circuits and Systems*, pages 242–252, February 2000.

[8] Virtual component interface standard. http://www.vsi.org, April 2001.

[9] M. Birnbaum and H. Sachs. How VSIA answers the SoC dilemma. *IEEE Computer*, pages 42–49, June 1999.

[10] The open core protocol international partnership.

[11] H. Zimmermann. OSI reference model - the ISO model of architecture for open systems interconnection. *IEEE Transactions on Communications*, 28(4):425–532, April 1980.

[12] E. Salminen, V. Lahtinen, K. Kuusilinna, and T. Hämäläinen. Overview of bus-based system-on-chip interconnections. In *Proceedings of ISCAS*, Scottsdale, Arizona, USA, May 2002.

[13] P. Guerrier and A. Greiner. A generic architecture for on-chip packet-switched interconnections. In *Proceedings of DATE 2000*, Paris, France, March 2000.

[14] K. Goossens, J. van Meerbergenm, A. Peeters, and P. Wielage. Networks on silicon: Combining best-effort and guaranteed services. In *Proceedings of DATE*, 2002.

[15] M. Sgroi, M. Sheets, A. Mihal, K. Keutzer, S. Malik, and A. Sangiovanni-Vincentelli. Addressing the system-on-a-chip interconnect woes through communication-based design. In *Proceedings of DAC*, Las Vegas, USA, June 2001.

[16] S. Kumar. On packet switched networks for on-chip communication. In A. Jantsch and H. Tenhunen, editors, *Networks on chip*, chapter 5, pages 85–106. Kluwer Academic, 2003.

[17] I. Saastamoinen, D. Sigüenza-Tortosa, and J. Nurmi. Interconnect ip node for future system-on-chip designs. In *Proceedings of DELTA 2002*, Christchurch, New Zealand, January 2002.

[18] D. A. Sigüenza-Tortosa and J. Nurmi. Proteo: A new approach to network-on-chip. In *Proceedings of IASTED Communication Systems and Networks*, Benalmádena, Málaga, Spain, September 2002.

[19] D. A. Sigüenza-Tortosa and J. Nurmi. Vhdl-based simulation environment for proteo NoC. In *Proceedings of IEEE HLDVT*, Cannes, France, October 2002.

[20] I. Saastamoinen, D. Sigüenza-Tortosa, and J. Nurmi. An IP-based on-chip packet-switched network. In A. Jantsch and H. Tenhunen, editors, *Networks on chip*, chapter 10, pages 193–213. Kluwer Academic, 2003.

[21] IEEE. *The Scalable Coherent Interface*, March 1992.

[22] I. Rubin and Ho-Ting Wu. Performance analysis and design of cqbt algorithm for a ring network with spatial reuse. *IEEE/ACM Transactions on Networking*, 4(4), August 1996.

[23] W. J. Dally. Performance analysis of k-ary n-cube interconnection networks. *IEEE Transaction on Computers*, 39(6):775–785, 1990.

Chapter 10

ARBITRATION AND ROUTING SCHEMES FOR ON-CHIP PACKET NETWORKS

Heikki Kariniemi and Jari Nurmi
Institute of Digital and Computer Systems, Tampere University of Technology, Finland

1. INTRODUCTION

The operation of arbitration and routing algorithms has a significant effect on the performance of the packet switched networks. Switch arbiters, which are responsible for scheduling the internal resources of the switches for packet transfers from input ports to output ports, determine the throughput of the switch nodes. Their operation can be modeled with a bipartite graph matching problem where each maximum matching corresponds to a valid schedule with maximum number of simultaneous transfers. Arbitration algorithms can be classified to maximum size matching (MSM) and maximum weight matching algorithms (MWM). The MWM algorithms, which are able to take into consideration the filling of the input buffers, are usually able to produce higher throughputs than the MSM algorithms. In spite of this, most of the present arbiters implement the MSM algorithm or its approximation, because the MWM algorithms do not basically have fast and simple hardware implementations. For this reason, the beginning of this chapter, which concerns arbitration schemes, focuses mostly on the MSM algorithms. However, recent research results show such a progression that it can be expected that practical implementations for the MWM algorithms and their approximation may also soon become usable.

Routing algorithms are responsible for controlling the routing of the packets to their desired destinations. The network topology, which can be either regular or irregular, affects the complexity of routing and restricts the set of usable algorithms. Generally speaking, routing in the networks with irregular topologies is more complex and requires extra resources for the processing and storing of the routing information, which is due to arbitrary connections between switches and processing nodes. In computer networks with regular topology the connections between the switches and computers follow some particular rules, which can be utilized by the routing algorithms.

There are multiple redundant routing paths between the source and destination nodes in typical computer and on-chip networks, which makes it possible to route traffic flexibly past congested switches. Adaptive routing algorithms are able to utilize redundant routing resources and adapt their routing decisions to the prevailing traffic conditions and balance the traffic load over the whole network in such a way that the possibility of congestions decreases, which also improves the performance of the network. Unlike adaptive algorithms, non-adaptive routing algorithms route packets deterministically along the specified routing paths from the sources to the destinations. Switches are not able to prevent congestion when, for example, two or more packets simultaneously request the same output port or the requested output port is already reserved, even if free ports would be available. Non-adaptive routing algorithms produce typically worse performance than adaptive, although load balancing could be used for generating routing paths randomly so as to distribute the traffic load over the whole network. The performance of the non-adaptive routing can also be improved by a smart allocation of network resources. A new approach, which aims at allocating link capacities to virtual circuits (or routing paths) in such a way that the network throughput would be maximized, is called oblivious routing.

Most of the solutions presented in this chapter have originally been designed for computer networks. They will also be used in future on-chip network implementations which will resemble current computer and telecommunications networks [1, 2]. Because also many of the System-on-Chip (SoC) circuits are used in computing intensive applications, the focus is on such solutions which have been destined for high-performance computing systems. This chapter is organized as follows. Section 2 discusses a switch architecture, which has become usual in telecommunication switches and is suitable for on-chip packet switched networks also. The objective of this section is to introduce the operation of the switches before presenting the arbitration and routing schemes. In Section 3 different arbitration schemes are presented. In the beginning of this section arbitration is examined from graph theoretic point of view in order to explain the scheduling problem. In Section 4 different routing schemes are presented. In the beginning of this section general features of routing algorithms are described. This section has been divided into two parts where routing in the networks with irregular and regular topologies are concerned separately. In Section 5 the arbitration and routing in extended generalized fat trees (XGFT) is concerned. This is used as a simple case study to show how the routing can affect the throughput of the network and how the arbitration could be implemented. The last section summarizes the contents of each of the sections briefly.

2. GENERIC SWITCH MODEL FOR ON-CHIP PACKET SWITCHED NETWORKS

Crossbar switches with different buffering architectures are commonly used as basic building blocks in different switch architectures in packet switched networks [3, 4, 5]. There are at least two good reasons for that. Firstly: crossbar switches have simple architecture and operation. Secondly: they have excellent performance. One of their disadvantages is that their size is proportional to the number of ports and grows fast as the port count is increased. Therefore, larger switches with high port count are usually constructed from smaller crossbar switches and their architectures resemble small communication networks. For example, multistage interconnection networks (MIN), which are also used as switches, consist of multiple levels of crossbar switches. Because in typical on-chip networks switches have relatively small number of ports, the crossbar switches can be considered the most probable choice for the switch architecture in most of the cases. In addition, crossbar switches with different buffering architectures will also be used in many future on-chip networks [6, 7, 8]. For this reason, in the following sections it is assumed that the switch nodes are crossbar switches.

2.1 Buffering schemes for crossbar switches

Different buffering schemes have been developed for crossbar switches. Most commonly used are output-buffered, crosspoint-buffered, and input-buffered switch fabrics. Output buffered switches have ideal performance with 100% throughput, but it presumes that the operation rate is speeded up in proportion to the number of the input ports. For instance, if N input ports would simultaneously have packets destined for the same output port, the transmission rate should be speeded up by N so that all of the N packets could be transferred without stalling the transfer to any of the input ports from the other network nodes. Crosspoint-buffered switches can achieve as high performance as the output-buffered switches without speeding up their operation rate, but the size of the crossbar grows quickly as the number of ports increases.

Input buffered crossbar switches are smaller than crosspoint-buffered switches, but a head-of-line (HOL) blocking decreases considerably their performance. This is illustrated in the left side of Figure 10.1 where empty white circles denote requested but not granted transfers, and black circles requested and granted transfers. Although input buffers contain packets for all of the output ports, only two requests have been granted, because the first packets in multiple buffers require the same output ports. The right side of Figure 10.1 illustrates the same situation when the queuing discipline has been changed from the first in – first out (FIFO) to another which allows the packets to be transferred out of the buffers in an arbitrary order which is different from the order of their arrivals. This can be implemented by replacing FIFO buffers with virtual output-queuing (VOQ) buffers or dynamically-allocated multi-queue (DAMQ) buffers [9]. The right side of Figure 10.1 illustrates how packets can be picked up form the VOQs or DAMQs in a suitable order so as to avoid HOL blocking and how more transfers can be performed simultaneously through the crossbar, which improves the performance.

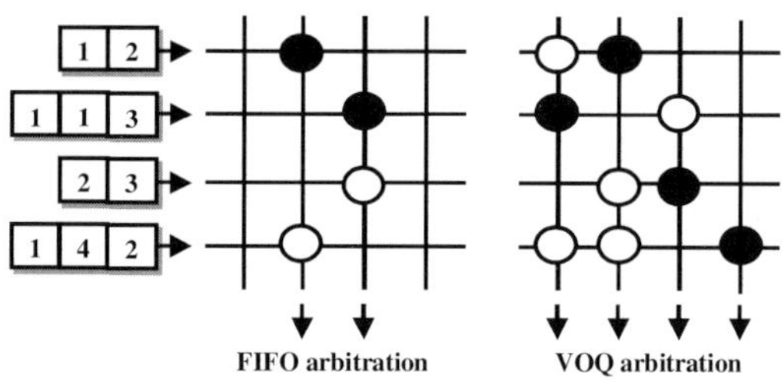

Figure 10.1. Example arbitrations with FIFO and VOQ buffering.

2.2 Switching strategy

The switching strategy determines how the channels and buffer resources are allocated to packets as they are transferred forward in the network and as the switch becomes temporarily congested. Cut-through routing tries to forward packets immediately after their headers have been received and the output ports can be determined. The cut-through routing strategy can be further classified to virtual cut-through routing [10] and wormhole routing [11] depending on how the packets are forwarded. Both the virtual cut-through and the wormhole switches determine the output port number according to the information stored into the packet headers and forward the packets immediately after their headers have been received, if the appropriate output port is free. If the appropriate output port is not free, wormhole switches keep the input port and the corresponding communication channel reserved until the desired output port becomes free.

Therefore, wormhole switches usually do not need to have buffer space for storing the whole packet and the transfer must be stopped. Unlike wormhole switches, virtual-cut through switches store the whole packets in the case that they can not be forwarded immediately after they have arrived. The virtual cut-through routing provides usually the best throughput, but the wormhole switches have the advantages of a small buffer space requirement and small switching latency.

Wormhole routing can be used with virtual channel flow control [11], which was originally designed for avoiding deadlocks in k-ary n-cubes and which resembles VOQ buffering. In this buffering scheme each virtual channel has its own buffer, but unlike the VOQs, the number of buffers per input port can also be smaller than the number of output ports. However, it has been shown that the higher the number of input buffers per input port is the higher is the network throughput [12]. In typical packet switched networks the maximum packet size is limited, unless the packet size is fixed. Therefore, when variable packet size is used, the input buffers could be dimensioned, for instance, in such a way that a couple of packets of maximum size fit into them.

3. ARBITRATION SCHEMES

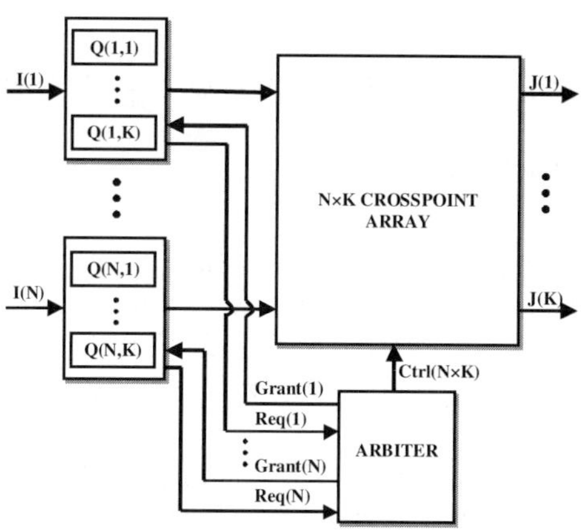

Figure 10.2. An input-buffered crossbar switch architecture with virtual output queues (VOQ) and centralized arbiter.

All arbitration algorithms solve generally the same problem when scheduling the transfers of the packets from the input ports to the output ports of the switch. The scheduling task is similar independently of the switch architecture, although here it is assumed that the switch is an input-buffered crossbar switch with VOQs depicted in Figure 10.2. This switch has N input ports and K output ports. Each of the input ports i (1≤j≤N) has K buffers for virtual output queues Q(i,j) which store the packets until the destination output port j (1≤j≤K) becomes free for the transfer. Input ports can request multiple transfers simultaneously and an arbiter is able to schedule several transfers at the same time during one arbitration cycle. For instance, the arbiter should be able to schedule a maximum of K transfers at the same time, if all of the input ports would have at least one packet in each of their VOQs and output ports would be requested correspondingly.

3.1 Modeling arbitration with a bipartite graph matching problem

The scheduling problem can be modeled with a bipartite graph matching problem. A bipartite graph is an undirected weighted graph G(V,E) where V is a union of two distinct sets of vertices I and J, and E is a set of edges which connect the vertices. Any pair of two vertices of graph G can be connected to each other with an edge belonging to set E only if they belong to different sets. In addition, each of the vertices can be connected to multiple other vertices. This is illustrated in Figure 10.3A.

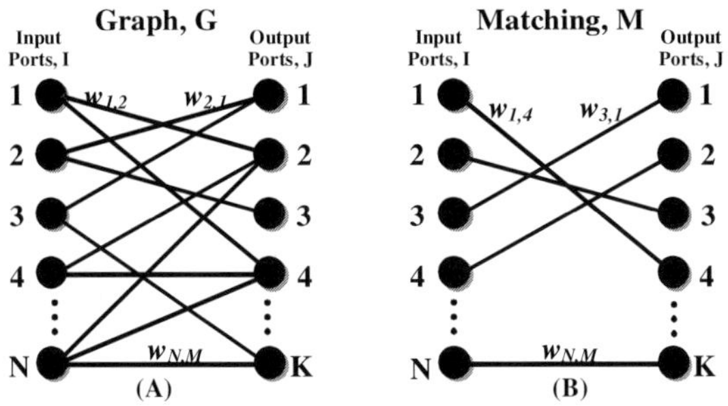

Figure 10.3. Graph G (A) and matching M (B) on it.

When modeling the scheduling problem vertices of set I correspond to the input ports and vertices of set J correspond to the output ports in such a way that there is one vertex for each of the ports in the sets. In Figure 10.3A edges between vertices correspond to requests that input ports send to the arbiter. For example, input port two has two packets one of which should be transferred to output port 1 and the other to output port 3. Matching M on G is any subset of E where none of the two edges in M have a common vertex. Figure 10.3B illustrates one possible matching M on graph G. Matching M_{max} with the highest possible number of edges is called the maximum matching. When solving the bipartite matching problem the objective is to find the maximum matching. This is what the arbiters are doing when they are trying to schedule as many simultaneous transfers as possible from the input ports to the output ports. It is also possible to associate other weight values $w_{i,j}$ than one with the edges so as to model, for instance, the occupancy of the buffers.

In the case that the weights associated with the edges are equal to one the maximum matching M_{max} is the matching with highest number of edges. If the weights are larger than one, the maximum matching M_{max} is the matching with the highest total sum of the weights. These matchings are called maximum size matching (MSM) and maximum weight matching (MWM) respectively. Below is a list of general properties which are desired for arbiters which will be used in high-speed switches [13].

1. **Provides a high throughput.** Arbiters should be able to find such schedules that the throughput of the switch will be maximized, which requires finding good matchings. In addition, the algorithm should be stable, i.e. the expected occupancy of the input buffers (VOQs) should be finite.
2. **Offers a starvation free service.** Arbiters should serve all of the input buffers (VOQs) fairly so that these would not become starved, i.e. the buffers do not remain unserved indefinitely.
3. **Achieves a high operation speed.** Arbiters should find the matchings as fast as possible so that they would not become a bottleneck for the system performance. This usually requires a simple and small hardware implementation.

The above list could also be supplemented with a requirement of a support for traffic classification to different service classes by prioritization, but it is not currently a commonly supported feature. Because the number of different choices is quite high, in the following subsections only some of the most commonly known MSM and MWM arbiters are reviewed. However, they give a view to the present state of the development of the arbiters.

3.2 Maximum size matching arbiters

The best time complexity of iterative algorithms that compute maximum size matchings (MSM) for bipartite graphs is $O(N^{5/2})$. These algorithms have not simple hardware implementations and they require too much time for computing the matchings, which is a problem. In addition, they do not produce necessarily such matchings which ensure starvation free service and stability. For this reason, typical hardware implementations of MSM-arbiters perform heuristic parallel algorithms which are able to compute only suboptimal matchings by approximating the behavior of an exact MSM-algorithm with smaller time complexity. These algorithms meet at least partially the requirements presented in the previous subsection. For instance, heuristic algorithms like iterative round-robin with SLIP (iSLIP) [13, 14] and iterative dual round-robin matching (iDRRM) [5, 15] only approximate MSM-algorithms. They are able to provide 100% throughput under uniformly distributed traffic, but under non-uniformly distributed load their performance is worse. Both of the aforementioned arbiters consist of multiple concurrently operating round-robin arbiters. Although they have practical small and fast implementations they require multiple operation cycles for computing the matching, which increases the delay. This problem is solved in pipeline-based maximal size matching (PMM) [16] arbiters where multiple subschedulers, which can be either iSLIP or iDRRM arbiters, operate in parallel.

Symmetric crossbar arbiters [9, 17] are another approach for solving the maximum matching problem. These arbiters consist of an array of arbitration cells. There is one arbitration cell for each of the cross-points in the array. These cells are competing for the transfers. Unlike iterative arbiters, the symmetric crossbar arbiters need only one operation cycle for computing the matching, which results from the increased parallelism.

3.2.1 Iterative arbiters

The operation of the iSLIP arbiter consists of three steps which can be iterated repeatedly in the following way. In this arbitration scheme both input and output ports have their own arbiters that communicate with each other so as to solve the maximal size matching.

1. **Request.** Each unmatched input port sends a request to every output for which it has a packet in one of its buffers.
2. **Grant.** If an unmatched output arbiter j ($1 \leq j \leq K$) receives multiple requests, it chooses the one that appears next in a fixed cyclic order (round-robin) starting from the highest priority element. The output arbiter j ($1 \leq j \leq K$) notifies each input arbiter whether or not its request was granted. The grant pointer g_j to the highest priority element is incremented (modulo N) to one location beyond the granted input port if, and only if, the grant is accepted in the next step (3).
3. **Accept.** If an input arbiter i ($1 \leq i \leq N$) receives a grant, it accepts the one that is next in a fixed cyclic order (round-robin) starting from the highest priority element. The accept pointer a_i to the highest priority element is

incremented (modulo K) to one location beyond the accepted output port. The accept pointers a_i are updated only in the first iteration.

In this scheme the grant and accept pointers are updated only if the grants are accepted and a match becomes completed. In this way the most recently matched pair of input and output ports gets the lowest priority, which ensures starvation free service of all of the ports. Furthermore, all of the ports are serviced fairly, because the pointers are updated one after another in a cyclic order like round-robin does. In this presentation the arbitration is implemented with multiple distinct arbiters placed at input and output ports. Therefore, iSLIP can also be constructed out of multiple small functional blocks instead of one large block so as to reduce the operation delays and to achieve as high operation speed as possible.

The operation of the iDRRM [5, 15] arbiter differs slightly from that of the iSLIP arbiter, but its operation consists of only request and grant steps. Unlike with iSLIP, with iDRRM input ports send their requests only to the output port which they have given the highest priority, not to all of the output ports for which they have packets in their VOQs. Both iDRRM and iSLIP can do multiple repeated iterations so as to find a better matching. The iDRRM arbiter serves input ports fairly, because the output arbiters select the requests to be granted in a cyclic order one after another like round-robin, and because the input arbiters do not update their request pointers before their requests become granted. This ensures starvation free service of all of the input ports.

In the previously presented arbitration schemes multiple iterations may be required for completing the arbitration process and each of the iterations takes more than one operation cycle. This may limit the system performance when the number of ports grows high unless the arbitration time can be reduced. One method of avoiding this problem is to perform several arbitrations concurrently so as to allow more time to be used for single arbitrations, which requires the usage of multiple arbiters. In pipelined-based maximal size matching (PMM)[16] arbiters this is implemented with multiple subschedulers. The subschedulers operate in parallel in a pipelined manner and it takes 3 operation cycles for each of them to complete the matching. In this way more computing time can be given for the scheduling of the transfers while the number of complete schedules per arbitration cycle can be increased. The exact implementation of the PPM arbiter depends on the subschedulers which can be either iSILP or iDRRM arbiters.

3.2.2 Symmetric crossbar arbiters

Symmetric crossbar arbiters like wave front arbiters (WFA) and wrapped wave front arbiters (WWFA) [9] have originally been designed to be used with dynamically-allocated multi-queue (DAMQ) input buffers. The DAMQs store a separate queue for all of the output ports like VOQs except that the queues share the same input buffer at each of the input ports. The left side of Figure 10.4 depicts an arbitration cell array which forms a core of wave front arbiters. There is one row of arbitration cells for each of the input ports and one column for each of the output ports in the cell array. Figure 10.4B depicts an arbitration cell. It has inputs for request (R) and grant (G) signals. The input N-signal indicates whether there are granted cells above in the same column, i.e. whether the output port has already been scheduled, and the input W-signal indicates whether there are granted cells to the left in the same row, i.e. whether the input port has already been scheduled. Outputs S and E, which are connected to inputs N and W of the next cells on the same columns and rows respectively, forward this information from a cell to another. Each of the arbitration cells has also signals which are used for determining which cells have the highest priority and form the "wave front". In Figure 10.4A arbitration cells along the "wave front" diagonals 1 and 5 can be given the highest priority at the same time, because they belong to different columns and rows. For this same reason, arbitration cells along diagonals 2 and 6 can be connected as well as diagonals 3 and 7. The result is a wrapped wave front arbiter (WWFA) where all of the cells of the "wrapped wave fronts" are always on different rows and columns.

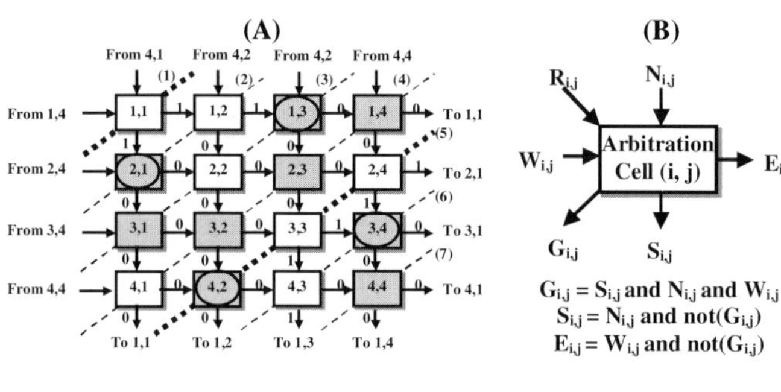

Figure 10.4. The operation and structure of the wrapped wave front arbiter (WWFA) (A) [9] (© [1993] IEEE) and an arbitration cell (B).

Figure 10.4A illustrates also how the WWFA-arbiter operates. Arbitration cells along diagonals 1 and 5 on thick dashed diagonal lines have the highest priority and they belong to the "wrapped wave front" which is moved diagonally over the arbitration cell array from the top left corner towards the bottom right corner during successive arbitration cycles. Arbitration cells with dark backgrounds show which output ports the input ports request. Ellipses within the cells show which of the requests have been granted while the values beside the arrows show the values of corresponding S-signals and E-signals (or N-signals and W-signals). As can be seen all of the input signals of the granted cells must have logical value one unless they have the top priority.

Both the WFA and the WWFA arbiters have the drawback that they are not able to provide starvation free schedules for all of the input queues. The starvation can be prevented if the input queues (VOQs) with the highest priority are allowed to maintain their priority until their requests become granted. This requires, however, modifications. Another drawback is that the WFA and WWFA arbiters have cyclic signal paths through combinational logic blocks, which is difficult to synthesize and test. This problem is solved in a diagonal propagation arbiter (DPA) [17]. In the DPA the arbitration cell array of the WWFA is extended by replicating the cells that belong to the same "wrapped wave fronts" during different arbitration cycles. The result is a (2×N-1)×N-array of cells. Arbitration cells have been connected directly together without feedback loops in such a way that the top most row of every block of N consecutive cell rows corresponds to one "wrapped wave front" of the WWFA with the highest priority. The area of active arbitration cells is determined with a window of N rows, which is moved over the cell array. This implements the round-robin rotation scheme and ensures the fairness of the service.

3.3 Maximum weight matching arbiters

Various scheduling algorithms can be designed by assigning different values to the weights of the edges of graph G. For example, weights can be queue lengths or maximum waiting times of packets. The time complexity of the best iterative algorithms which compute exact maximum weight matchings (MWM) is $O(N^3\log(N))$. Pure MWM arbiters like longest queue first (LQF) and oldest cell first (OCF) have impractically complex hardware implementations and high time complexities. For this reason, heuristic algorithms like iterative longest queue first (iLQF) and oldest cell first (iOCF) [13], which compute only approximations of the maximum weight matchings, have been developed. Another heuristic MWM algorithm is iterative longest port service (iLPF) [18] which actually finds at first a set of maximum size matchings from which it chooses the matching with the largest total weight. In this way it is able to take advantage of the high speed of the MSM algorithms while it is also able to produce a more optimal scheduling like the MWM algorithms. In modified longest queue first serviced (mLQF) [19] arbiters sums of the queue length and the maximum waiting time are used as the weights. It assigns value zero to weights before the next arbitration cycle each time after the request of the VOQ has become granted.

3.4 Summary of arbitration schemes

In typical broadband networks switches have either a small number of high-speed ports or a high number of slower ports and the total input packet rates are high. In broadband networks a fixed small packet size is also usually used, which further increases the packet rate and sets harder requirements for the operation speed and performance of the arbiters. In a typical on-chip networks switches have smaller number of ports. In addition, the input packet rate may also be smaller because of the usage of larger variable sized packets. Therefore, it can be assumed that MSM arbiters presented earlier are suitable for the switches of the on-chip networks without any special changes. Generally speaking, MWM arbiters are able to produce higher throughput, but they are not useful, if wormhole routing strategy is used, because wormhole switches have usually small input buffers which have space for only at most few packets. In addition, MSM arbiters can also be easily supplemented to support traffic prioritization by allowing only requests with highest priority to take part in arbitration.

4. ROUTING SCHEMES

Routing is a process which is performed in each switch node of the network for every packet that arrives at them. It determines the paths along which the packets traverse the network from their source to their destination. The operation of the routing is specified by a routing algorithm which can be either adaptive or non-adaptive. Adaptive routing algorithms are able to route packets past congested switch nodes. Therefore, they typically produce better throughput than non-adaptive algorithms which are not able to do so. Furthermore, they can provide more fault tolerant and reliable routing, because they are usually able to change the routing paths if connections between switches are broken. Non-adaptive routing algorithms route packets always deterministically to the destination according to the routing information, and switches are not able to change the routing path although the next switch along it would be congested. A simple example of a non-adaptive routing algorithm is a source routing algorithm. It attaches the routing information directly to the packet headers in the form of output port numbers of the successive switches that are along packet's routing path. For this reason, the switches do not need to do any complex and time consuming computations, which is an advantage. Another advantage of the source routing is that the load balancing can be partially implemented in it, because packet sources can give randomly generated routing paths for successive packets, which spreads the traffic over the whole network, reduces the probability of the congestions, and improves the performance of the network.

Routing algorithms that will be used in on-chip networks should generally satisfy requirements of the following list [20].

1. **Routing algorithms must route packets correctly.** This includes also that algorithms should provide deadlock-free routing so that it would be impossible that multiple packets are waiting in a cycle for each other to be routed forward, although they will never be routed forward. The routing algorithms should also be livelock-free, which ensures that all of the packets will always finally arrive at their destinations.
2. **Routing algorithms should consist of simple computations.** Simple computations have usually small and fast hardware implementation unlike complex computations which require more time and hardware resources.
3. **Routing algorithms should be adaptive.** Adaptive routing algorithms are able to route packets along alternative routing paths to the destinations and avoid using congested network nodes. They are also able to distribute and balance the traffic load in the network. In addition, they usually operate correctly also in such conditions where the network is partially out of action.
4. **Routing algorithms should primarily choose the shortest routing paths for the packets and use only secondarily the other paths.** The routing delays are smallest along the shortest routing paths, if they are not congested. This property would also eliminate the possibility of the livelock.
5. **Routing algorithms should service fairly all network users.** This eliminates the problem of an indefinite postponement which occurs when the packets are waiting for an event (routing) which can happen but never does.

Network topology, which can be either regular or irregular, has a significant effect on the implementation of the routing algorithms. In the networks with regular topology connections between network nodes follow certain rules, which can be taken advantage of. Because of the regularity of the connections, the routing algorithms are simpler and they have smaller and faster hardware implementations. In the networks with irregular topology the connections does not follow any special rules and the routing is usually based on the usage of routing tables which must be configured before the system is completely ready for operation. Because on-chip networks can have topologies of both of the types, this section concerns routing in the networks with both irregular and regular topologies separately.

4.1 Routing in networks with irregular topologies

Computer networks with irregular topology consist of arbitrarily connected switches and computers as Figure 10.5 depicts. The advantage of the irregular topology is that the network can be expanded with new computers or switches in a flexible manner. In addition, if the system size will be fixed and communication patterns are known, the system architecture can be optimized in order to minimize communication latencies and to maximize the network throughput as well as the system performance. As Figure 10.5 illustrates the network topology and the number of computers and switches can be chosen flexibly. Furthermore, it would be possible to use switches of different sizes in different parts of the system, and the number of transmission links between nodes and channels within the links could also vary.

Examples of computer networks with irregular topologies are Autonet [21] and Myrinet [22]. Autonet is a self-configuring network, which uses a distributed Up*/Down* routing algorithm. Each switch has a processor for running an algorithm which determines the topology of the network as a spanning tree and the position of the switch within it. In the spanning tree hosts are at the leaves at level zero and switch nodes at higher levels of the tree. After the spanning tree has been determined each switch node is able to compute and load its forwarding table according to the structure of the spanning tree. At each switch the input port number and the destination addresses of the packets are used for addressing the forwarding table which contains the output port numbers. Because it is not possible to adapt routing decisions to the varying traffic conditions in the network, this form of the Up*/Down* routing is non-adaptive. Routing in Myrinet is also based on a similar usage of spanning trees as in Autonet, but instead of distributed routing Myrinet uses source routing where packet headers contain routing information for steering the routing of the packets through the switches. The hosts perform address-to-route transformation for each packet before transmission to the network. The performance of the Up*/Down* routing algorithms depends on the way the spanning tree and routing tables are computed [23]. They are not able to balance traffic load on the network either as fully adaptive algorithms like, for instance, Adaptive Trail Routing [24], which is a disadvantage.

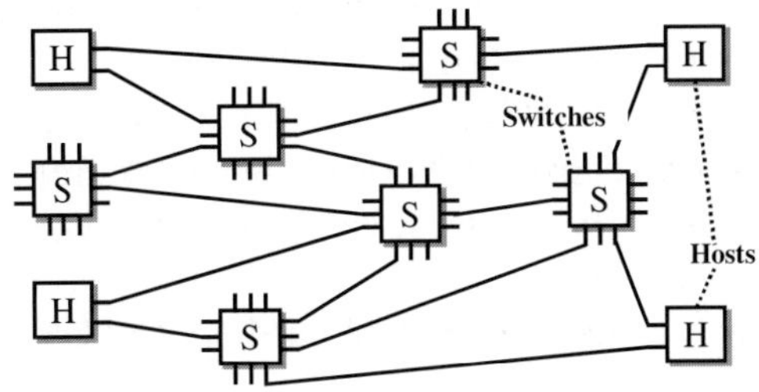

Figure 10.5. A part of a network of workstations with four hosts (H) and six switches (S).

Online oblivious routing algorithms [25, 26, 27, 28] can be used for routing virtual circuits optimally on the networks minimizing the probability of congestion. Oblivious routing algorithms specify routing paths between every communicating pair of source and destination nodes in such a way that the load of each transmission link is maximized and the probability of congestion is minimized. This routing scheme is basically non-adaptive and it could be used in all networks independently of the topology. Usable polynomial time algorithms for solving the online oblivious routing problem have also been developed.

4.2 Routing in networks with regular topologies

Communication networks with regular topologies like different trees, multistage interconnection networks (MIN), n-dimensional meshes, and k-ary n-cubes are the most common in multiprocessor systems [20]. One of the basic non-adaptive routing algorithms used in meshes and k-ary n-cubes is called dimension order routing (DOR). In this non-adaptive routing algorithm, each packet is routed along one dimension at a time until it arrives at a node the local coordinate of which is equal to that of the destination processor node. After this the packet is routed to the next dimension. E-cube is a similar basic routing algorithm used in k-ary n-cubes. In 2-ary n-cubes the coordinates of the destination node can be represented as a binary number which is stored as a destination address into the header of the packet to be transmitted. At each node the e-cube computes the position of the next coordinate or address bit of the local node which differs from that of the destination node, and the packet is forwarded to the direction of the corresponding coordinate.

Adaptive deadlock free routing algorithms can be designed systematically for meshes and k-ary n-cubes. One of the methods is based on restricting the number of turns along the routing paths [29]. Figure 10.6 illustrates sets of turns of tree different turn model routing algorithms. Figure 10.6A shows all of the eight possible turns and the simple cycles these turns can form. In Figure 10.6B two turns are eliminated from both of the cycles, which corresponds to the deterministic DOR in 2-dimensional meshes. Dashed arrow lines denote the eliminated turns. Only one turn is allowed, which makes this non-adaptive algorithm deadlock-free. In Figure 10.6C only turns to the west have been eliminated from both of the cycles. Therefore, if it is necessary to route packets to the west, it should be done at first, because no turns to the west can be used later. After that the packet can be routed adaptively to the north, east, and south. In Figure 10.6D the two turns have been eliminated in such a way that the packet is not allowed to be turned to the other direction after it has been once turned to the north. This corresponds to the north-last routing algorithm.

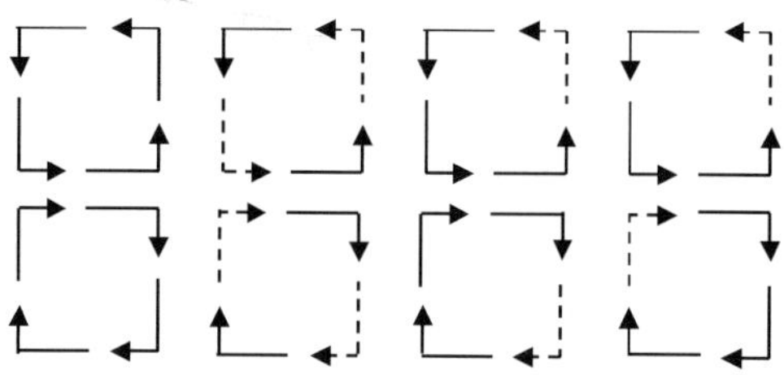

Figure 10.6. Allowed turns of the minimal turn model routing in two dimensions. (A) all of the eight possible turns. (B) four turns allowed in 2-dimensional dimension order routing. (C) six turns allowed in west-first routing. (D) six turns allowed in north-last routing.

The operation of the west-first routing algorithm in a two dimensional mesh is depicted in Figure 10.7. Arrows show the routing directions and walls illustrate unavailable channels which can be either reserved or faulty. The route in the bottom right corner will be blocked until the wall breaks, because the west-first routing algorithm is not able to do turns to west after it has once turned to some other direction. Let Δx and Δy denote the number of hops between the source and destination nodes along x and y dimensions. Then the number of shortest paths usable for fully adaptive routing algorithms from the source node to the destination node is equal to $(\Delta x+\Delta y)!/(\Delta x!\Delta y!)$. For west-first algorithm there are also $(\Delta x+\Delta y)!/(\Delta x!\Delta y!)$ shortest paths available when the x-coordinate of the destination node is greater than or equal to that of the source node, otherwise only 1. Therefore, it is only a partially adaptive routing algorithm like all the other turn model routing algorithms also are. It is also a non-minimal algorithm, because it does not necessarily route packets along the shortest paths only. Two other non-minimal adaptive routing algorithms are static dimension reversal routing and dynamic dimension reversal routing algorithms [30] which provide the same number of routing paths to both of the directions between the source and destination processor nodes. Fully adaptive routing algorithms have also been developed for meshes and k-ary n-cubes [31, 32].

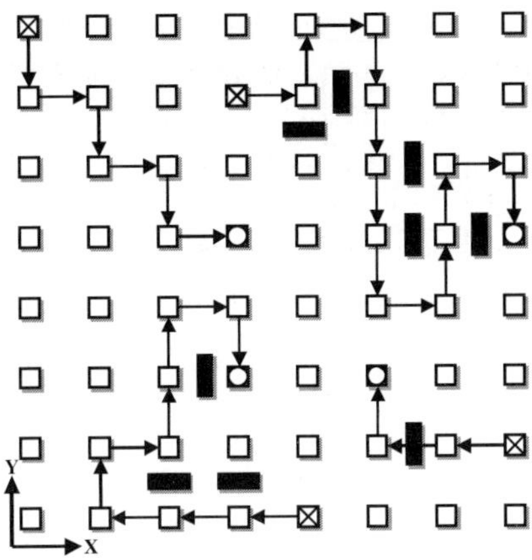

Figure 10.7. The operation of the west-first routing in a 2-dimensional 8×8 mesh [29] (© [1992] IEEE).

Oblivious routing algorithms are also used in networks with regular topologies. The operation of the simplest oblivious routing algorithms consists of only two steps [33]. In the first step they send the packets to a randomly chosen node in the network. In the second step they route them to their correct destinations. Simple oblivious routing algorithms like IVAL and 2TURN produce clearly better average-case and worst-case throughputs than DOR in 2-dimensional torus networks [34]. In the first step of IVAL packets are routed to a randomly chosen intermediate node using DOR. The second step uses also DOR, but reverses the order of dimension traversal. The 2TURN algorithm does not have a closed form description like IVAL. It has only a closed form description of the possible routing paths which can be any such paths, which contain at most two turns except "u-turns" within the same dimension. The paths of 2TURN are a superset of those of IVAL. Because of this, 2TURN has better performance.

Multistage interconnection networks like butterfly networks have regular topologies which consist of multiple stages of switch nodes. Bi-directional multistage network (BMIN) is made by connecting two butterflies back to back and by folding one of the halves on the other half. The switch nodes must also be modified respectively in such a way that they are able to route the packets at each stage over bi-directional links instead of unidirectional links. This is illustrated in Figure 10.8 where transmission links between switches are bi-directional.

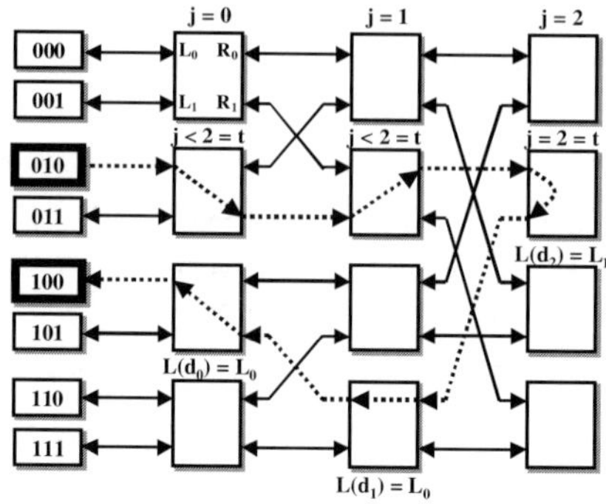

Figure 10.8. Turnaround routing in a butterfly BMIN.

Typically minimal (shortest-path) routing algorithms like Turnaround [35] are used in BMINs. Assume that the source address S and the destination address D are k-ary numbers $s_{n-1}...s_1s_0$ and $d_{n-1}...d_1d_0$ respectively. In addition, define that function FirstDifference(S, D) returns the first position t ($0 \leq t \leq n-1$) starting from the left, where digits d_t and s_t are not equal. The Turnaround algorithm operates in a switch at stage j in the following way. In this description the left and right ports of the switches are denoted with letters L and R respectively.

1. t = FirstDifference(S, D).
2. If j = t, then turnaround the packet and send it forward through port $L(d_j)$.
3. If j < t and the packet comes from an input port L(m) ($0 \leq m \leq k-1$), then send the packet forward through any available output port R(i) , for $0 \leq i \leq k-1$.
4. If j < t and the packet comes from an input port R(m) ($0 \leq m \leq k-1$), then send the packet forward through the output port $L(d_j)$.

It can be proved that the Turnaround algorithm routes packets always along one of the shortest paths from the source to the destination node. Figure 10.8 illustrates one of four possible routing paths drawn with dashed arrow lines from source $(s_2s_1s_0)_2 = 010_2$ to destination $(d_2d_1d_0)_2 = 100_2$ in Butterfly BMIN. The route from the source node to the turning switch at stage 2 has been chosen randomly whereas the routing path from that same switch to the destination node is determined by the destination address.

4.3 Summary of routing schemes

Both adaptive and non-adaptive routing algorithms can be used in networks with both regular and irregular topologies. For the moment most of the on-chip networks have regular topologies, which is probably because of the more complex and expensive routing algorithms required by the irregular topologies. In addition, the flexibility of the configuration of the irregular topologies can not be effectively utilized in System-on-Chip (SoC) circuits. Because adaptive algorithms do not necessarily require more complex computations especially in the networks which have a regular topology, can they be preferred to non-adaptive algorithms like, for example, source routing. However, non-adaptive routing like source routing can be used for forcing the packets to traverse the network through particular switches and links, which can be utilized in fault diagnosis and for routing virtual circuits. For this reason, the non-adaptive source routing could be used in the networks side by side with adaptive routing algorithms. Furthermore, source

5. CASE STUDY: ARBITRATION AND ROUTING IN EXTENDED GENERALIZED FAT TREE ON-CHIP NETWORKS

The previous sections presented arbitration and routing schemes and concerned their usage in on-chip networks. This section presents a simple case study to show how these schemes are applied in extended generalized fat tree (XGFT) [36] interconnection networks which are also BMINs. The following subsections illustrate how the switch nodes and their arbiters operate. They also show how the operation of the routing algorithm can affect the performance and scalability of the network. According to results of simulations performed with various XGFT configurations the performance of the XGFTs can be improved substantially by using a new Turn Back When Possible (TBWP) [37, 38] routing algorithm instead of usually used shortest-path routing algorithms like Turnaround algorithm [35]. In addition, it can be estimated that the improved performance is achievable with only a negligible increase of costs. These results were achieved with such XGFTs where packets were routed upwards and downwards with separate networks which were connected to each other with one link within each switch node. These two networks are called up-routing network and down-routing network, respectively.

5.1 Network topology

Fat trees [39] have several advantageous properties such as scalability, reliability, a regular and simple topology, and a good performance. In addition, fat trees can be proved to be area-universal interconnection networks, which means that they can simulate any other interconnection network with the same silicon area with only a logarithmic slowdown. Usually only a polynomial slowdown is achieved, when networks are simulated with other networks. Fat tree networks are usually implemented with switches of a constant size. The usage of switches of constant size limits the possible number of processor leaf nodes, which is a problem. For instance, if 2×4-switches (2 parents, 4 children) would be used, the number of processor leaf nodes could be only some integer equal to 4^h where h is the height of the fat tree. This problem has been solved in XGFTs [36] by allowing switches of different sizes to be used in different levels of the XGFT. For example, the number of processor leaf nodes could be 32 or 48 if 2×2-switches or 2×3-switches would be used at the first level of the XGFT depicted in Figure 10.9A. This same property can also be used for changing the number of routing resources of the network in a flexible manner. For example, the number of redundant routing paths is higher in the XGFT illustrated in Figure 10.9B, because the 2×4-switches of the two topmost levels have been replaced with 4×4-switches.

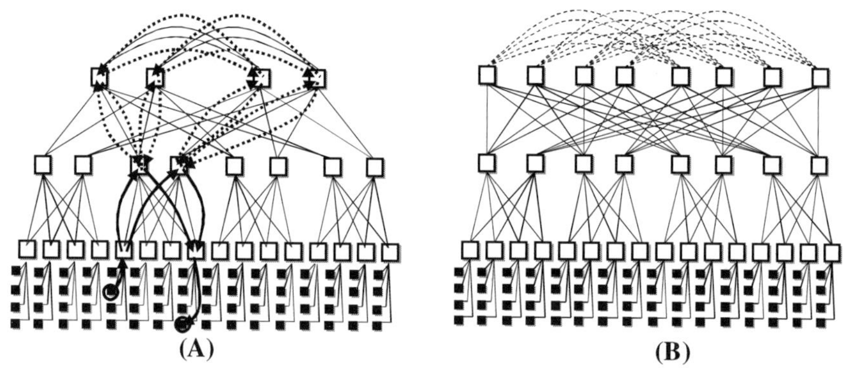

Figure 10.9 The XGFT with connected top most roots and the operation of the Turnaround and TBWP routing algorithms [38] (A) and the XGFT with inserted additional root switches [38] (B) (© [2003] IEEE).

In typical fat tree implementations the top most roots have been connected only with their child nodes through their down-routing ports and the up-routing ports have been left unused [7, 8]. XGFTs illustrated in Figure 10.9 have also been modified by connecting their topmost switches to each other with new additional links which have been drawn with dashed lines in Figure 10.9B. This modification increases the number of connections between the up-routing and down-routing networks and improves the throughput of the XGFT, which will be shown by simulation results later in this same chapter.

5.2 Switching and arbitration in XGFT

There are two separate networks for routing packets upwards and downwards in the modified XGFT [38]. This has been arranged by dividing network nodes into two distinct switches, one for switching packets in the up-routing network and the other in the down-routing network. In each network node unidirectional channels connect the two networks with each other [38]. These channels are called turn back channels. They are used for switching packets from the up-routing network to the down-routing network, if the switch node is a root of the sub XGFT the leaves of which the destination and source nodes are. The architecture of the switch nodes implemented with two smaller switches instead of one large switch provides several advantages including more flexible placement and routing of the switch logic. Furthermore, the switch halves can be different, which allows simpler and more optimal implementations.

In the simulated system wormhole routing strategy and input-output buffered crossbar switch architecture were used in both of the switch halves of the switch nodes. The operation of the simulated arbiter was not exactly similar to that of any of the arbiters presented in subsection 3.2. However, it was able to schedule multiple transfers during a single arbitration cycle and serve requests of different input ports fairly, which are the most important requirements. Transfers were scheduled slightly differently in different switch halves. In the up-routing switches the requests of the input ports were serviced in a cyclic order. The input port with the highest priority was allowed to maintain its priority until its request became granted. Before the next arbitration cycle the highest priority was given in a cyclic order to the next input port the request of which was not granted yet. In the up-routing switches the next output port was also selected in a cyclic order from the set of free ports. As the packet was to be turned back towards the leaves, the input ports requested all of the output ports including the port connected to the turn back channel. Otherwise, they request only the output ports which were connected to the parent switches. In the down-routing switches the input ports were serviced like in the up-routing switches, but the destination addresses of the packets determine unambiguously which output ports are requested. This kind of arbitration scheme corresponds approximately that of earlier presented MSM arbiters. In the simulated system the operation of the switches was pipelined in such a way that it was possible to continue the transfer of the packets unbrokenly during the arbitration. Therefore, the transfer of the packets could be continued unbrokenly through the switch if the requested output ports were free.

5.3 Operation of TBWP routing algorithm

The Turn Back When Possible (TBWP) routing algorithm was originally developed to improve the performance of modified fat tree networks where the top most switches are connected to each other with additional links [37]. It has been proved that the TBWP is also able to route packets correctly in XGFTs modified in a similar way [38]. The TBWP operates clearly differently from Turnaround routing [35] algorithm which routes packets always along one of the shortest paths to the destination. Minimal (shortest-path) routing algorithms like Turnaround are most commonly used in computer systems where fat trees [40, 41] (TMC CM-5, Meiko CS-2) or bi-directional multistage interconnection networks (BMIN) [42] (IBM RS/6000 SP) are used for communication. These algorithms route packets back towards the leaf nodes immediately after they have arrived at such a switch node which is a common ancestor of both the source and destination nodes. Unlike the Turnaround, the TBWP algorithm is able to search for free routing paths through upper switch levels, if the turn back channels have already been reserved. In the topmost switch level the TBWP routes packets from one root to another over the inserted additional transmission links.

In the following description of the TBWP routing algorithm [38] it is assumed that there are only two unidirectional channels within each bidirectional transmission link. The configuration of the XGFT is defined here with vectors $M = (m_0, m_1, m_2, \ldots, m_h)$ and $W = (w_1, w_2, \ldots, w_h)$ which hold the number of ports to child (m_i) and parent nodes (w_i) of the switches in level i ($1 \leq i \leq h$) of the XGFT. The first element m_0 of vector M, which is always equal to one, has been inserted so that the second part of the TBWP algorithm would operate correctly also in the first level of the XGFT. In the description, destination (D) and source (S) nodes are addressed with a pair of numbers (0, D) and (0, S), where $0 \leq D, S \leq m_0m_1m_2\ldots m_h-1$, when parameter h is the height of the fat tree and product $m_0m_1m_2\ldots m_h$ is the total number of processor nodes. Number zero denotes the level of the leave nodes. Switch nodes are addressed with a pair of numbers (L, X), where $1 \leq L \leq h$ and $0 \leq X \leq w_1w_2\ldots w_{L-1}m_{L+1}\ldots m_h-1$, when L denotes the level and product $w_1w_2\ldots w_{L-1}m_{L+1}\ldots m_h$ is the total number of the switch nodes in level L. In the following description the up-routing switches perform the first part of the description and the down-routing switches perform the second part. DIV and MOD operations produce integer quotient and remainder of the division of two positive integers.

TBWP routing algorithm:

- **Part 1 (Up-routing):** *If the packet comes to the node (L, X) from one of the children and if S DIV $m_1m_2.m_L$ = D DIV $m_1m_2.m_L$, then route the* *packet to the turn back channel, else route it to one of the parent nodes. If the turn back channel is already reserved, then route the packet to one of the parent nodes.*

- **Part 2 (Down-routing):** *If the packet comes to the node (L, X) from one of the parents or from the turn back channel, then route it to a child node through output port C[(D DIV* $m_0 m_1 m_2 .. m_{L-1}$*) MOD* m_L*].*

In the case that transmission links would contain several asynchronous channels there would be a group of ports instead of only one port for routing packets to some particular switch from the switches. This would not make the implementation of the TBWP much more complex, because the first part would choose and request output port groups as was described earlier instead of output ports while the second part of the TBWP would compute the address of the port group from which the free port would be chosen. However, because of the higher number of ports the switch arbiters would be larger, which would be a disadvantage.

The difference between the operation of the Turnaround and the TBWP algorithms is illustrated in Figure 10.9A. Solid thick arrow lines show how the Turnaround would route packets from leaf node 17 to node 31. Dashed thick arrow lines show additional routing paths that the TBWP algorithm could use for routing the packets. Unlike the Turnaround algorithm, the TBWP is able to switch packets to one of the parents also in the case that the Turn Back channel is already reserved. For this reason, the TBWP is able to use higher number of alternative routing paths than the Turnaround routing as Figure 10.9A illustrates. As a consequence, the possibility of congestion is smaller and the throughput is higher. It is also able to use additional links between the up-routing and down-routing networks. It would also have been possible to connect switch halves of the same top most root switches of the XGFT instead of different root switches in order to achieve similar results.

5.4 Performance

In the simulated system transmission links were bi-directional and 32 bits wide. There were two unidirectional asynchronous channels within each of the links and one asynchronous turn back channel in each of the switch nodes. The switch nodes were input-output buffered crossbar switches with input and output buffers of eight words. The transfer was performed asynchronously over the channels in such a way that the receiving side was able to stop the transmission, when the input buffer became full and the words were not able to be forwarded. The transmitting side had to wait until the stopped transmission was allowed to be continued. During simulations traffic sources, i.e. the computing leaf nodes, could be in three different states which were transmitting, waiting, and being idle. The simulation time was divided into time slots the length of which was equal to the time it takes to transfer one word of data over asynchronous channels. It was also equal to the length of one operation cycle (or a clock cycle) of the switch nodes. The operation of the switches was pipelined and the switching latency of five time slots was used in the simulations.

Simulations were done with 32 bits wide packets, which consist of a header of three words, a payload field of a variable number of words, and a tail of one word [38]. The first word of packets contains 16 bits wide source and destination addresses. The next two words contain fields for a packet length, a priority number, a sequence number, and other relevant information needed for controlling the transfer of packets. In the XGFTs the sequence number is necessary for ordering the packets, because usually the packets are routed upwards along randomly chosen paths and they may arrive at the destination node in disorder. A packet tail carries a frame check sequence field which is used for detecting bit errors that may occur during the transfer.

Simulations were performed using randomly generated traffic which was produced in the following way. During each time slot a uniformly distributed pseudo random number *prn* was generated from real number interval [0.0, 1.0]. This pseudo random number *prn* was compared to a quotient $\rho/plen$, where parameter ρ was a load factor and parameter *plen* was the length of a single packet or a burst of packets to be transmitted in words. The transmission was performed, if the generated random number *prn* was smaller than or equal to the quotient $\rho/plen$. If the transmission was still unfinished, the total number of packets *tlen* to be transmitted was accumulated by one or by the length of the burst of packets, and the transmission was continued without any suspensions. Every time a packet was sent to the network the value of *tlen* was decreased by one. In the simulated system traffic sources generated new traffic only during the time slots they were transmitting or idle, i.e. during the time which was available for transmitting. During simulations the utilization of the available time corresponded quite accurately the value of parameter ρ and each source loaded the network equally according to ρ's value. As ρ's value was increased the actual throughput also increased as well as the waiting time until the network became saturated. Near the saturation point the increase in ρ had only a negligible effect on the actual throughput and only the waiting time increased, although the proportion of the transmission time to the available time further increased according to ρ's value and was almost equal to it. Uniformly distributed pseudo random numbers were also used for addressing the destinations. As a consequence of this the packets were distributed uniformly over all of the leaf processing nodes.

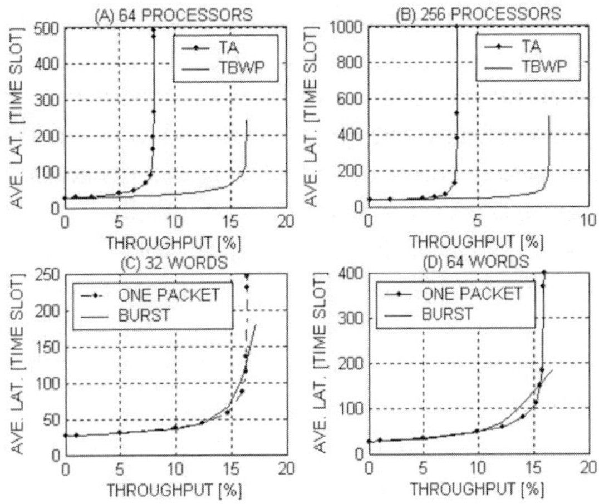

Figure 10.10. Throughputs produced by the Turnaround and the TBWP with different systems sizes (A, B), and when large packets and burst of small packets are used with the TBWP (C, D). The simulated system is illustrated in Figure 10.9A.

The first two simulations were performed in order to compare the performance of the Turnaround (TA) and the TBWP routing algorithms with two different system sizes. In the simulated systems the topmost switches were connected like Figure 10.9A illustrates, the packet size was 32 words, and systems were constructed with constant sized 2×4-switches. These simulations were accomplished with 64 and 256 processing nodes. The simulation results are presented in Figures 10.10A and 10.10B where the throughputs (THROUGHPUT) are given in percentages and they show the proportion of the average number of time slots the various processor leaf nodes have been able to use for transmission to the total number of time slots. The average latencies (AVE. LAT.) show the average number of time slots between the time instants of the transmission and reception of the first word of the packets. These results show that TBWP produces approximately two-fold maximum throughput per computing node compared with the Turnaround and that the maximum throughput halves as the system size is quadrupled.

The next two simulations were performed with the system illustrated in Figure 10.9A in order to study the effect of the packet size on the throughput. As Figure 10.10C shows with packets of 32 words (ONE PACKET) network became saturated at the point of 16.4% and with bursts of 4 packets of 8 words (BURST) at the point of 17.2%. However, the effective throughputs were 14.9% and 10.8% respectively, when only the transferred payload data was taken into account excluding the headers of three words and tails. Simulations with packets of 64 words and bursts of 8 packets of 8 words were also accomplished in order to investigate the effect of the change of the packet size on the throughput. Results in Figure 10.10D are quite similar to those in Figure 10.10C. The average throughputs saturated at the points of 15.9% and 16.7% while the effective average throughputs were 15.2% and 10.4% in a respective order. Because the usage of a large packet size seems to results in higher effective throughput, the next simulations were performed using only one packet size of 32 words.

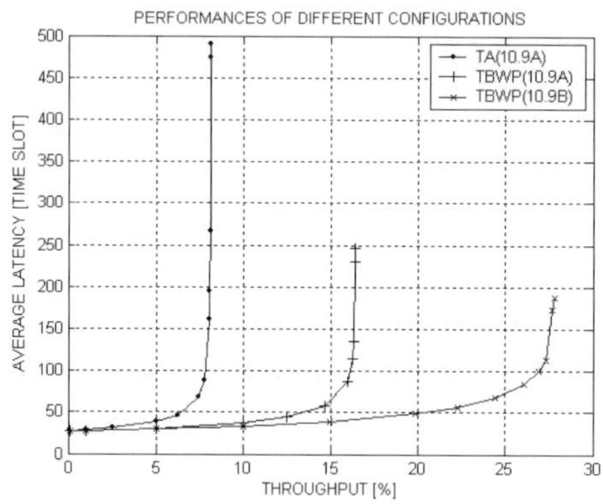

Figure 10.11. The performances of different network configurations [38] (© [2003] IEEE).

Figure 10.11 illustrates results of simulations performed with different network configurations. For the purpose of comparison, results of two earlier simulation were transferred from Figure 10.10A to Figure 10.11 where they are denoted with TA(10.9A) and TBWP(10.9A). In addition, simulations were also performed with a configuration illustrated in Figure 10.10B where two topmost switch layers consist of 4×4-switches. The results of this simulation is denoted with TBWP(10.9B) in Figure 10.12. The maximum throughputs of different configurations were 8.1%, 16.4%, and 27.8% in a respective order. These results show that the throughput of the second configuration (TBWP(10.9A)) is approximately 102% higher than that of the first configuration (TA(10.9A)) where Turnaround was simulated with unconnected topmost root switches. Because this improvement can be achieved simply by connecting the roots and by changing the routing algorithm to the TBWP, it can be assumed that the circuit area does not increase in the same proportion. The throughput of the third configuration (TBWP(10.9B)) is approximately 70% higher than that of the second configuration, which is also a significant improvement taking into consideration that the circuit area does not increase in the same proportion. Comparing the throughputs of the first (TA(10.9A)) and the third configuration (TBWP(10.9B)) the improvement is 243%. These results show how the operation of the routing algorithm can affect the performance and the cost of the system. It can be assumed that the systems size grows proportionally much less than the performance as the system is modified.

Packet traffic, which consists of packets of only one size, is quite artificial. However, measurements with e.g. real local area networks (LAN) show that their traffic consists of packets of only a relatively small number of different sizes [43]. In such networks where the minimum and maximum packet sizes are limited the first packets are always of the maximum possible size and only the last packet is shorter, when large amount of data is transferred over the network. Because the packet size will also be limited in the on-chip networks, the presented simulation results are usable in evaluating the throughputs of the different network configurations and in comparing the different routing algorithms. Moreover, these simulations were not primarily performed in order to compute accurate estimates of the performance, but for the purposes of comparison of different alternatives. The total length of each of the simulations was little longer than 500×10^3 time slots. For example, with a throughput of 27.8% this corresponds to a transmission time of approximately 139×10^3 time slots per traffic source on average. In this special case a total of 278×10^3 packets of 32 words were transmitted during the simulation when there were 64 traffic sources in the system. In addition, the presented results are averages of results of three different simulations.

6. SUMMARY

In this chapter various arbitration and routing schemes have been reviewed. The arbitration was concerned from both theoretic and practical point of view. The arbitration was at first modeled as a maximum matching problem before the different arbitration algorithms were presented. Because current maximum weight matching algorithms have still quite complex and slow hardware implementations, the focus was on maximal size matching algorithms like iSLIP, iDRRM, and WWFA which have smaller and simpler hardware implementations.

Different routing schemes were reviewed after arbitration schemes. Because the operation of the routing algorithms varies along with the network topology, multiple different topologies were also presented briefly. Basically, the routing algorithms are either adaptive or non-adaptive depending on whether they are able to take into consideration the prevailing traffic conditions in the networks when making decisions of the routing paths. Non-adaptive deterministic routing is usually simpler to implement than adaptive routing which requires distributed decision-making processes at the switch nodes along the routing paths. However, adaptive routing produces typically better performance and for networks with regular topology it has realizable implementations which are not necessarily much more complex than those of non-adaptive routing algorithms.

In the case study the earlier presented arbitration and routing schemes were applied to practice in the modified extended generalized fat tree (XGFT) interconnection network. A new adaptive TBWP routing algorithm was introduced and its performance was compared to the performance of adaptive Turnaround routing algorithm. Presented simulation results show that the TBWP routing algorithm can utilize more efficiently all of the available free resources of the network than the Turnaround routing algorithm with slightly increased circuit area, which demonstrates how the operation of the routing algorithm affects the system performance and costs.

REFERENCES

[1] L. Benini, and G. De Micheli, Networks on Chips: A New SoC Paradigm, Computer, 35(1): 70-78, 2002.

[2] S.K. Tewksbury, M. Uppuluri, and L.A. Hornak, Interconnections/Micro-Networks for Integrated Microelectronics, Proceedings of GLOBECOM'92, Orlando, FL, USA, 1992, pages 180-186.

[3] F.A. Tobagi, Fast Packet Switched Architectures for Broadband Integrated Services Digital Networks, Proceedings of the IEEE 78(1)(1990): 133-167.

[4] H. Ahmadi, and W.E. Denzel, A Survey of Modern High-Performance Switching Techniques, IEEE Journal of Selected Areas in Communications 7(7)(1989): 1091-1103.

[5] H.J. Chao, Next Generation Routers, Proceedings of the IEEE, 90(9)(2002): 1518-1558.

[6] E. Rijpkema, K.G.W. Goossens, A. Radulescu, J. Dielissen, J. van Meerberg, P. Wielage, and E. Waterlander, Trade Offs in the Design of a Router with Both Guaranteed and Best-Effort Services for Networks on Chip, Proceedings of the Design, Automation and Test in Europe Conference and Exhibition (DATE 2003), Messe Munich, Germany, 2003, pages 350-355.

[7] P. Guerrier, and A. Greiner, A Generic Architecture for On-Chip Packet-Switched Interconnections, Proceedings of the Design, Automation and Test in Europe Conference and Exhibition (DATE 2000), Paris, France, 2000, pages 250-256.

[8] P.P. Pande, C. Grecu, A. Ivanov, and R. Saleh, Design of a Switch for Network on Chip Applications, Proceedings of International Symposium on Circuits and Systems (ISCAS 2003), Bangkok, Thailand, 2003, pages 217-220.

[9] Y. Tamir, and H.-C. Chi, Symmetric Crossbar Arbiters for VLSI Communication Switches, IEEE Transactions on Parallel and Distributed Systems 4(1) (1993) 13-27.

[10] P. Kermani, and L. Kleinrock, Virtual Cut-Through: A New Computer Communication Switching Technique, Computer Networks 3(4) (1979): 267-286.

[11] W.J. Dally, and C.L. Seitz, Deadlock-Free Message Routing in Multicomputer Interconnecion Networks, IEEE Transactions on Computers C-36(5) (1987): 547-553.

[12] W.J. Dally, Virtual-Channel Flow Control, Proceedings of 17th Annual International Symposium on Computer Architecture, Seattle, WA, USA, 1990, pages 60-68.

[13] N. McKeown, Scheduling Algorithms for Input-Queued Cell Switches, Ph.D. Thesis, University of California at Berkley, 1995.

[14] N. McKeown, The iSLIP Scheduling Algorithm for Input Queued Switches, IEEE/ACM Transactions on Networking 7(2)(1999): 188-201.

[15] H.J. Chao, Saturn: A Terabit Packet Switch Using Dual Round-Robin, Proceedings of Global Telecommunications Conference (GLOBECOM 2000), San Francisco, CA, USA, 2000, pages 487-495.

[16] E. Oki, R. Rojas-Cessa, and H.J. Chao, PMM: A Pipelined Maximal-Sized Matching Scheduling Approach for Input-Buffered Switches, Proceedings of Global Telecommunications Conference (GLOBECOM 2001), San Antonio, TX, USA, 2001, pages 35-39.

[17] J. Hurt, A. May, X. Zhu, and B. Lin, Design and Implementation of High-Speed Symmetric Crossbar Schedulers, Proceedings of IEEE International Conference on Communications, Vancouver, BC, Canada, 1999, pages 1478-1483.

[18] A. Mekkittikul, and N. McKeown, A Practical Scheduling Algorithm to Achieve 100% Throughput in Input-Queued Switches, Proceedings of the 17th Annual Joint Conference of IEEE Computer and Communications Societies (INFOCOM 1998), San Francisco, CA, USA, 1998, pages 792-799.

[19] Z. Lisheng, and H. Chengdu, A Practical Scheduling Algorithm for Input Buffered Switch, Proceedings of International Conference on Communication Technology (ICCT 2000), Beijing, China, 2000, pages 1059-1064.

[20] D.E. Culler, J. Pal Singh, and A. Gupta, Parallel Computer Architecture: A Hardware/Software Approach, Morgan Kaufmann Publishers Inc. San Francisco, USA, 1999.

[21] M.D. Schroeder, A.D. Birrel, M. Burrows, H. Murray, R.M. Needham, T.L. Rodeheffer, E.H. Satterthwaite, and C.P. Thacker, Autonet: A High-Speed, Self-Configuring Local Area Network Using Point-to-Point Links, IEEE Journal on Selected Areas in Communications, 9(8) (1991): 1318-1335.

[22] N.J. Boden, D. Cohen, R.E. Felderman, A.E. Kulawik, C.L. Seitz, J.N. Seizovic, and S. Wen-King, Myrinet: A Gigabit-per-Second Local Area Network, IEEE Micro 15(1)(1995): 29-36.

[23] J.C. Sancho, A. Robles, and J. Duato, On the Relative Behaviour of Source and Distributed Routing in NOWs Using Up*/Down* Routing Schemes, Proceedings of the 9th Euromicro Workshop on Parallel and Distributed Processing, Mantova, Italy, 2001, pages 11-18.

[24] W. Qiao, and L.M. Ni, Adaptive Routing in Irregular Networks Using Cut-Through Switches, Proceeding of the International Conference on Parallel Processing, Ithaca, NY, USA, 1996, pages 52-60.

[25] H. Räcke, Minimizing Congestion in General Networks, Proceeding of the 43rd Annual IEEE Symposium on Foundations of Computer Science, Vancouver, Canada, 2002, pages 43-52.

[26] Y. Azar, E. Cohen, A. Fiat, H. Kaplan, and H. Räcke, Optimal Oblivious Routing in Polynomial Time, Proceedings of the 35th ACM Symposium on Theory of Computing, San Diego, California, USA, 2003, pages 383-388.

[27] N. Bansal, A, Blum, S. Chawla, and A. Meyerson, Online Oblivious Routing, Proceedings of the 15th Annual Symposium on Parallel Algorithms and Architectures, San Diego, California, USA, 2003, pages 44-49.

[28] M. Bienkowski, M. Korzeniowski, and H. Räcke, A Practical Algorithm for Constructing Oblivious Routing Schemes, Proceedings of the 15th Annual ACM Symposium on Parallel Algorithms and Architectures, San Diego, California, USA, 2003, pages 24-33.

[29] C.J. Glass, and L.M. Ni, The Turn Model for Adaptive Routing, Proceeding of the 19th Annual International Symposium on Computer Architecture, 1992, pages 278-287.

[30] W.J. Dally, and H. Aoki, Deadlock-Free Adaptive Routing in Multicomputer Networks Using Virtual Channels, IEEE Transactions on Parallel and Distributed Systems 4(4)(1993): 466-475.

[31] G.D. Pifarre, L. Gravano, S.A. Felperin, and J.L.C. Sanz, Fully Adaptive Minimal Deadlock-Free Packet Routing in Hypercubes, Meshes, and Other Networks: Algorithms and Simulations, IEEE Transactions on Parallel and Distributed Systems, 5(3)(1994): 247- 263.

[32] D.H. Linder, and J.C. Harden, An Adaptive and Fault Tolerant Wormhole Routing Strategy for k-ary n-cubes, IEEE Transactions on Computers 40(1)(1991): 2-12.

[33] L.G. Valiant, and G.J. Brebner, Universal Schemes for Parallel Communication, Proceedings of the 13th Annual ACM Symposium on Theory of Computing, Milwaukee, Wisconsin, USA, 1981, pages 263-277.

[34] B. Towles, W.J. Dally, and S. Boyd, Throughput Centric Routing Algorithm Design, Proceedings of the 15th Annual ACM Symposium on Parallel Algorithms and Architectures, San Diego, California, USA, 2003, pages 200-209.

[35] L.M. Ni, Y. Gui, and S. Moore, Performance Evaluation of Switch Based Wormhole Networks, IEEE Transaction on Parallel and Distributed Systems, 8(5)(1997): 462-474.

[36] S.R. Ohring, M. Ibel, S.K. Das, and M.J. Kumar, On Generalized Fat Trees, Proceedings of 9th International Parallel Processing Symposium, Santa Barbara, CA, USA, 1995, pages 37-44.

[37] H. Kariniemi, and J. Nurmi, New Routing Algorithm for Improving the Throughput of Fat Tree Interconnection Networks, Proceedings of the IASTED International Conference on Computer Science and Technology, Cancun, Mexico, 2003, pages 375-381.

[38] H. Kariniemi, and J. Nurmi, New Adaptive Routing Algorithm for Extended Generalized Fat Tree On-Chip, Proceedings of International Symposium on System-on-Chip 2003, Tampere, Finland, 2003, pages 113-118.

[39] C.E. Leiserson, Fat Trees: Universal Networks for Hardware-Efficient Supercomputing, IEEE Transactions on Computers C-34(10)(1985): 892-901.

[40] C.E. Leiserson, Z.S. Abuhamdeh, D.C. Douglas, C.R. Feynman, M.N. Ganmukhi, J.V. Hill, W.D. Hillis, B.C. Kuszmaul, M.A. St. Pierre, D.S. Wells, M.C. Wong, S.-W. Yang, R. Zak, The Network Architecture of the Connection Machine CM-5, Proceedings of the 4th Annual ACM Symposium on Parallel Algorithms and Architectures, San Diego, California, USA, 1992, pages 272-285.

[41] J. Beecroft, M. Homewood, and M. McLaren, Meiko CS-2 Interconnect Elan-Elite Design, Parallel Computing 20(10-11) (1994): 1627-1638.

[42] H. Sethu, C.B. Stunkel, and R.F. Stucke, IBM RS/6000 SP Interconnect Network Topologies for Large Systems, Proceedings of IEEE International Conference on Parallel Processing, Minneapolis, MN, USA, 1998, pages 620-627.

[43] K.M. Khalil, K.Q. Luc, and D.V. Wilson, LAN Traffic Analysis and Workload Characterization, Proceedings of Conference on Local Computer Networks, Minneapolis, USA, 1990, pages 112-122.

III

DESIGN METHODOLOGY AND TOOLS

Chapter 11

SELF-TIMED APPROACH FOR NOISE REDUCTION IN NOC

Pasi Liljeberg, Johanna Tuominen, Sampo Tuuna, Juha Plosila and Jouni Isoaho
University of Turku, Finland

1. Introduction

In this chapter, we present how self-timed circuit design can be used for reduction of crosstalk and switching noise in high performance NoC circuits. In this approach, *time-interleaving* is used to reduce the effect of interconnect signal coupling and high current peaks. Time-interleaving is implemented by dividing the system into partitions which are *de-synchronized* internally and with respect to each other. The focus is on interconnects within and between components. Impact of asynchronous signaling techniques, such as dual-rail and 1-of-4 encoding, on overall noise levels and signal timing is studied in several contexts.

As technology scales down into the deep submicron regime and the size of chips grows larger noise immunity will be one of the most important design metrics in system design. It is more difficult to manage different types of noise, such as power supply, crosstalk, and leakage noise, because of the continuous reduction of supply and threshold voltages. If noise is not handled properly it will introduce additional signal transition delays and might even cause false switching leading to unreliable operation of the circuit [1]. This chapter will focus on minimizing power supply noise and crosstalk which both have increasingly important impact as the size of the chips gets larger with a higher circuit density.

Power supply noise or unwanted fluctuation of the supply voltage within a digital ULSI chip mainly originates from simultaneous clock-induced switching of CMOS circuits which causes high peak current draws from the power source. The total power supply noise is the sum of two major components: the resistive voltage drop IR and the inductive switching noise $L\Delta I/\Delta t$ [2, 3]. Here I is current, R and L are the effective power supply wire resistance and inductance, respectively, and ΔI is the total current change during the rise or fall time Δt of the concurrently transitioning signals. Hence, dealing with high current peaks, rather than average current, is a key issue to decrease switching noise and an important factor for power distribution network design. Typically, large on-chip decoupling capacitors are needed to provide a stable power supply for every system module on a chip [4, 5]. The area required by these capacitors naturally increases with the size and complexity of the system. In order to decrease the need for decoupling capacitance, current peaks must be lowered. This can be accomplished by decreasing the number of simultaneous switching events in the system. Such an approach, based on de-synchronization of the system, is presented in this chapter.

Crosstalk noise caused by capacitive and inductive coupling between on-chip signal wires is another fundamental problem in modern high-speed system-on-chip design [1]. Capacitive coupling is currently dominant, but inductive coupling becomes more and more significant as signal frequencies and on-chip wire lengths increase. Crosstalk has two major detrimental effects [5]. First, if the magnitude and duration of the coupled noise is sufficient, a signal may temporarily assume an erroneous logic value which in turn may lead to a logical failure. Secondly, crosstalk also affects timing. The delay of a wire not only depends on the properties of the wire itself but also on how the wires that are capacitively or inductively coupled to it are switching. If a wire and another wire coupled to it switch simultaneously in opposite directions, crosstalk increases the delay of the wires because twice as much charge must be transferred across the coupling capacitance. On the other hand, if the coupled wires switch in the same direction, the delay is reduced.

Crosstalk can be kept within appropriate limits by sufficient wire spacing and shielding. These requirements can be significantly eased, resulting in a more compact bus implementation, by controlling transmission of signals in such a way that crosstalk is minimized. The noise reduction methodology described in this study combines time-interleaving techniques with appropriate asynchronous data encoding schemes to achieve this goal in on-chip bus design.

In this chapter, the self-timed approach for noise reduction is illustrated using two case studies. One employs time-interleaving and asynchronous signaling techniques to reduce noise in high performance bus segments. The other exploits the advantages of the de-synchronization and time-interleaving methods applied for the path metric unit of the Viterbi decoder. For the case studies we have used 0.13, 0.18 and 0.35 μm CMOS technologies.

2. Self-Timed Techniques for Noise Reduction

In many designs noise is caused to a large extent by the synchronous operation of the circuit. As chips grow larger and the energy consumption increases, the part of the noise induced by the synchronous operation will increase. The clock dictated operation forces a great deal of gates and flip-flops in the chip to change their states nearly at the same moment. As a consequence, the current profile of the circuit is dominated by the clock induced high peaks, which are the main source for power supply noise. Utilizing asynchronous techniques these peaks can be folded to a longer period of time by adjusting the timing of the circuit. Furthermore, in a synchronous design a large number of capacitively and inductively coupled interconnects along the chip also switch simultaneously. Crosstalk between the interconnect wires can be reduced by employing self-timed protocols and encoding methods.

2.1 Self-timed design

An asynchronous circuit is a self-timed device, whose operation is not tied to a global clock signal [6], as is the case in the conventional synchronous design approach. Instead, the execution and synchronization of the system is controlled locally, which allows truly modular design and easy re-usage of existing components. Because there is no global clock to sequence the events in the system, the operation relies on strictly controlling the order of transitions on control wires. These control wires can be used to create handshake communication channels which are utilized to manage the state transitions and data transfers in the system. Hence, the problems of distributing the clock signal and consequences of clock dictated operation can be avoided.

2.2 Asynchronous signaling protocols

In contrast to synchronous signaling protocols, asynchronous protocols do not rely on any external timing sources, such as clock signals. Instead, a transaction between parties involved in a certain communication relies on events in a set of asynchronously driven control signals. Almost all protocols are based on *requests* and *acknowledgements* between the initiator and the responder of a transaction. A request is used to initiate a transaction while the acknowledge is used to signal the completion of a transaction or indicate when the next one may begin. These control signals include all of the necessary timing information for a reliable data transmission and ordering the events in a self-timed system. There are several possibilities how state changes are encoded to the control wires and how many transitions are used. However, asynchronous signaling schemes usually require explicit alternation between events on request and acknowledge signals. In the following paragraphs we discuss two of the most commonly used asynchronous protocols, *4-phase* and *2-phase protocols.*

4-phase protocol [7], illustrated in Figure 11.1a, is also sometimes referred to as *level-signaling* protocol. It is a return to zero protocol, as the signaling levels at the end of each communication cycle will be the same as in the beginning. It requires four transitions on handshake lines, two on both request and acknowledge wires. A 4-phase transaction between two parties is initiated when the sender sets the request high. The receiver responds by setting the acknowledge high which will be answered eventually by the sender taking the request low. Finally the receiver sets the acknowledge low indicating that a new cycle may begin. Notice that the time between distinct events can be arbitrarily long, making it attractive for communication between system modules with different operating speeds.

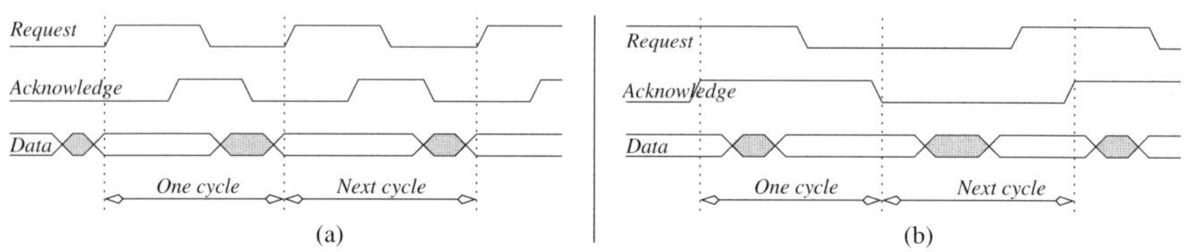

Figure 11.1. Self-timed signaling, (a) 4-phase protocol (b) 2-phase protocol.

In the 2-phase protocol [8], illustrated in Figure 11.1b, all transitions, either rising or falling, have the same meaning. In contrast to the 4-phase protocol it is a non-return to zero protocol, as the signal levels at the end of each communication cycle are the opposite as they were in the beginning. Hence, the absolute state of control wires is not important, only the events, i.e., transitions either to a low or a high state are important. For that reason the 2-phase protocol is also called *transition signaling.* As suggested by its name the 2-phase protocol requires two transitions on handshake lines, one on both the request and acknowledge wire per communication cycle. A transaction between two parties is initiated when the sender produces an event on the request wire, sets the line to a high or a low state. The receiver completes the transaction simply by producing an event in the acknowledge wire. The 2-phase signaling is an especially attractive choice for a communication protocol over long on-chip wires, with high parasitic properties causing a considerable delay, due to the minimal number of signal transitions.

2.3 Data encoding with self-timed protocols

A common choice for self-timed data encoding is the *bundled* data encoding [8], which can be used either with the 4-phase or the 2-phase protocol. Some texts use the term *single-rail* for this encoding. The bundled data uses one wire for each bit, thus transmitting n-bit data from the sender to the receiver requires $n+2$ wires, n wires for data and one wire for both request and acknowledge. The bundled data approach assumes that the propagation delays of control signals are larger than or equal to the propagation delay of data signals. Hence, the control signals have to be bundled with data, routed so that they experience the same parasitic properties and physical wire length as data bits. Furthermore, the control signals could be slightly delayed so that a large enough safety margin is achieved. If these constraints are not met, the receiver could start its operation with incorrect data, or even worse, if it reads the input at the moment when data is changing the data lines in the receiver could enter into a metastable state.

There are three different data validity schemes that can be used with the bundled 4-phase protocol; *early, broad*, and *late* [9]. In the first, probably the most commonly used, the sender issues valid data before setting request high and can remove data at will after receiving a rising acknowledge signal. In the broad scheme data is valid during the whole cycle, from request high to acknowledge low. In the late scheme data is valid in the down going part of the handshake, from request low to acknowledge low.

An alternative to the bundled data method is *dual-rail* encoding which uses separate requests for each data bit [10]. Each data bit is encoded onto 2 wires in such a way that data itself acts as a request. Therefore, transmitting n-bit data requires $2n + 1$ wires, $2n$ for data and one for acknowledge. An often used 4-phase dual-rail encoding has four states: '00' for idle, '10' for valid zero, '01' for valid one and '11' is a not used illegal state. Transmission of a bit requires a transition from the idle state to either the valid 0 or valid 1 state. After the sender has received the acknowledge, it must initialize data back to the idle state. In the 2-phase dual-rail encoding there are no idle and illegal states. The transmitted value is encoded into events so that only one of the two wires is allowed to make a transition during a cycle. A transition on one of the two wires indicates the sending of a zero while a transition on the other indicates sending of a one. After the receiver has acknowledged the data, a new transfer cycle may start immediately by a transition in either of the two wires. Notice that in this case the value of the code word is not important, only the mutually exclusive events matter. The dual-rail encoding is insensitive to the wire delays and hence there is no need for any timing assumptions like in the bundled data signaling. This is advantageous when one is using automated routing particularly for long on-chip interconnects.

So far we have assumed that we have a *push channel* where the sender is the active party that initiates the transaction. However, it is also possible to have a *pull channel* so that the receiver is the active party, requesting data from the sender.

In addition to the above commonly used encoding methods there are plenty of other techniques [11]. A particularly interesting one is the 1-of-4 delay insensitive data encoding scheme [12]. It bears a resemblance to the dual-rail encoding and it can use both transition and level signaling protocols, even though the decoding of the transition signaling is quite a lot of more complex. In addition to that, the number of required wires is the same. In the 1-of-4 data encoding the information of the two-bit symbol is transmitted by using four wires. A two-bit code symbol, '00', '01', '10', or '11', is transmitted by changing the signal level on just one of the four wires. 1-of-4 encoding as well as all the other 1-of-N encoding methods are delay insensitive [11].

3. Method for Power Supply Noise Reduction

The purpose of the method is to decrease the number of simultaneous switching events so that the current peaks will be lower, and consequently to decrease power supply noise and electromagnetic interference (EMI) [13]. The method is based on the idea of dividing the system into partitions which are de-synchronized internally and with respect to each other. By doing so, simultaneous clock related events are distributed over a larger time frame while keeping functionality and correctness of the design intact. The technique is based on re-tuning the timing of the system by facilitating self-timed design so that the current draw during a time interval will be below a certain level set by a designer. The method for switching noise reduction can be applied separately at three different levels; chip, sub-system, and component levels as shown in Figure 11.2.

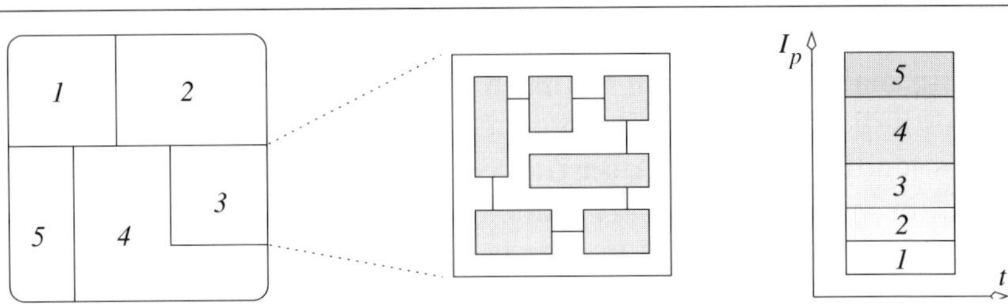

Figure 11.2. System partitioning for decreasing concurrent events

At the chip level the method is applied to adjust the timing between system modules, such as processing elements, IP-blocks, customized blocks, and memories. At the second level, sub-system level, the effort is concentrated on the relative timing of modules inside a sub-system, such as arithmetic units, caches, and combinational logic blocks. At the lowest level the method is used to tune the internal timing of the components in the sub-systems. This includes for example manipulating the timing of the flip-flops in a large FIFO queue, sequencing the activation of interconnect drivers, and controlling the internal calculation process in an arithmetic unit.

It is not allways necessary to utilize the method at all above mentioned levels, one could only focus on one or two levels. Hence, by only concentrating on the two lowest levels, or only on the third level, a local hot spot in the power supply network can be fixed without inserting local decoupling capacitors or re-designing a system component. On the other hand, by exploiting the first level of the method the problems caused by a high system level peak current draw can be solved without need for increasing the number of power supply pins on the chip or enlarging the size of the chip-level power supply network decoupling capacitors.

Even though the goal is the same at all levels of the method, reduction of simultaneous switching activity, the means to achieve this objective somewhat differ from each other. One reason for that is the difference between the wire delays, skew, which has to taken into account when operating at the chip level with long distances whilst it can be disregarded when operating at the register level with negligible wire delays. Furthermore, the effect of data dependencies on the ordering of operations is rather different in each level.

3.1 De-synchronization at different levels

Level 1 At the first level, chip level, the system built around a shared communication medium is de-synchronized so that distinct system modules receive the clock edge at different times. This can be achieved for instance by adding appropriate delays to the clock lines or by using a separate clock source for each module. Consequently, the clock induced current draw of the system will spread over a longer time period and the current profile will become flatter. This kind of clocking leads to a globally asynchronous locally synchronous (GALS) [14] system where the system is divided into several clock domains. Self-timed handshake signaling provides a reliable way of implementing inter-module communication in such a system. This architecture enables flexible use of stoppable clocks providing automatic power down of idle system modules.

Furthermore, the clock skew-related timing problems are restricted only to relatively small locally synchronous islands. The inter-module communication in such a system is out of the scope of this chapter, but can be found in [15–17]. Notice that it is not necessary to de-synchronize all modules, there can be modules operating and interacting synchronously while the others follow the GALS based approach.

Level 2 The second level of the method concentrates on adjusting the relative timing of components inside a sub-system. The method at this level does not require re-design of the entire sub-system or components, it is necessary only to add a self-timed controller, re-route existing clock lines, and in some cases add register levels. Five different phases of the method at this level can be identified, as follows:

(1) Sources of the highest current peaks need to be identified. This can be done by investigating switching activity of the circuits and locating power consuming parts of the sub-system such as arithmetic units, long synchronous pipelines with a large bit width, interconnect drivers, etc. A more reliable way is to use some place and route tool with a support of graphical representation of voltage drops in different locations of the chip. As before, it is not necessary to apply the method to all components in a sub-system. There is no point to put any design effort to de-synchronize components with a low current draw, unless the number of those components is considerable. Instead, in many cases it is enough to focus only on a few energy consuming ones.

(2) In this phase the timing requirements and data dependencies of the components in the sub-system need to be analyzed. In addition to that, one has to take into account also the timing requirements of the external interface of the sub-system. For each component the worst-case duration of its operation is recorded, including the setup and hold times of its external interface. In a synchronous system the clock period has to be chosen according to the slowest path in the design. Therefore, the clock period of a sub-system with several components is selected according to the slowest component, or if the clock is shared with other parts of a chip, it has to be chosen in accordance with the slowest chip part. Hence, the activation of components which have delays shorter than the clock cycle can be easily time-interleaved. In the best case, there are components whose operation time is short enough, i.e. the sum of operation times is less than a clock cycle, so that they can be executed completely in a sequential fashion. However, one has to take into account input data dependencies between distinct components before arranging sequential operations. This is for example if a component with a operation time less than a half clock cycle is activated by the ris-

ing clock edge in the synchronous design and its output data is assumed to be available for some other component at the falling clock edge.

(3) The clock lines of the components that are selected for the de-synchronization are replaced with asynchronous communication links connected to the self-timed controller of the sub-system, illustrated in Figure 11.3. These links are used to control the operation of the components so that the number of simultaneous events in the sub-system is decreased. A synchronous component considers a pulse arriving from the link as a clock signal, it has the same frequency and duration as the original clock. The only difference is that distinct components receive the pulse at different times. Hence, from the component point of view it makes no difference if the control signal is a clock or a self-timed pulse as long as the duration of the pulse is sufficient. In the case that a more complex protocol is applied to the communication link, a circuit that responds to the self-timed controller has to be added to the component.

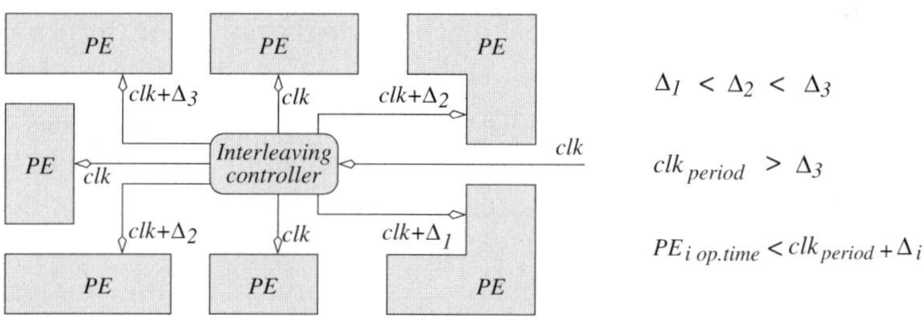

Figure 11.3. Concept of de-synchronization applied to a sub-system.

(4) A self-timed controller is inserted into the sub-system. Even though its operation is based on asynchronous logic, it uses the clock signal as a reference point to sequence the operation of the sub-system. This is necessary since the external timing of the sub-system needs to be kept intact. The controller distributes time-interleaved pulses to the components that are selected for de-synchronization. There are two design approaches how the time-interleaved pulses can be generated. The first implementation employs delay lines to produce pulses whose rising edges are spread over a certain time window. In this approach the implementation will be fast, especially suitable for high-frequency systems. However, producing long delays between pulses is impractical, due to the need for implementing extremely long delay lines. The other possibility is to employ an asynchronous protocol between the controller and components. In this approach, the operation of the sub-system can be easily kept under control with handshake signals. It is possible to sequence the activation of distinct components so that some components are ac-

tivated after the others have completed their operation, provided that the timing margin of the system allows. However, this approach is not well-suited for a high-frequency system or a design where all components have a tight timing margin.

(5) The last phase of the method in the second level is optional. As explained above, each component selected for the de-synchronization process has an independent self-timed driven clock signal. Hence, it is easy to stop the clock just for one or more components of the sub-system without affecting to the clock signals of the other components. This requires only minor changes to the self-timed control unit. Notice that this provides a flexible way to stop clocks of distinct components in a distributed manner, allowing a finer granularity than conventional clock stopping of larger modules. Even though the stopping of the clock of some components occasionally does not decrease the worst-case current peaks, it will decrease average energy consumption over a longer period of time.

Level 3 The third level of the method deals with the internal timing of a component. Opposite to the first two levels, this level focuses directly on the ultimate destination the clock signal, flip-flops. The time-interleaving in this level is carried out without inserting any asynchronous logic to the design. Instead, delays are added to the selected clock lines inside the component to make sure that distinct sets of flip-flops will not switch simultaneously. As before, a timing analysis has to be performed before inserting delays to find out the possible amount of time for interleaving. Attractive targets for this are for example bus or interconnect drivers, large register banks and all parallel operating circuits. In the case that there are input data dependencies or the data originates from two or more already time interleaved sources, a register level can be inserted in front of a combinational logic block. With this arrangement possible redundant switching of combinational logic caused by different data arrival times can be avoided.

Discussion The height of current peaks can be adjusted by changing the granularity during partitioning the system at distinct levels into smaller blocks. With a coarse grain granularity the system is partitioned at each level only in a few time interleaved blocks and the size of the self-timed control will be small. When a smoother current profile is required the system needs to be partitioned into several blocks. On the other hand, with a finer granularity the timing analysis and verification of correct operation will be more complex and the size of the self-timed control increases.

4. Crosstalk

Crosstalk has become a major source of noise in high-speed integrated circuits because of the non-proportional scaling of vertical and horizontal dimensions of interconnects and decreasing wire pitch. The effects of crosstalk noise depend on the properties of interconnects and the signals propagating on them. In the following sections their influence and methods to model and reduce crosstalk are addressed.

4.1 Influence of Design Parameters

An IC designer cannot directly set the electrical properties of interconnects, but he or she can affect wire width, length and separation distances that contribute to the electrical properties. In the following the influence of the physical dimensions of two coupled wires on crosstalk noise is demonstrated. A change in a single physical dimension always affects several electrical properties. For instance, an increase in the width of a wire affects its resistance, inductance and capacitance.

The effect of rise times and wire length on crosstalk noise is illustrated in Figure 11.4. The wires used in the simulation are 0.30 μm wide and 0.40 μm thick, and separated at 0.30 μm from each other. The length of the wires was varied between 0.2 mm and 8 mm, and signal rise time between 20 ps and 1000 ps. Crosstalk noise increases as rise times become shorter since the induced current and voltage are dependent on the speed of the transition on the switching line. Noise increases also with wire length because the two wires run in parallel to each other, causing the total coupling between them to increase with length.

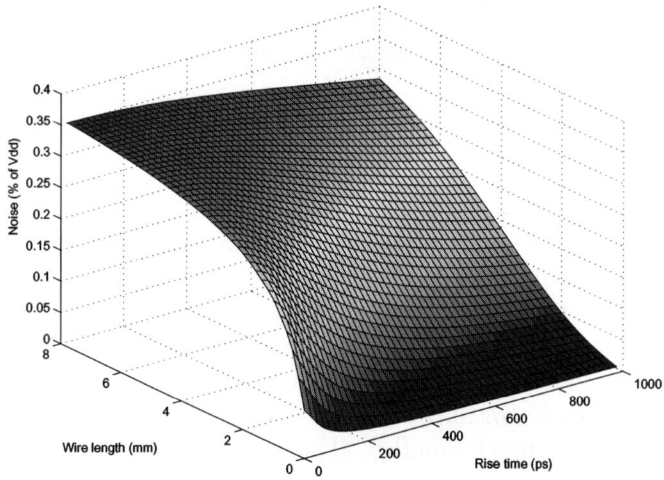

Figure 11.4. Crosstalk noise on a quiet wire at different wire lengths and rise times.

It can be observed that when the rise time is above 700 ps the increase in noise is nearly linear with wire length. However, when the rise time is below 200 ps the noise increases much more rapidly with wire length. The rate of increase slows down as the length of the wires increases. This is caused by the inductance of the wires whose effects become significant at high speeds. The importance of inductance diminishes as the length of the wires increases due to resistance that increases with wire length.

Crosstalk noise can be reduced by limiting the amount of coupling, which is accomplished by increasing the distance between adjacent wires. Crosstalk noise induced onto a quiet wire at different separation distances and lengths is depicted in Figure 11.5. The width and thickness of the wires are 0.3 μm and 0.4 μm, respectively. The rise time of the aggressor is 250 ps.

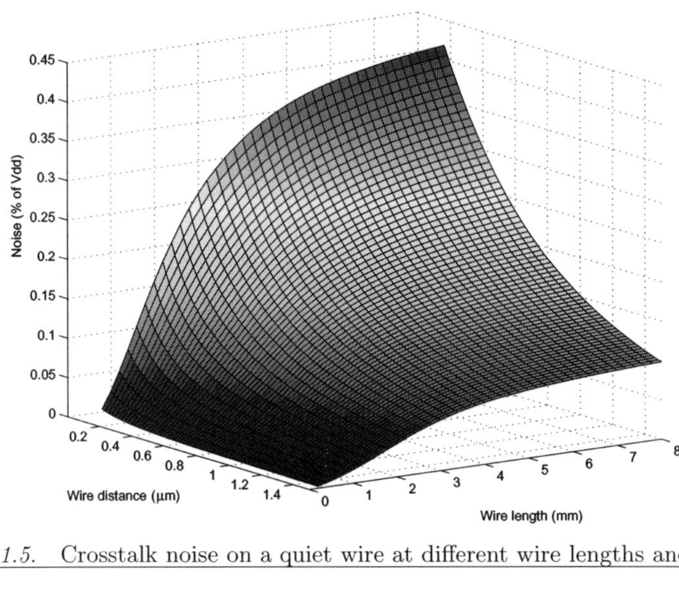

Figure 11.5. Crosstalk noise on a quiet wire at different wire lengths and distances.

It should be noticed that the wire distance axis has been reversed to improve the readability of the figure. It can be seen that increasing the distance between wires helps to reduce crosstalk noise. The rate of reduction is greatest initially and it diminishes as the distance between the wires grows. It can also be seen that crosstalk noise increases almost linearly with wire length up to about four millimeters. After that the increase in noise is much smaller. The same behavior was also seen in Figure 11.4 for different rise times.

4.2 Influence of Switching Patterns

The amount of crosstalk noise induced on a wire is affected by the activity of neighboring wires. Coupled wires switching in the same direction reduce the propagation delay, while switching in the opposite direction increases the propagation delay. The amount of crosstalk noise in an 8-bit planar bus for different switching patterns is shown in Table 11.1. The wires are numbered from left to right and noise is measured on the fourth wire to obtain the maximum noise. The wires are 3 mm long and their width and height are 0.6 μm and 1.2 μm, respectively, and the rise time is 100 ps.

The maximum amount of noise induced onto the quiet wire is 37 percent of the operating voltage. This situation arises when all wires except the fourth one are switching. If the two closest wires on both sides of the fourth wire are switching, the noise is reduced only slightly to 35 percent. However, if only the third and fifth wire are switching the noise is reduced to 28 percent of the operating voltage. The third and fifth wire that are the nearest ones are the main source of noise, but the second and sixth wire still have a noticeable effect on noise. It can also be seen from Table 11.1 that a wire switching in opposite direction can effectively cancel the noise caused by other wires. In the sixth switching pattern crosstalk noise is canceled altogether.

Table 11.1. Crosstalk noise on the fourth wire as a percentage of V_{dd} in an 8-bit bus for different switching patterns.

Wire number	1	2	3	4	5	6	7	8	Crosstalk noise
Pattern 1	↑	↑	↑	-	↑	↑	↑	↑	37 %
Pattern 2	↑	↑	-	-	-	↑	↑	↑	9 %
Pattern 3	↑	-	-	-	-	-	↑	↑	2 %
Pattern 4	-	-	↑	-	↑	-	-	-	28 %
Pattern 5	-	↑	↑	-	↑	↑	-	-	35 %
Pattern 6	↑	↑	↑	-	↓	↓	↓	↓	0 %

The influence of switching patterns on propagation delay is presented in Table 11.2. The delay of the fourth wire is given since it is affected most by coupling to the other wires in the bus. The delay of the wire varies between 30 ps and 136 ps. The shortest propagation delay is obtained when all wires are switching in the same direction and the longest delay when the wires coupled to the fourth wire are switching in the opposite direction. In case the fourth wire is the only one switching, the delay is 67 ps. The same delay is obtained when the wires on one side are switching in the opposite direction and on the other side in the

Table 11.2. Propagation delay of the fourth wire in an 8-bit bus for different switching patterns.

Wire number	1	2	3	4	5	6	7	8	Propagation delay
Pattern 1	↑	↑	↑	↑	↑	↑	↑	↑	30 ps
Pattern 2	↓	↓	↓	↑	↓	↓	↓	↓	136 ps
Pattern 3	-	-	-	↑	-	-	-	-	67 ps
Pattern 4	↓	↓	↓	↑	↑	↑	↑	↑	67 ps
Pattern 5	-	-	↑	↑	↑	-	-	-	38 ps
Pattern 6	-	-	↓	↑	↓	-	-	-	131 ps
Pattern 7	-	↑	↑	↑	↑	↑	-	-	32 ps
Pattern 8	-	↓	↓	↑	↓	↓	-	-	136 ps

same direction, since both sides cancel each other's effects. Patterns 5–8 demonstrate the influence of adjacent wires on propagation delay. Wires three and five constitute the majority of delay variation.

4.3 Crosstalk Modeling and Reduction

Noise analysis and avoidance is nowadays a critical factor in the design of integrated circuits. Design tools need to take into account the effect of crosstalk and perform interconnect optimizations at various design stages. The amount of crosstalk noise can be estimated with crosstalk models. These models need to be both accurate and efficient in order to be used to model today's complex integrated circuits. Simulation techniques such as SPICE and model order reduction [18] can be used, but they are in several cases too time-consuming [19]. Over the last decade, several fast crosstalk models have been proposed. Analytical models with closed-form formulas are very efficient, but they often lack in accuracy and versatility. Analytical formulas have been derived for lumped interconnects using Laplace transform in [20, 21], while telegrapher equations for distributed interconnects have been used in [22].

There are several ways to reduce the adverse effects of crosstalk in integrated circuits. One method is to reduce the amount of coupling. This can be achieved by increasing the spacing between wires, avoiding long parallel runs of wires and by routing wires on adjacent layers in perpendicular directions. Other methods are to shield sensitive signals using ground and V_{dd} shield wires, and making signal rise times as long as possible.

In long wires, buffers can be used to reduce delay and the effects of crosstalk noise. Coding can be used to reduce delay in buses by eliminating transitions that cause a large delay [23]. It is also possible to reduce delay in buses by avoiding simultaneous transitions by skewing signal transition timing of adjacent wires [24]. Uneven signal timings are an inherent part of asynchronous signaling. In the following case studies the influence of signal timing and coding on noise is discussed.

5. Case Studies

The methods and techniques to reduce on-chip noise, introduced in the previous sections, are applied to two different case studies. In the first case study both crosstalk and switching noise on a high performance bus will be studied. The bus in question is a pipelined bus architecture which is segmented into independent parallel operating sections. The second case study concentrates on switching noise reduction for a Viterbi decoder. The effort is focused on the path metric unit, which is the core of the computation in the Viterbi decoder and dominates the power consumption.

5.1 Case study 1: Minimizing noise in a high performance on-chip bus

In this case study both crosstalk and switching noise characteristics are analyzed on an advanced pipelined bus structure [17, 25] targeted for GALS based system-on-chip design. The bus architecture is a distributed organization based on self-timed communication. Unlike in the case of a conventional shared bus, Figure 11.6.a, the pipelined bus can be simultaneously accessed by all the attached processing elements. The physical wires that implement the bus are divided into N-1 segments, where N is the number of modules connected to the bus. The segments are isolated from each other by N *transfer stages*, one attached to each module. The arbitration and control are evenly distributed among the transfer stages which contain internal FIFO queues for pipelining the data flow. A bus segment between adjacent stages consists of two separate unidirectional point-to-point interconnects which transfer data asynchronously between the stages in opposite directions. These two links of a segment can operate in parallel, and due to pipelining, all segments of the system bus can transfer data concurrently. The pipelined bus architecture is illustrated in Figure 11.6.b.

In the following paragraphs crosstalk and switching noise in a segment of the pipelined bus are investigated. Different signaling methods are compared in order to minimize noise, and their effect on the performance is analyzed. The pipelined bus was implemented by using a 0.13 μm technology with 1.2 V supply voltage. The wires in the bus segment in question are placed 0.6 μm apart from each other and they have length

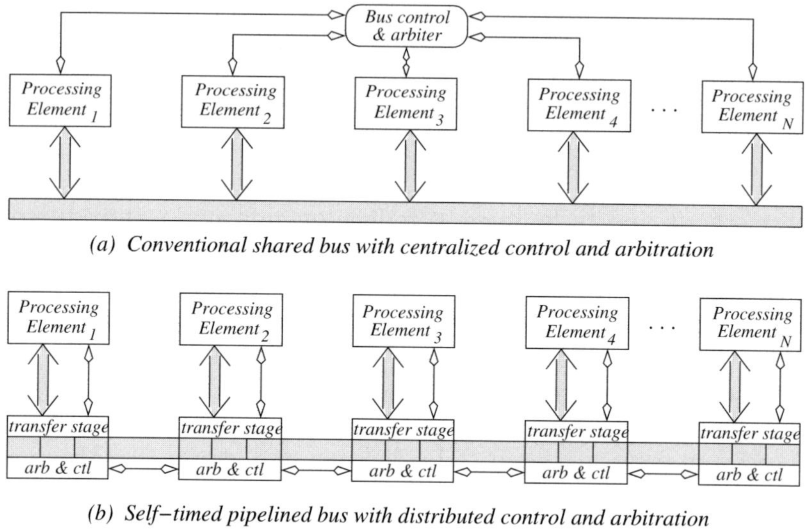

Figure 11.6. Shared (a) and pipelined (b) bus architectures

of 2 mm, width of 0.6 μm and thickness of 0.32 μm. The size of the transmitted messages is 32-bit in all the cases. The rise and fall times of each signal are 150 ps.

5.1.1 Noise characteristics of the bus segment.

In the first phase noise characteristics on a bus segment are analyzed when employing the bundled data convention with the two-phase signaling protocol. This protocol was chosen in order to minimize the signaling events, and thereby the delay, in the rather long interconnect with considerable parasitic properties. However, in this case the signal protocol could be as well synchronous from the noise analysis point of view. This is due to the fact that all wires may switch at the same time and no encoding techniques are used. This signaling method serves as a reference point of the noise comparison towards the other methods.

The worst-case voltage coupling between adjacent wires, crosstalk, in such a bus is shown in Figure 11.7, where three transfer cycles are considered. It occurs when all the wires in the bus except one in the middle switch simultaneously from zero to one while the intended value for the wire in the middle is zero. The maximum crosstalk voltage that is coupled to that wire is 144 mV or 12 % of the supply voltage. One might think 12 % is not a significant amount of noise but it could easily be vital when added to the other noise sources such as switching noise, electromagnetic interference, receiver and transmitter offset and so on.

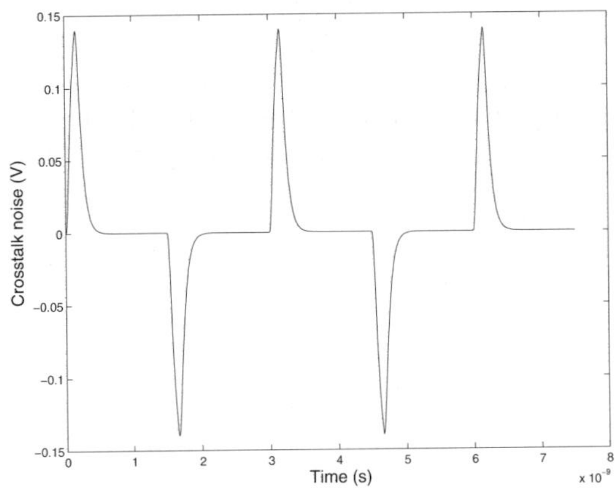

Figure 11.7. Crosstalk voltage on a bus segment.

The current draw of the bus drivers is shown in Figure 11.8 by curve a, the upper one. As it can be seen, the current profile is clearly dominated by the clock induced current peaks.

5.1.2 Asynchronously time-interleaved transmission.

In this approach data bits are sent in a time-interleaved fashion. A message is divided into bit groups which are transmitted at slightly different times with respect to each other. Hence, the power hungry bus drivers will not switch exactly at the same moment of time, instead only a set of drivers will switch simultaneously. This considerably reduces the peak current draw and therefore the switching noise. Each of these groups is formed so that a group does not contain adjacent bits. For example if a 32-bit message is divided into four groups, the first group contains bits 0, 4, 8, 12, 16, 20, 24, 28. This reduces crosstalk since neighboring bus wires do not switch exactly at the same moment. In addition to attenuated crosstalk characteristics, such a bit division subsequently reduces the switching noise, since the bus drivers involved in transmitting a certain group of bits are not located next to each other. Therefore the concurrent current draw of the drivers is spread into a larger area, reducing the local peak current. The number of wires in the bus are kept same as before, n-bit messages require n wires for data and two wires for the handshake signals. Obviously this method sacrifices a part of the performance to the increased noise margin. However, by keeping the delays between bit groups relatively small, considerable reduction of noise can be achieved with a minor performance loss.

The current profile caused by the bus segment drivers with time-interleaving is shown in Figure 11.8 by curve b. The messages were divided into four 8-bit groups which were transmitted in interleaved fashion. The time interval between groups was 80 ps, obtained by using an average size driver. As can be seen, the peak value is considerably lower when compared to the synchronous case. Furthermore, the profile is significantly smoother while the peaks are reduced by 43 %.

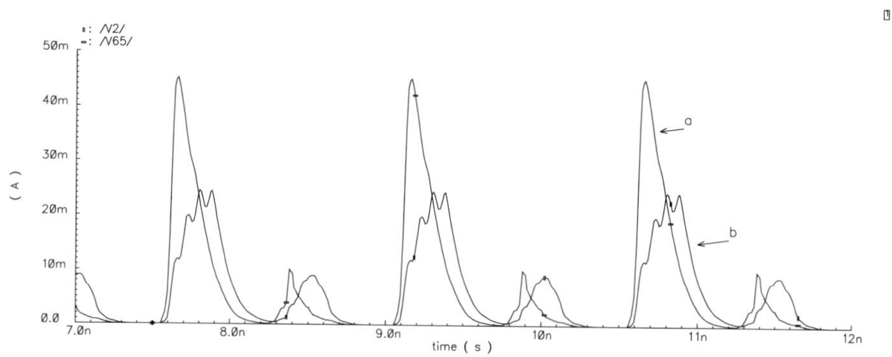

Figure 11.8. Current profiles of a bus segment a) synchronous, b) time-interleaved.

In addition to the above analysis, the effect of different interleaving times on crosstalk and current peaks were studied. Furthermore, partitioning the data transmission into two and three groups was also considered. The crosstalk and peak current values with the three different partitioning techniques are shown in Figure 11.9, where crosstalk values are illustrated as solid lines and peak current values are presented as dashed lines. The analysis was performed with different interleaving times up to 200 ps. This was chosen to be the upper bound of the analysis since the meaning was not to sacrifice the performance entirely. The effect of data partitioning into two, three and four groups is illustrated with curves noted with letters a, b and c, respectively. Observe that, with 200 ps time interleaving and partitioning data into two or three groups the reduction in both values can be thought saturated.

As can be expected, the amount of noise decreases when the time between bit groups increases, and the rate of reduction is faster when more groups are used. To make the comparison fair between different bit partitionings one has to take into account the total increase in the transfer time. For example, reduction in noise that can be achieved by using 80 ps intervals with division into four groups, has to be compared against 120 ps time-interleaving when using three groups since both are causing a 240 ps increase to the total transfer time. The lowest noise characteristics can be obtained when the data transfer is partitioned into

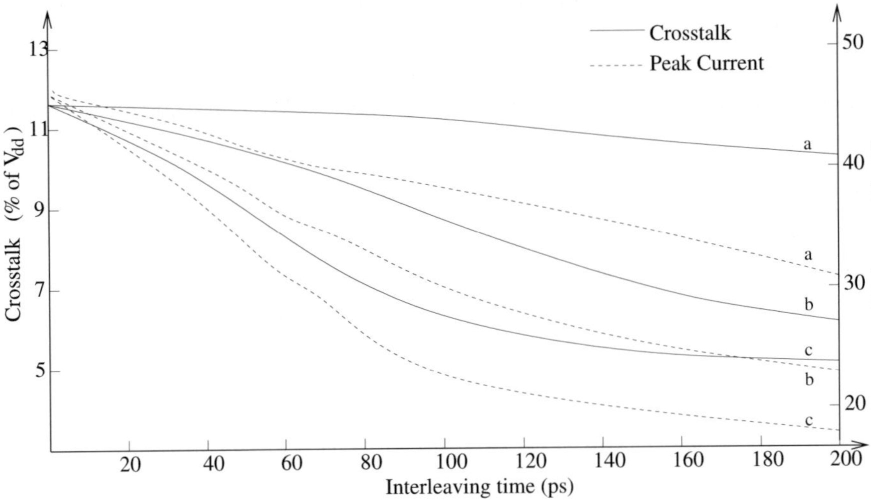

Figure 11.9. Crosstalk and peak current as a function of time-interleaving

four 8-bit groups, shown by curve c. Applying this partitioning scheme with 100 ps intervals the transfer time is increased by 300 ps but at the same time the crosstalk is reduced by 50 %, from 12 % to 6 %, and the peak current has decreased by 49 %. Similar behavior but with a smaller decrease in noise characteristics can be seen when the data is divided into three or two groups. However, the decrease in crosstalk is rather small when the data transfer is divided into two groups.

5.1.3 Dual-rail and 1-of-4 data encoding.

The effect of the noise was analyzed with two encoding schemes, four-phase dual-rail and 1-of-4 data encoding. For the dual-rail encoding the current peaks are not reduced since similar amount of switching occurs as in the 32-bit synchronous bus. In the synchronous bus all the four closest wires have an effect on the victim line in contrast to the dual-rail implementation where three out of four have an impact to the victim line. This is due to the dual-rail encoding as explained in section 11.2.3. Hence, the crosstalk is reduced only by 10 % compared to the synchronous case, to 11 % or 130 mV of the supply voltage. However the crosstalk should not be as detrimental to performance with the four-phase dual-rail as it is with the two-phase one, since the adjacent wires switch at the same direction. Furthermore, the four-phase dual-rail has an automatic error detecting mechanism as a default, due to the illegal state '11'. For instance if the codeword should be '10', but due to crosstalk the last bit assumes temporarily an erroneous logic value '1'.

This will lead to illegal state and the transmission system can be build so that it observes it.

The current profiles of the synchronous and 1-of-4 encoded bus are shown in Figure 11.10. The current peaks in the latter one are significantly lower compared to the synchronous bus, while the effect on crosstalk is rather small. The peak current is decreased by 50 %, while the crosstalk is 120 mV or 10.4 % of the supply voltage. Similarly as with the dual-rail encoding, the crosstalk should not be as detrimental to the performance for a 1-of-4 encoded interconnect as it is for synchronous implementation, since any crosstalk that does occur will be between wires switching at same direction. Furthermore, the likelihood that two adjacent wires will switch simultaneously is quite small with 1-of-4 encoding [12]. The 1-of-4 encoded bus is attractive in the low-power perspective, since it transmits two bits of information producing a transition only on one wire compared to the synchronous one where each bit requires transition on a corresponding wire. The average current of the 32-bit bus is 7.5 mA for the synchronous one and 3.5 mA for the 1-of-4 encoding interconnect, the reduction being 53 %. This supports the low-power behavior of the 1-of-4 encoding.

In addition to the above mentioned advantages, it is also possible to employ the time-interleaving with both the dual-rail and 1-of-4 encoding techniques. This way the noise coefficients can be even more decreased as illustrated in the Figure 11.9.

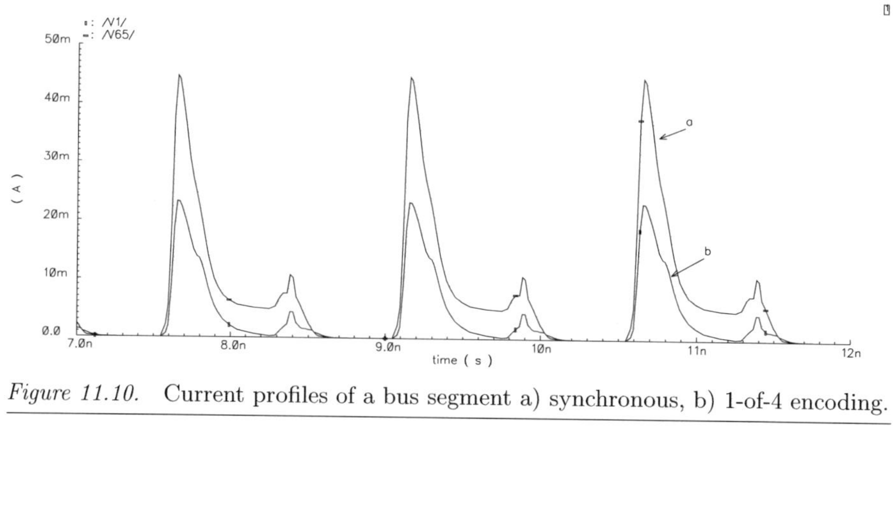

Figure 11.10. Current profiles of a bus segment a) synchronous, b) 1-of-4 encoding.

5.2 Case study 2: On-chip switching noise reduction for Viterbi decoder

A Viterbi decoder is used to decode convolutionally encoded data [26]. This kind of encoding is widely used in a large portion of digital transmission and digital recording systems, including mobile phones and digital TV broadcast. Convolutional codes are one of the major families of error correcting codes. Error correcting codes are commonly used in the transmission of digital data due to their ability to reduce the error probability [27]. These codes operate by adding redundancy into a signal in a way that some errors that might occur during transmission can be eliminated in the decoder. The path metric unit (PMU) is the core of the computation in the Viterbi decoder and therefore it dominates the overall power consumption. For instance, with a random input bit sequence the power dissipation of the PMU can be as high as 90 % of the overall power consumption [28]. This motivates to concentrate design efforts on the PMU in order to smoothen the current profile of the circuit.

5.2.1 Implementation strategies.

The synchronous decoder is implemented using hard-decision decoding [29]. It uses one bit quantization on the received bits and Hamming distance [27, 26] to update the path metrics. The architecture of the decoder is illustrated in Figure 11.11 a. The branch metric unit calculates the branch metrics according to the Hamming distance method. These metrics are sent to the path metric unit, which consists of eight add-compare-select (ACS) processing elements and a control unit [29]. Each of those processing elements contains two ACS cells which all have similar functionality with three different basic operations: addition, comparison and selection. One ACS cell contains two adders, a comparator, a path metric out unit, and a selector. The path metric out unit is used to store the results of the comparison at the current counting cycle and to transfer data. The selector chooses the smallest path metric value and counts the corresponding state. The interconnects between processors are used to transmit every clock cycle the previous path metric values which are needed to calculate the new ones. At every rising clock edge each of those processing elements calculates new path metric values and decision values. The survivor memory unit is used to store the decision values, which are the results of the ACS counting. The trace back operation reads these values backwards and outputs the original bit sequence.

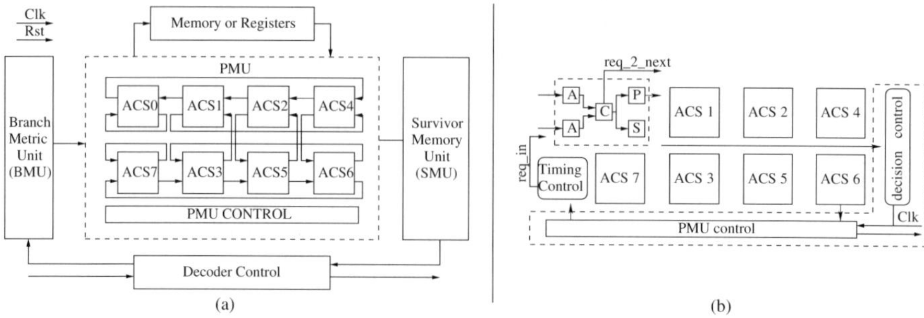

Figure 11.11. a) The structure of synchronous Viterbi decoder b) self-timed PMU

The Viterbi decoder was de-synchronized following the methodology given in Section 3. The external interface and the original timing restrictions of the self-timed PMU are equal to the synchronous one. Furthermore, the number of processing elements and the inter-processor connections were the same as those in the synchronous implementation. The structure of the self-timed PMU is shown in Figure 11.11 b. The synchronous decoder operates with a 25 MHz clock which sets the limit for the duration of one counting cycle in the self-timed structure. The counting cycle is the time period that it takes to decode and output one decoded bit. The number of required counting cycles depends on the number of states used. One decoded bit requires 10 cycles for a 64 state decoder and 38 cycles for a 512 state decoder.

The self-timed PMU can be divided into two parts: synchronous and asynchronous. The synchronous part consists of PMU control and decision control modules. The PMU control unit acts as an interface between synchronous and asynchronous domains. It controls the handshake logic that activates the calculation and communicates with other parts of the decoder. The decision control unit is used to collect the decision values that arrive asynchronously. The reading operation of the decision values is clocked because it sends the decision values to the synchronous survivor memory unit.

The asynchronous part consists of the eight ACS processing elements and a timing control. Timing control is used to interleave the operation of the ACS processing elements via self-timed communication channels. Each of those elements are divided to four separate stages. The stages are controlled in a self-timed manner so that their operations are time interleaved. The counting cycle starts by activating two adders, the first two stages, at a slightly different time. After the completion of both additions the comparator and the next processing element are activated. Hence, comparison, third stage, and additions in the next processing element are executed in parallel. Last stage, the selection and transfer of the new path metric value, is activated immediately after comparison. Meanwhile, stages at the different processing elements are activated in a domino like fashion. After all ACS processing elements have completed their tasks the timing control is acknowledged and a new counting cycle can be started.

The interprocessor communication was implemented with three different methods. The purpose was to find a method that reduces most the peak current values caused by driving the interconnects between the processing elements. The asynchronous interface between the processing elements was implemented according to the guidelines given in Section 3. The first case is asynchronous interconnect which is implemented utilizing the four-phase handshake protocol. In the second one the data transmission is divided into four 4-bit groups which are transmitted in the time-interleaved fashion. The third one is implemented using four-phase dual-rail encoding.

5.2.2 Current profile.

The motivation for reducing the peak current values is illustrated in Figure 11.12. During the active phase of the processing elements in the synchronous design the clock related current peaks are as high as 1.2 A. Furthermore the idle time between the calculation causes peak current values around 1 A. Hence in synchronous systems many flip-flops switch without having an actual input to process because they are connected to the clock.

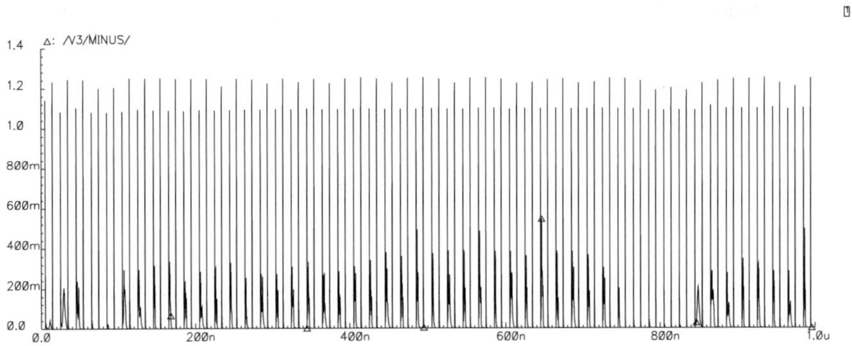

Figure 11.12. Current profile of the synchronous 512 state PMU

The average capacitive loads of the interconnects are from 100 fF to 240 fF in the 0.35 μm technology. The corresponding average resistive loads are from 5 Ω to 10 Ω. Above are the minimum and the maximum values obtained from analysis performed after place and route which are a function of the interconnect length. Therefore the values used to model the interconnects in the current profile analysis are spread between the above values. The 64 state PMU was analyzed with a 0.35 μm technology using three different inter-processor communication methods between the processing elements as explained earlier. Each bit of the 16-bit wide interconnects was modeled separately with its actual capacitive and resistive load.

A corresponding analysis was performed with a 0.18 μm technology. The average capacitive and resistive loads of the interconnects were from 37 fF to 72 fF and from 39 Ω to 75 Ω respectively. The separate interconnect analysis was performed to illustrate the effect of technology scaling. The effect of the technology scaling was a decrease in capacitance values and an increase in resistance values when compared to the 0.35 μm technology, as can be expected. The summary of the simulation results for both technologies are shown in Table 11.3. These results were measured as peak to peak values.

The current profile of the 64 state PMU with 0.35 μm technology is shown in Figure 11.13. During the active phase of the processing elements, the current peaks are around 550 mA which is only half of those in the corresponding 512 state implementation. Similarly with the 512 states PMU the decrease is minor during the idle period of the processing elements. With the self-timed implementation using asynchronous interconnects the current peaks are reduced by 75 % during the active phase. If applied to the 512 state implementation the peak values during counting should decrease from 1.2 A to 300 mA.

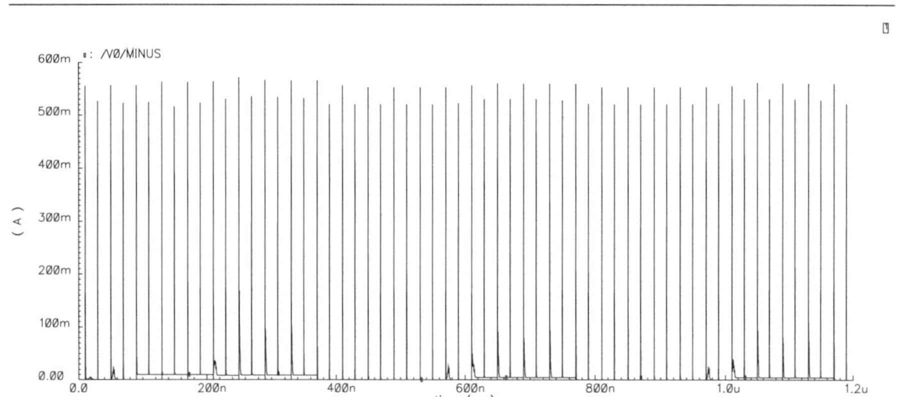

Figure 11.13. Current profile of the synchronous 64 state PMU

The peak current values are reduced most when the self-timed PMU is implemented with the time-interleaved interconnect method. This current profile is shown in Figure 11.14. The reduction is 87 % during the counting cycle. One of the advantages of the self-timed method can be seen during the idle period between the counting cycles. This is illustrated in Figure 11.14 as significantly lowered current peak values, the reduction is 97 %. The above values result from the partitioning of data transfer and the natural support of the power down during the idle period in the self-timed implementation.

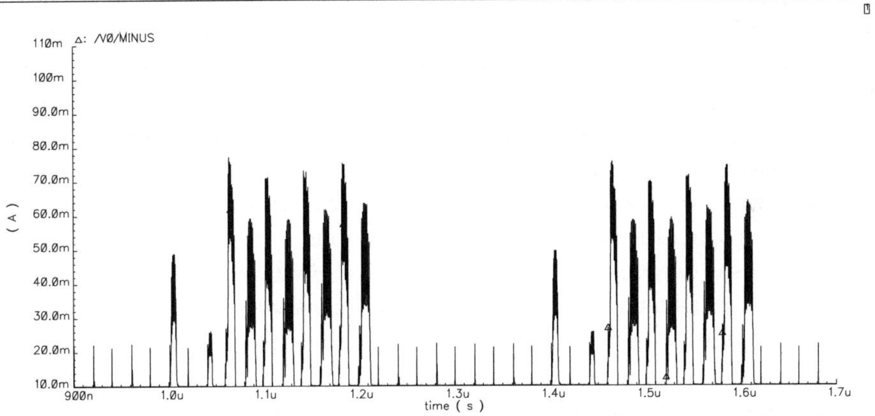

Figure 11.14. Current profile of the asynchronous 64 state PMU with 4-bit interconnects

The results of the current profile analysis for the 0.35 μm and the 0.18 μm technologies are collected in Table 11.3. Separate peak current values are presented for interconnects with utilizing the loads from the 64 state and 512 state implementations. It shows that all asynchronous implementations provide lower current peaks than the corresponding synchronous ones.

Table 11.3. Peak current values with different technologies.

	0.35μm			0.18μm	
Design	PMU(64)	Int.co.(64)	Int.co.(512)	Int.co.(64)	Int.co(512)
Sync.	550 mA	36 mA	39 mA	15 mA	16.5 mA
Async.	140 mA	4.5 mA	6.2 mA	1.9 mA	3.2 mA
Interleaved	70 mA	3.2 mA	3.2 mA	1.3 mA	1.3 mA
Dual-Rail	80 mA	8.4 mA	9.3 mA	4.7 mA	4.8 mA

The optimal interconnect implementation is the asynchronous time-interleaved data transfer where the message is partitioned into four 4-bit groups. The current peaks are reduced by 92 % regardless of the number of states used. In this implementation the interconnects were relatively short. Therefore the effect of the interconnects on the current profile was rather small. All of the three design methods reduce the current peaks considerably compared to the synchronous one. The self-timed PMU implemented with the time-interleaved interconnects gives the most optimal result regardless of the technology. The above results show that a large amount of the switching noise can be reduced by using self-timed design method.

5.2.3 Area.

The area comparisons between the synchronous and asynchronous implementations were made according to the results gained from the synthesis. The differences in areas between the two different design approaches with 0.35 μm technology are shown in Table 11.4. In order to make the area comparison between implementations straigthforward, relative areas are used where synchronous 64 state PMU serves as a reference point. In the asynchronous 64 state implementation the area is 21 % larger than the area of the corresponding synchronous system. As the size of the decoder increases the area penalty decreases, with 512 state implementation the difference in total area is 2 %. A similar comparison

Table 11.4. Relative areas with different technologies.

	0.35μm			0.18μm		
Number of states	Comb.	Seq.	Total	Comb.	Seq.	Total
Sync. 64	1	1	1	1	1	1
Async. 64	1.13	1.42	1.28	1.14	1.22	1.18
Sync. 512	4.07	5.63	4.85	3.67	5.37	4.55
Async. 512	4.48	5.38	4.93	4.38	4.70	4.55

made with the 0.18 μm technology is shown in Table 11.4. The relative area decreases in every design compared to the corresponding areas from 0.35 μm technology. The asynchronous 64 state implementation is 16 % larger than the corresponding synchronous one. Similarly as with the 0.35 μm technology, the area penalties decrease when the size of the decoder increases. In fact the total area of the asynchronous 512 state implementation is the same as it is in the synchronous one.

The absence of clock circuitry in the asynchronous implementations saves area that can be used for sequential logic. However, the amount of combinatorial logic increases due to the self-timed control circuitry. In the 0.35 μm and 0.18 μm technologies the asynchronous implementations of the 512 state case have about the same total area as the corresponding synchronous ones. When the size of the decoder increases the amount of sequential area increases significantly. This leads to the situation where the area needed for the clock circuitry compensates the area needed for the self-timed control circuitry.

6. Conclusion

A self-timed approach for minimizing crosstalk and switching noise was presented in this chapter. It was based on self-timed circuit design techniques, system partitioning and time-interleaved communication. In this approach, time-interleaving was used to reduce the effect of interconnect signal coupling, current peaks and the probability of erroneous states. Time-interleaving was implemented by dividing the system into partitions which were de-synchronized internally and with respect to the other within the same partition. The technique was based on re-tuning the timing using self-timed design. The impact of asynchronous signaling techniques on overall noise levels and signal timing was studied in several contexts. The method was illustrated using two case studies. The first case study employed time-interleaving to reduce noise in high performance on-chip bus segments. The second one exploited the advantages of the de-synchronization method applied to the path metric unit of the Viterbi decoder.

Case study 1: Time-interleaving and asynchronous encoding schemes were applied to 32-bit bus segments. Time-interleaved bit groups guarantee that all the interconnect drivers do not switch simultaneously. This reduces both crosstalk and switching noise considerably. According to simulations, the peak current was decreased by 43 % and the crosstalk was reduced by 37 % compared to the synchronous segment, when the bus transactions were divided into four bit groups with only 70 ps time-interleaving. In addition to the above, a particularly interesting case was the 1-of-4 delay insensitive data encoding scheme. It is attractive in the low-power design perspective — the average current was reduced by 53 % compared to the synchronous bus.

Case study 2: The eight synchronous processing elements of the path metric unit were de-synchronized so that their internal operation was time-interleaved. Furthermore, three different self-timed inter-processor communication schemes were studied in order to decrease current peaks caused by interconnect drivers between the processing elements. The impact of the self-timed approach was significant, since the reduction of current peaks was 87 % compared to the synchronous version. This came with an area penalty of 21 %. However, the area penalty decreases as the size of the decoder increases.

Reducing noise has an immediate impact, boosting both reliability and performance. The study considered in this chapter revealed the possibility to decrease the crosstalk and the power supply noise by utilizing the self-timed design approach.

References

[1] H. Bakoglu. *Circuits, Interconnections, and Packaging for VLSI.* Addison-Wesley, 1990

[2] H. H. Chen and J. S. Neely. "Interconnect and Circuit Modeling Techniques for Full-Chip Power Supply Noise Analysis", in *IEEE Transactions on Components, Packaging, and Manufacturing Technology* - part B, Vol. 21, No.3, August 1998.

[3] K. L. Shepard and V. Narayanan. "Noise in Deep Submicron Digital Design", In *Proc. of International Conference on Conputer Aided Design (ICCAD)*, 1999.

[4] H. H. Chen and D. D. Ling. "Power Supply Noise Analysis Methodology for Deep-Submicron VLSI Chip Design", in *Proc. of Design Automation Conference*, Anaheim, California, 1997.

[5] W. J. Dally and J. W. Poulton. *Digital Systems Engineering*, Cambridge University Press, 1998.

[6] C. L. Seitz. "System timing". Chapter 7 in Mead and Conway, editors, *Introduction to VLSI Systems*, Addison-Wesley, 1980.

[7] A. Davis and S. M. Nowick. "Asynchronous circuit design: motivation, background and methods." In G. Birtwistle and A. Davis, editors, *Asynchronous Digital Circuit Design*, Springer, 1995.

[8] I. E. Sutherland. "Micropipelines. The 1988 Turing Award Lecture", *Communications of the ACM*, Vol.32, No.6, pp.720-738, June 1989.

[9] A.M.G. Peeters. Single-Rail Handshake Circuits. Ph.D. Thesis, Eindhoven University of Technology, The Netherlands, 1996.

[10] J. Sparso and S. Furber. *Principles of Asynchronous Circuit Design - A System Perspective.* Kluwer Academic Publishers, 2001.

[11] Tom Verhoeff. "Delay-insensitive codes - an overview", *Distributed computing*, 3(1):1-8, 1988.

[12] W. J. Bainbridge and S. B. Furber. "Delay Insensitive System-on-Chip Interconnect using 1-of-4 Data Encoding", in *Proc. of International Symposium on Asynchronous Circuits and Systems (ASYNC)*, March 2001.

[13] S. Hayashi and M.Yamada. "EMI-Noise Analysis Under ASIC Design Environment", *IEEE Transactions on Computer-Aided Design of Integrated Circuits and Systems*, Vol.19, No.11, November 2000.

[14] D. M. Chapiro. *Globally-Asynchronous Locally-Synchronous Systems.* Ph.D. thesis, Stanford University, October 1984.

[15] P. Liljeberg, J. Plosila and J. Isoaho. "Asynchronous Interface for Locally Clocked Modules in ULSI Systems", in *Proc. of International Symposium on Circuits and Systems (ISCAS)*, Sydney, Australia, May 2001.

[16] J. Muttersbach, T. Villiger and W. Fichtner. "Practical Design of Globally-Asynchronous Locally-Synchronous Systems", in *Proc. of International Symposium on Advanced Research in Asynchronous Circuits and Systems (ASYNC)*, April 2000.

[17] P. Liljeberg, J. Plosila and J. Isoaho. "Self-Timed Ring Architecture for SoC Applications" in *Proc. of IEEE International SoC Conference*, Portland, OR, USA, September 2003.

[18] L. T. Pillage and R. R. Rohrer. "Asymptotic Waveform Evaluation for Timing Analysis", in *IEEE Transactions on Computer-Aided Design of Integrated Circuits and Systems*, Vol.9, No.4, pp.352-366, April 1990.

[19] A. Devgan. "Efficient Coupled Noise Estimation for On-Chip Interconnects" in *IEEE/ACM International Conference on Computer-Aided Design*, San Jose, CA, USA, 1997.

[20] M. Becer and I. N. Hajj. "An Analytical Model for Delay and Crosstalk Estimation in Interconnects under General Switching Conditions", in *the 7th IEEE International Conference on Electronics, Circuits and Systems*, Kaslik, Lebanon, 2000.

[21] S. Tuuna and J. Isoaho. "Estimation of Crosstalk Noise for On-Chip Buses" in *Proc. of 13th International Workshop on Power and Timing Modeling, Optimization and Simulation*, Turin, Italy, September 2003.

[22] H. Kawaguchi and T. Sakurai. "Delay and Noise Formulas for Capacitively Coupled Distributed RC Lines" in *Proc. of Asia and South Pacific Design Automation Conference*, Yokohama, Japan, 1998.

[23] P. Sotiriadis and A. Chandrakasan. "Reducing Bus Delay in Submicron Technology Using Coding", in *Proc. of IEEE Asia and South Pacific Design Automation Conference*, 2001.

[24] K. Hirose and H. Yasuura. "A Bus Delay Reduction Technique Considering Crosstalk", in *Proc. of the Design Automation and Test in Europe*, 2000.

[25] J. Plosila, P. Liljeberg and J. Isoaho. "Pipelined On-Chip Bus Architecture with Distributed Self-Timed Control", in *Proc. of IEEE International Symposium on Signals, Circuits and Systems (SCS)*, Iasi, Romania, 2003.

[26] M. S. Ryan and G. R. Nudd. "The Viterbi Algorithm", *Warwick Research Report*, No.238, University of Warwick, England, February 1993.

[27] T. K. Moon and W. C. Stirling. *Mathematical Methods and Algorithms for Signal Processing*, Prentice Hall, 2000.

[28] P. A. Riocreux, L. E. M. Brackenbury, M. Cumpstey and S. Furber. "A Low-Power Self-Timed Viterbi Decoder", in *Proc. of International Symposium on Asynchronous Circuits and Systems (ASYNC)*, March 2001.

[29] M. Savolainen. *Reusable Viterbi Decoder Implementation*, M.Sc. Thesis, Tampere University of Technology, October 2001.

Chapter 12

FORMAL COMMUNICATION MODELING AND REFINEMENT

Juha Plosila, Tiberiu Seceleanu
University of Turku, Department of Information Technology, Turku, Finland
{juha.plosila,tiberiu.seceleanu}@utu.fi

Kaisa Sere
Åbo Akademi University, Department of Computer Science, Turku, Finland
kaisa.sere@abo.fi

1. Introduction

Nowadays we are used with terms like *system-on-chip* (SOC) or *platform-based design*. One of the characteristics of such designs is the dramatically increased amount of interconnect, communication issues starting to have a more and more important impact on all others. In a somehow "classical" style, usual design process starts with a description of functionality, as specified by concurrent processes, communicating via a variety of mechanisms, such as procedure calls, message passing, shared variables, etc. The most effort in design is consumed in specifying behavior and, towards the end of the process, the elements are to be placed together and hopefully provide the intended global behavior. Lately, communication related analysis started to move to earlier stages of the design and even drive the further process decisions, hence the apparition of the *communication-based design* (CBD) concept [1], in the context of platform-based design. Basically, the system design is split along functional and communication characteristics.

Certain goals of CBD may be pointed as (i) protocol identification, (ii) level(s) of abstraction, (iii) combination between different models of computation and architectures, (iv) development of procedures to move from one level of abstraction to another one, usually trying to obtain a correct derivation that would ensure the compliance of the implementation with the initial specification.

One way to approach such aspects of design is to work within, or close to a formal environment, where design steps can be either verified (*after* the decision) or controlled (*during* the transformation). The Action Systems formalism is one possible framework that may help us reach the mentioned goals. It provides a mathematical, rigorous framework for defining constraints for design specifications, reasoning about concurrency and communication issues, logical properties of system descriptions, and proving correctness of taken design steps.

The Action Systems framework has been applied in the design of asynchronous [2] and synchronous [3] systems. Therefore, it offers a powerful unifying basis for hardware design generally. It offers the possibility to work in a stepwise manner with systems that contain both synchronous and asynchronous components, leading to correctly derived descriptions and implementations. This is a feature not usually found in other formal approaches. They mainly address either synchronous or asynchronous design aspects, or, in the case when they also cover both design styles, they provide distinct rules for each of them.

Chapter overview. We start by presenting basic notions of the Action Systems formalism. In Section 12.3, we discuss techniques on how to formally model communication channels using Action Systems, starting from the assumption that the target system is viewed as a composition of master and slave modules, communicating on possibly multiple channels, using possible multiple protocols. Asynchronous and synchronous communication schemes are presented. In the following section, we analyse the procedures used to bring an abstract model of communication closer to implementation levels. This is continued with a specification example in Section 12.5. Final remarks on the present work are discussed in Section 12.6.

2. Action Systems

The *Action Systems* formalism initially proposed by Back and Kurki-Suonio [4] and extended in other studies [5, 6] is a framework for specification and correctness-preserving development of reactive systems. Based on an extended version of the guarded command language of Dijkstra [7], Action Systems is a state-based formalism. It has the *Refinement Calculus* [8] as the mathematical basis for disciplined stepwise derivation.

2.1 Actions

An *action* A is defined by

$$
\begin{array}{rcll}
A & ::= & abort & \text{(abortion, non-termination)} \\
 & | & skip & \text{(empty statement)} \\
 & | & x := x'.R & \text{(non-deterministic assignment)} \\
 & | & x := e & \text{((multiple) assignment)} \\
 & | & A_1 \parallel \ldots \parallel A_m & \text{(non-deterministic choice)} \\
 & | & A_1; \ldots ; A_m & \text{(sequential composition)} \\
 & | & A_1 * \ldots * A_m & \text{(simultaneous composition)} \\
 & | & A_1 // \ldots // A_m & \text{(prioritized composition)} \\
 & | & P \rightarrow B & \text{(guarded action)}
\end{array}
$$

where $A_j, j = 1, \ldots, m$, and B are actions, P and R predicates (boolean conditions), x a variable or a list of variables and e an expression or a list of expressions.

Semantics of actions. An action A is considered *atomic*, that is, only its input-output behavior is of interest. This indicates that only the *initial* and *final* state of an action can be observed. Atomic actions may be represented by simple assignments or by more complex action compositions, such as the atomic sequence.

The *total correctness* of an action A with respect to a precondition P and a postcondition Q is denoted $P\ A\ Q$ and defined by

$$P\ A\ Q \mathrel{\widehat{=}} P \Rightarrow \mathrm{wp}(A, Q)$$

where $\mathrm{wp}(A, Q)$ stands for the *weakest precondition* [7] of an action A to establish the postcondition Q.

We define for example:

$$
\begin{array}{rcl}
\mathrm{wp}(x := x'.R, Q) & \widehat{=} & \forall x'.R \Rightarrow Q[x'/x] \\
\mathrm{wp}(x := e, Q) & \widehat{=} & Q[e/x] \\
\mathrm{wp}(A_1 \parallel \ldots \parallel A_m, Q) & \widehat{=} & \mathrm{wp}(A_1, Q) \wedge \ldots \wedge \mathrm{wp}(A_m, Q),
\end{array}
$$

where the notation $Q[e/x]$ represents the replacement of x by e in the predicate Q. The *guard* gA of an action A is defined by $gA \mathrel{\widehat{=}} \neg\mathrm{wp}(A,$ *false*). Considering a guarded action $A \mathrel{\widehat{=}} P \rightarrow B$ we have that $gA = P \wedge gB$. An action A is said to be *enabled* in some state, if its guard is *true* in that state. Otherwise A is *disabled*. Observe that any action A can be written in the form *true* $\rightarrow A$, and thus each action can be considered a guarded action, even though a non-trivial guard does not exist. If gA is invariantly *true*, the action is *always enabled*.

Non-atomic actions. Atomic action compositions do not always allow efficient modeling of complex system behavior. For example, when describing a communication sequence between systems, an atomic sequential composition cannot be used as communication events take place in turns in separate atomic actions. In other words, the execution of a sequence is temporarily stopped in some system state, and resumed later in some other state. This kind of behavior is reflected by *non-atomic* action compositions, which allow the observation of intermediate states. In such a construct, the component actions may be atomic constructs of their own, but the composition in itself is not. A non-atomic composition is referred to as a *non-atomic action*.

We use four different non-atomic constructors: ; (*sequential composition*, each action is executed once in the specified order), $\|$ (*parallel composition*, each action is executed once in any order), $\oplus$ (*exclusive composition*, selection of one component action disables the others until the selected one has been completed) and ∇ (*synchronous composition*, actions are executed simultaneously in two successive phases: read phase and write phase). All of these operators are defined in terms of the non-deterministic choice $[\!]$ and a set of auxiliary local variables, *program counters*. As an example, the non-atomic sequential composition of two actions A_1 and A_2 can be defined by

$$A_1 \; ; \; A_2 \mathrel{\widehat{=}} ((p = 0 \rightarrow A_1 \; [\!] \; p = 1 \rightarrow A_2); \; p := p + 1 \; \mathbf{mod} \; 2)$$

where the program counter p is assumed to have the initial value of 0.

Quantified composition. Any of the above atomic and non-atomic composition operators, denoted below as '$\bullet$', can be *quantified*. The notation is defined for a *parametrized* action $A(i)$ as follows

$$[\; \bullet \; 1 \leq i \leq n : \; A(i) \;] \mathrel{\widehat{=}} A(1) \bullet \ldots \bullet A(n)$$

Note that, in general, the parameter i does not have to run through a range of values. The set of values may be defined explicitly, for example: $i \in \{5, 2, 6, 9\}$. The leftmost value is considered first, the rightmost value last.

2.2 Hierarchical Action Systems

An *action system* $\mathcal{A}$ is an iterative composition of actions executing on a set of local and global variables. A simplified hierarchical Action Systems model has the following form

```
sys A (interface list) [parameter list] ::
|[ var local variable list
   expressions expression list
   subsys subsystem instance list
   actions action list
   init initialization of variables
   exec
      do action composition od
      SysOp subsystem composition
]|
```

The *interface list* defines the global variables through which $\mathcal{A}$ communicates with other systems. The *parameter list* defines the generic parameters used within the system, such as the range of certain variables or the dimensions of some arrays. The **var** clause defines the local variables, visible only within $\mathcal{A}$. The **expressions** clause defines local shorthand notations for some expressions used within $\mathcal{A}$, for example in the guards of the actions. The items declared here are evaluated every time they are met during the execution of the system. The **subsys** clause defines, in the case of a hierarchical Action Systems model, unique instance names for the subsystem components specified in detail elsewhere, describing how the formal interface and parameter lists of the subsystems are replaced with actual variables and parameters within the system $\mathcal{A}$. The **actions** clause describes the atomic actions present in the system. A unique name is given for each action. An action can also be, partly or completely, composed of other actions defined in the **actions** clause. In the **init** clause, all the local and interface variables are initialized.

An underscore '_' signals that the current line is continued on the next line.

The **exec** clause contains both a **do-od** loop and a composition of subsystem instances defined in the **subsys** clause, or either of them. The loop encloses a composition of actions defined in the **actions** clause. This construct is realized using the atomic action composition operators $[\!]$ and $//$, and the non-atomic operators **;** , $\|$, $\oplus$, and ∇. The system composition operators $SysOp \in \{\|, \nabla, //\}$ are used both within the subsystem composition and between the **do-od** loop and the subsystem composition. The semantics of composing action systems is discussed separately in Section 12.2.3 below.

Behavior. An action system $\mathcal{A}$ operates as follows. After the initialization of the variables in the **init** clause, enabled actions in the **do-od** loop and in the subsystem instances are selected for execution one by one. Selection between simultaneously enabled actions is nondeterministic. Parallel behavior is modeled by simultaneously enabled actions which can be *interleaved*, i.e., executed in any order. This is the case, if the actions in question do not write onto same variables, and can neither disable nor enable each other. When all actions both in the **do-od** loop and in the subsystem instances become disabled, the computation stops temporarily until a system communicating with $\mathcal{A}$ enables one of the actions again via the interface variables. If the *interface list* is empty, i.e., $\mathcal{A}$ is a closed system, the computation terminates when there are no more enabled actions left.

Invariants. A predicate P is an *invariant* over an action A, if $P \Rightarrow wp(A, P)$. At the system level, P is an invariant of an action system $\mathcal{A}$ if it is established by the initialization, i.e. $true \Rightarrow wp(Init, P)$, and it is an invariant of each of the component actions of $\mathcal{A}$.

2.3 Composing Action Systems

Consider two action systems $\mathcal{A}_1$ and $\mathcal{A}_2$ which do not include any subsystems. The *parallel composition* of such systems, denoted $\mathcal{A}_1 \parallel \mathcal{A}_2$, is defined to be an action system whose **do-od** loop has the form **do** $A_1 \mathbin{[\!]} A_2$ **od**, where A_1 and A_2 represent the actions of the constituent systems $\mathcal{A}_1$ and $\mathcal{A}_2$, respectively. This system merges the interface variables of the components $\mathcal{A}_1$ and $\mathcal{A}_2$ keeping the local identifiers distinct. The components interact or communicate via their shared interface variables.

Assume now that $\mathcal{A}_1$ and $\mathcal{A}_2$ are instantiated as unique subsystem instances I_1 and I_2 within some enclosing system $\mathcal{B}$, so that the **exec** clause of $\mathcal{B}$ has the form **do** B **od** $\parallel I_1 \parallel I_2$.

Then all local identifiers (i.e. constants, variables, action names etc.) within the components $\mathcal{A}_1$ and $\mathcal{A}_2$, denoted here collectively by li_1 and li_2, respectively, are considered local identifiers of $\mathcal{B}$ and are implicitly renamed to $I_1.li_1$ and $I_2.li_2$ to ensure that they are distinct within the enclosing system $\mathcal{B}$. Hence, the above **exec** clause of $\mathcal{B}$ is considered to be equivalent to **do** $B \mathbin{[\!]} I_1.A_1 \mathbin{[\!]} I_2.A_2$ **od**.

This systematic naming scheme can be easily extended to cover systems with multiple hierarchy levels. For example, if a system $\mathcal{C}$ is composed of a loop **do** C **od** in parallel with an instance I of the system $\mathcal{B}$, we can *flatten* the hierarchy into the equivalent form:

$$\textbf{do}\ C \mathbin{[\!]} I.B \mathbin{[\!]} I.I_1.A_1 \mathbin{[\!]} I.I_2.A_2\ \textbf{od}$$

In addition to the parallel composition $\mathcal{A}_1 \parallel \mathcal{A}_2$ discussed above, we also use the *prioritized composition* $\mathcal{A}_1 // \mathcal{A}_2$ [6] and the *synchronous composition* $\mathcal{A}_1 \nabla \mathcal{A}_2$ [3], where the loops of the composed systems have the forms **do** $A_1 // A_2$ **od** and **do** $A_1 \nabla A_2$ **od**, respectively. Since both the prioritized operator $//$ and the synchronous operator ∇ between actions are defined in terms of the choice operator $[\!]$, a prioritized or a synchronous composition of action systems can always be viewed as a special kind of parallel composition.

From a *flat*, non-hierarchical description, such as the above one, performing the reverse operation, we can *decompose* the original system loop into dedicated, coherent parts. We can then *extract* these parts as separate subsystems for further reuse.

2.4 Refinement of Action Systems

Action systems are intended to be developed in a stepwise manner within the refinement calculus framework. The system transformations performed at each step taking us from a high, abstract level towards a more concrete one are guaranteed to be correct. In this section, we revise briefly the basic concepts of the refinement calculus. Comprehensive studies on the refinement calculus can be found elsewhere [8].

Refinement of actions. In weakest precondition terms, an action A is said to be (correctly) refined by an action C, denoted $A \leq C$, if

$$\forall Q.(wp(A,Q) \Rightarrow wp(C,Q))$$

holds. This is equivalent to the condition

$$\forall P,Q.((PAQ) \Rightarrow (PCQ)),$$

which means that the concrete action C preserves every total correctness property of the abstract action A.

Data refinement. *Data* refinement [9, 8] is the technique we use for transforming an abstract specification into a more concrete one. Assume that A is an action on the program variables a and u, and C is an action on the variables c and u, respectively. Let $R(a,c)$ be a boolean relation between the variables a and c. Then the *abstract* action A is data refined by the *concrete* action C using the abstraction relation $R(a,c)$, denoted $A \leq_R C$, if

$$\forall Q.(R \wedge wp(A,Q) \Rightarrow wp(C, \exists a.R \wedge Q))$$

holds. Note that Q is a predicate on the variables a, u, and $(\exists a.R{\wedge}Q)$ is a predicate on the program variables c and u. The data refinement $A \leq_R C$ replaces the variable a with the variable c preserving the variables u.

Refinement of action systems. When refining systems, we base our reasoning on the evaluation of observable behaviors, *traces*. A concrete action system $\mathcal{C}$ *trace refines* another system $\mathcal{A}$, denoted $\mathcal{A} \sqsubseteq \mathcal{C}$ if we can prove certain relations on the traces of the two systems. Details about this technique are preseneted in [5].

3. Modeling Communication Channels

In this section, we present techniques on how to formally model communication channels using Action Systems, starting from the assumption that the target system is viewed as a composition of master and slave modules. They communicate on possibly multiple channels. A channel can be either a point-to-point link or a resource shared by several master and slave modules. Both asynchronous and synchronous communication is considered. A classification of communication ports and channels is presented based on the direction of data flow and the type of the module producing the data transferred through the channel in question. We also show how to create new communication channels using refinement techniques.

3.1 Asynchronous Communication

Asynchronous interfacing provides a viable approach to construct modern complex systems-on-chip, composed of several subsystem units, in a modular and reliable manner. In asynchronous communication, a data transfer event between two system components consists of two phases: *request* and *acknowledge*. Depending on the application, the duration of each phase may be either arbitrary or up-bounded. Asynchronously communicating modules form an asynchronous system architecture in which a component module, taken separately, can internally be implemented as either an asynchronous (self-timed) or a synchronous (clocked) circuit block, or a piece of software running in a standard or an application-specific processor core.

Communication channels. Asynchronous interaction between Action Systems models of subsystems is arranged via communication channels each composed of a set of interface variables of the involved modules. Note that some of the interface variables may not directly belong to any communication channel but are used as auxiliary status or control signals between modules. In our formal design framework, a communication channel $c\langle d \rangle$ or a channel c in short, is defined to be a tuple (c, d), where c is a *communication variable* and d is the list of those *data variables* whose values are transferred from a system module to another by communicating via c. When referring to a single party of communication, we talk about *communication ports* rather than channels. If the list d is empty, c is called a *control channel*. Otherwise we have a *data channel* $c\langle d \rangle$. Notice, however, that here d does not necessarily represent actual computational data only but it can represent any kind of control information as well, for example a memory address, an operation code, or an error message.

Generally, a communication variable c is of the enumerated type $com_{m,n}$ defined by

$$\textbf{type}\ \ com_{m,n} : (req_0, \ldots, req_{m-1}, ack_0, \ldots, ack_{n-1})$$

where $req_0, \ldots, req_{m-1}$ and $ack_0, \ldots, ack_{n-1}$ are request and acknowledge states (values), respectively. A variable $c : com_{m,n}$ is initialized to one of its acknowledge states ack_j. If $m = 1$ or $n = 1$, the default value is just *req* or *ack*, respectively. Hence, the simplest and the most usual type $com_{1,1}$ is equivalent to (req, ack) by default. We denote $com_{1,1}$ simply by *com*.

Renaming of the values of the type $com_{m,n}$ is denoted by $com_{m,n}(value\_list)$. For example, the expression $com_{2,1}(count, load, done)$ indicates that the default values req_0, req_1, and *ack* of $com_{2,1}$ are replaced with *count*, *load*, and *done*, respectively. Furthermore, when instantiating subsystems, a formal interface variable c_f of the type $com_{m,n}$ gets the corresponding values specified for an actual interface variable c_a. In some cases, only a part of the values of c_a is assigned to c_f. For instance, consider a library component $\mathcal{C}(c_f : com; \ldots)$ which is to be included in a system. Assuming that the actual variable c_a is of the type $com_{m,n}$, where $m, n > 1$, the **subsys**-clause of the system is provided with an instantiation of the form $Instance\_Name : \mathcal{C}\,(c_a(req_i, ack_j), \ldots)$, where the notation $c_a(req_i, ack_i)$ means that the default values *req* and *ack* of c_f are replaced with the actual values req_i and ack_j of c_a in the component instance *Instance_Name*.

Point-to-point. A *point-to-point* communication channel connects two action systems, one of which acts as the *master* and the other as the *slave* in a given data transfer event. The master side of a channel is called an *active* communication port, and the slave side is referred to as a *passive* communication port. A *communication cycle* on a channel $c\langle\, d\,\rangle$ consists of two phases. When the active party, the master, initiates the cycle by setting c to a request state req_i, the cycle is said to be in the *request phase*. Correspondingly, when the passive party, the slave, responds by setting c to an acknowledge state ack_j, the communication cycle on c is said to be in the *acknowledge phase*. The data d can be transferred either in the request phase from the master to the slave (*push channel*), in the acknowledge phase from the slave to the master (*pull channel*), or in both phases bidirectionally (*biput channel*) [10]. It is also possible that the roles of the communicating parties are switched at certain moments, so that the current master becomes the slave and the current slave becomes the master of the channel $c\langle\, d\,\rangle$. Such switching is controlled by another system module which has to ensure that the parties are never simultaneously in the master mode.

As a generic example of a point-to-point link, consider the below master module $\mathcal{M}_1$ and the slave module $\mathcal{S}_1$ which are instantiated, i.e. composed in parallel, as the subsystem instances *Master* and *Slave* in the system $\mathcal{P}\_to\_p$ also given below. They communicate via the biput channel $c\langle\, d\,\rangle$, where c is of the type $com_{2,1}$ and d is a data variable of any type symbolized by the generic type *data*. *Master* ($\mathcal{M}_1$) initiates a communication cycle in the action M_1 by assigning a value to d and setting c to one of its request values req_0 and req_1. *Slave* ($\mathcal{S}_1$) then responds by executing an appropriate action S_1 or S_2 in which the value of d is copied to a local buffer variable buf_0 or buf_1, a new value is assigned to d, and the communication variable c is set to the acknowledge state *ack*. *Master* detects the acknowledgement and stores the value of d to the local buffer *buf* in its M_2 action, sequentially composed with the action M_1. After this, a new communication cycle may begin.

```
sys M1 (c : com2,1; d : data) ::
|[ var buf : data
   actions
     M1 : d := d'. (d' ∈ data);
          c := c'. (c' ∈ {req0, req1})
     M2 : c = ack → buf := d
   init c := ack
   exec   do M1 ; M2 od
]|
```

```
sys S1 (c : com2,1; d : data) ::
|[ var buf0, buf1 : data
   actions
     S1 : c = req0 → buf0 := d;
          d := d'. (d' ∈ data); c := ack
     S2 : c = req1 → buf1 := d;
          d := d'. (d' ∈ data); c := ack
   init c := ack
   exec   do S1 [] S2 od
]|
```

```
sys P_to_p () ::
|[ var c : com2,1; d : data
   subsys
     Master : M1 (c, d)
     Slave : S1 (c, d)
   init c := ack
   exec   Master || Slave
]|
```

Point-to-multipoint. In the case of a *point-to-multipoint* channel a single master is connected to several slaves so that one slave is accessed at a time. Such operation can be generally modeled using a communication variable c of the type $com_{N,1}$, where N is the number of the involved slaves. In other words, c has a distinct request value req_i for each slave and a common acknowledge value *ack* shared by the slaves. Note that a master accessing N slaves *simultaneously* is modeled by N separate point-to-point channels, not by a single point-to-multipoint channel.

As an example, consider the below system *P_to_mp* which contains an instance *Master* of the module $\mathcal{M}_1$, already defined above, and two instances *Slave*(0) and *Slave*(1) of the module $\mathcal{S}_2$ given below. Similarly as in the point-to-point case, the subsystems communicate via the biput channel $c\langle d \rangle$, where c is of the type $com_{2,1}$. However, now the channel can be viewed as a combination of two mutually exclusive subchannels one of which connects *Master* and *Slave*(0), the other *Master* and *Slave*(1), correspondingly. This means in practice that *Slave*(i), $i \in \{0, 1\}$, uses the values req_i and *ack* of c, as specified in the **subsys** clause of the system *P_to_mp*. *Master* initiates a communication cycle with either *Slave*(0) or *Slave*(1) by setting c to req_0 or req_1, respectively.

```
sys S₂ (c : com; d : data) ::
|[ var buf : data
   actions
     S : c = req → buf := d;
         d := d'. (d' ∈ data); c := ack
   init c := ack
   exec  do S od
]|
```

```
sys P_to_mp () ::
|[ var c : com_{2,1}; d : data
   subsys
     Master : M₁ (c, d)
     Slave(i) : S₂ (c(req_i, ack), d)
   init c := ack
   exec  Master || Slave(0) || Slave(1)
]|
```

Bus. A *bus* is a shared communication resource with several masters competing for the access to the slaves connected to the bus. At any given time only one master can use the bus. Hence, *arbitration* is needed to decide which master is to be allowed to access the shared communication resource. Due to this control, it is convenient to model a whole bus as a separate Action Systems module which takes care of the arbitration procedure and data transfer between masters and slaves.

The below module $\mathcal{B}us$ is a generic model of a simple bus supporting N_M (> 0) masters, which have the form of the module $\mathcal{M}_2$ specified below, and N_S (> 0) slaves which are assumed to be instances of the abstract slave module $\mathcal{S}_2$ defined above in the point-to-multipoint case. The new master model $\mathcal{M}_2$ includes, in contrast to the earlier $\mathcal{M}_1$, an interface variable (a) used as an address to select one of the N_S slaves. Notice that the slave count N_S is a generic parameter of $\mathcal{M}_2$, and both N_S and the master count N_M are generic parameters of the bus model $\mathcal{B}us$. The system $\mathcal{B}us\_sys$, also given below, is an example bus system configuration with $N_M, N_S = 2$, composed of the instances *Master*(0) and *Master*(1) of $\mathcal{M}_2$, the instances *Slave*(0) and *Slave*(1) of $\mathcal{S}_2$, and the instance *Bus* of the component $\mathcal{B}us$.

```
sys Bus (m[N_M] : com; a[N_M] : 0..N_S − 1; dm[N_M] : data;
         s : com_{N_S,1}; ds : data) [N_M, N_S : natural] ::
|[ actions
     Req(i) : m[i] = req → ds := dm[i]; s := req_{a[i]}
     Ack(i) : s = ack → dm[i] := ds; m[i] := ack
   init m, s := ack
   exec  do [ ⊕ 0 ≤ i ≤ N_M − 1 : (Req(i) ; Ack(i)) ] od
]|
```

```
sys M₂ (c : com; a : 0..N_S − 1; d : data)
      [N_S : natural] ::
|[ var buf : data
   actions
     M₁ : d := d'. (d' ∈ data);
          a := a'. (0 ≤ a' ≤ N_S − 1);
          c := req
     M₂ : c = ack → buf := d
   init c := ack
   exec  do M₁ ; M₂ od
]|
```

```
sys Bus_sys () ::
|[ var m[2] : com; s : com_{2,1}; a[2] : 0..1;
       dm[2], ds : data
   subsys
     Master(i) : M₂ (m[i], a[i], dm[i]) [2]
     Bus : Bus (m, a, dm, s, ds) [2, 2]
     Slave(i) : S₂ (s(req_i, ack), ds)
   init m, s := ack
   exec  Master(0) || Master(1) || Bus
           || Slave(0) || Slave(1)
]|
```

The module $\mathcal{B}us$ communicates with the attached masters via N_M point-to-point channels $m[i]\langle\, a[i], dm[i]\,\rangle$, $i \in \{0, \ldots, N_M - 1\}$, where

$m[i]$ is a communication variable of the type *com*, $a[i] \in \{0, \ldots, N_S - 1\}$ is a slave address provided by the master i, and $dm[i]$ represents data transferred between the master i and $\mathcal{B}us$. In fact, $m[i]$ is a biput channel with respect to $dm[i]$, as in the request phase (action $Req(i)$) data moves from the master to $\mathcal{B}us$, and in the acknowledge phase (action $Ack(i)$) from $\mathcal{B}us$ to the master.

Communication between $\mathcal{B}us$ and the attached N_S slaves takes place via a biput-type point-to-multipoint channel $s\langle\, ds\,\rangle$, where the communication variable s is of the type $com_{N_S,1}$, and ds represents data transferred between $\mathcal{B}us$ and a slave in both directions. We assume in this model that ds, and thereby also $dm[i]$ at the master side, not only represents actual computational data but also various control information, such as a memory address, a function/mode selector, or a slave response message, exchanged between a master and a slave during a bus transaction. $\mathcal{B}us$ arbitrates between the requesting bus masters and forwards the request of the granted master i to the slave $a[i]$ by setting the communication variable s to $req_{a[i]}$ in the action $Req(i)$. Note that arbitration is here explicitly modeled by the exclusive action composition operator '$\oplus$' in the **do-od** loop of the module $\mathcal{B}us$. The idea is that only one of the N_M communication sequences modeled by the action compositions $Req(i)$; $Ack(i)$, where $i \in \{0, \ldots, N_M - 1\}$, can be enabled at any given time.

The presented abstract bus model does not contain any priority control. This means that selection between requesting masters is completely arbitrary, independently of how long each master has waited for access to the bus. One way to solve this problem is to use a *fairness assumption* by which we can conclude that each requesting master will be *eventually* served. The other approach would be to increase determinism by implementing an appropriate *priority scheme* which would give a higher priority to certain masters, for example to those ones that have been waiting longer. A simple priority scheme can be added to the above $\mathcal{B}us$ module by introducing a local priority array $pr[N_M]$ of the type *natural*, with all elements initialized to 0, and then modifying the action $Req(i)$ to

$$
\begin{aligned}
&(m[i] = req) \wedge (\forall j\,.\,(0 \le j \le N_M - 1) \wedge (j \ne i)\,.\,pr[j] \le pr[i]) \rightarrow \\
&\quad pr := pr'\,.\,(\forall j\,.\,0 \le j \le N_M - 1\,.\,P_1 \wedge P_2 \wedge P_3); \\
&\quad ds := dm[i];\;\; s := req_{a[i]}
\end{aligned}
$$

where

$$
\begin{aligned}
P_1 &\;\widehat{=}\; j = i \Rightarrow pr'[j] = 0 \\
P_2 &\;\widehat{=}\; (j \ne i) \wedge (m[j] = req) \Rightarrow pr'[j] = pr[j] + 1 \\
P_3 &\;\widehat{=}\; (j \ne i) \wedge (m[j] \ne req) \Rightarrow pr'[j] = pr[j]
\end{aligned}
$$

Now the action $Req(i)$ can be enabled, i.e., is able to actually participate in arbitration, only if there are no higher priority requests $m[j] = req$ $(j \neq i)$ present such that $pr[j]$ is larger than $pr[i]$. The priority array is updated every time the action $Req(i)$ (for any i) is executed, according to the predicates P_1, P_2, and P_3 defined above. In other words, $pr[i]$ is initialized back to 0 (P_1), the elements $pr[j]$ corresponding to the pending requests that lost arbitration are incremented by one to increase priority of the requests in question (P_2), and the elements $pr[j]$ corresponding to the inactive requests are not changed (P_3). This control mechanism ensures, without any fairness assumptions, that every requesting master will get the access to the bus, and that the longer the master has been waiting for its turn the better chance it has to be selected.

Network. A conventional bus structure serves as a platform for rapid construction of complex systems-on-chip. However, it introduces a performance bottleneck, as parallel communication between system modules is not possible. To enable flexible design of high-performance systems, a more parallel, *network-like* communication platform is needed. In such a platform model, each system module or *processing element* has a unique address, and all modules can simultaneously access the communication medium.

An abstract template model $\mathcal{P}e$ of a processing element is given below. It has, as a generic parameter, an identification number $id \in \{0, \dots, N-1\}$, where N is the total number of processing elements in the system.

```
sys Pe (co : com; ao : 0..N − 1; dout : data;
        ci : com; ai : 0..N − 1; din : data) [N : natural; id : 0..N − 1] ::
|[ var dbuf : data; abuf : 0..N − 1
   actions
     Snd : co = ack → dout := dout′. (dout′ ∈ data);
               ao := ao′. (0 ≤ ao′ ≤ N − 1) ∧ (ao′ ≠ id); co := req
     Rec : ci = req → dbuf, abuf := din, ai; ci := ack
   init co, ci := ack
   exec   do Snd ∥ Rec od
]|
```

The module is composed of two actions *Snd* ("send") and *Rec* ("receive") which can be enabled simultaneously. $\mathcal{P}e$ acts as a master when sending data to the network by executing the action *Snd*, and as a slave when receiving data from the network by executing the action *Rec*. It has the point-to-point output port $co\langle\, ao, dout\,\rangle$, managed by *Snd*, where co is a communication variable of the type *com*, ao is a destination address between 0 and $N-1$, and *dout* represents data of any type to be sent to the destination element through the network. In the case of the point-to-point input port $ci\langle\, ai, din\,\rangle$, managed by *Rec*, ci is again a variable of the type *com*, and ai is a source address, i.e., the identification number of the processing element that has sent the incoming data *din*. The two communication ports, output and input, can work in parallel enabling fast data exchange through the interface.

The communication platform itself, the network, can be modeled as the below generic Action Systems module $\mathcal{N}et$ supporting N processing elements which are assumed to be instances of the above template model $\mathcal{P}e$. An example configuration with $N = 4$ is given below as the system $\mathcal{N}et\_sys$ which is composed of the instance *Network* of the module $\mathcal{N}et$, and the four instances $Host(i)$, $i \in \{0, \ldots, 3\}$, of the module $\mathcal{P}e$.

In general, the module $\mathcal{N}et$ has N point-to-point host ports, one for each attached host processing element i, $i \in \{0, \ldots, N-1\}$. Each host port is composed of the input port $pi[i]\langle\, pia[i], pid[i]\,\rangle$ and the output port $po[i]\langle\, poa[i], pod[i]\,\rangle$. The former is connected to the output port $co\langle\, ao, dout\,\rangle$ and the latter to the input port $ci\langle\, ai, din\,\rangle$ of the corresponding host element $\mathcal{P}e$. Hence, $pi[i]$ and $po[i]$ are communication variables of the type *com*, $pia[i]$ is the destination address of data $pid[i]$, and $poa[i]$ is the source address of data $pod[i]$.

sys $\mathcal{N}et$ $(pi[N] : com;\ pia[N] : 0..N-1;\ pid[N] : data;$
$\quad po[N] : com;\ poa[N] : 0..N-1;\ pod[N] : data)\ [N : natural]$ ::
|[**var** $tr[N,N] : com;\ db[N,N] : data$
actions
$Snd(i) :\ (pi[i] = req) \wedge (tr[i, pia[i]] = ack) \rightarrow db[i, pia[i]] := pid[i];\ \_$
$\quad tr[i, pia[i]], pi[i] := req, ack$
$Rec(i,j) :\ (tr[j,i] = req) \wedge (po[i] = ack) \rightarrow pod[i], poa[i] := db[j,i], j;\ \_$
$\quad po[i], tr[j,i] := req, ack$
init $pi, po, tr := ack$
exec **do** $[\ \| \ 0 \le i \le N-1 :\ (\,Snd(i)\ \|\ \ [\ \|\ (0 \le j \le N-1) \wedge (j \ne i) :\ Rec(i,j)]\,)\ ]$ **od**
]|

sys $\mathcal{N}et\_sys$ () ::
|[**var** $co[4], ci[4] : com;\ ao[4], ai[4] : 0..3;\ com_{2,1};\ dout[4], din[4] : data$
subsys
$Host(i) :\ \mathcal{P}e\,(co[i], ao[i], dout[i], ci[i], ai[i], din[i])\ [4, i]$
$Network :\ \mathcal{N}et\,(co, ao, dout, ci, ai, din)\ [4]$
init $co, ci := ack$
exec $Network \,\|\, Host(0) \,\|\, Host(1) \,\|\, Host(2) \,\|\, Host(3)$
]|

$\mathcal{N}et$ contains two parametrized actions in its **actions** clause: $Snd(i)$ and $Rec(i,j)$. As specified in the **do-od** loop of the module, each host port i is managed by the action $Snd(i)$, which is responsible of the input port $pi[i]$, and $N-1$ actions $Rec(i,j)$ with $j \in \{0, \ldots, N-1\} \setminus \{i\}$, which take care of the output port $po[i]$. Here the parameter j refers to all possible source ports sending data to the destination port i. The actions $Snd(i)$ and $Rec(i,j)$ can be simultaneously enabled, so that data is received from and sent to the host element i in parallel through $pi[i]$ and $po[i]$. Furthermore, all actions $Snd(i)$, for $i \in \{0, \ldots, N-1\}$, can execute in parallel, and thereby several, even all of the actions $Rec(i,j)$

for a given i can become enabled at the same time. The execution order of such simultaneously enabled $Rec(i,j)$ actions is completely arbitrary, modeled by the nondeterministic choice, unless a priority scheme is implemented in a similar manner as in the bus model presented earlier.

Data transfer from a host port i to the destination port $pia[i]$ is modeled by the internal point-to-point communication channel $tr[i, pia[i]]$ $\langle\, db[i, pia[i]]\,\rangle$, where $tr[i, pia[i]]$ is of the type *com*, and $db[i, pia[i]]$ is a copy of the data variable $pid[i]$, assigned by the action $Snd(i)$. The request sent by the master action $Snd(i)$ via $tr[i, pia[i]]$ is detected by the corresponding slave action $Rec(pia[i], i)$. This copies the value of the intermediate data variable $db[i, pia[i]]$ and the source port identification number i to $pod[pia[i]]$ and $poa[pia[i]]$, respectively, and sends then a request to the destination module $pia[i]$ through $po[pia[i]]$, and an acknowledgement to the action $Snd(i)$ through $tr[i, pia[i]]$. Observe that as there is a distinct channel $tr[i, pia[i]]\langle\, db[i, pia[i]]\,\rangle$ for each possible destination $pia[i]$, the port i can communicate with all destination ports in parallel, even though the communication cycles in question are activated by $Snd(i)$ one at a time according to the sequence of addresses $pia[i]$ provided by the host processing element i via the input port $pi[i]$.

3.2 Synchronous Communication

The main characteristic of synchronous devices is the employment of a common signal, the clock, which continuously switches between the high and the low values. All updates are performed in some kind of relation with this signal: either on the rising/falling edge or on the high or low level of the clock signal. Intrinsically, there is a time tag associated with the clock activity, thus a period denoting the time between two similar consecutive transitions. We start by giving a short description of the synchronous design framework within the Action Systems formalism.

Synchronous Action Systems. Synchronous Action Systems [3] is a timeless approach towards the representation of synchronous hardware devices. Briefly, let us consider two initial always enabled assignments, $A \mathrel{\widehat{=}} w_A := w'.Q_A$, $B \mathrel{\widehat{=}} w_B := w'.Q_B$. Their synchronous individual representation is reached if we split them into read and write parts – R_A, R_B, with the help of intermediate variables, u_A, u_B:

$$R_{A,B} \mathrel{\widehat{=}} u_{A,B} := v'_{A,B}.Q_{A,B}[u_{A,B}/w_{A,B}],\ W_{A,B} \mathrel{\widehat{=}} w_{A,B} := u_{A,B}$$

Adding another variable that models the clock signal, the boolean p, we have the synchronous versions:

$$S_{A,B} = \neg p \rightarrow R_{A,B}; p := true \,\|\, p \rightarrow W_{A,B}; p := false$$

Now we can further synchronously compose S_A and S_B by simultaneously executing their read and write parts:

$$S_A \nabla S_B \quad = \quad \neg p \rightarrow R_A \star R_B; p := true \parallel p \rightarrow W_A \star W_B; p := false$$

Notice in the above the behavior of $S_A \nabla S_B$: the inputs are read and an intermediate result is produced, after which the real update is performed on the output variables. The two phases, read and write are run in turns, based on the value of the boolean variable p, modeling the clock. However, at the highest level, the synchronous operator is used to hide the details of the above composition: if certain actions A and B are (or can be transformed into) *always enabled* assignments, it is enough to write $A \nabla B$ and we do not have to expose the underlying read / write phases and the clock modeling variable p. The composition of two synchronous action systems is done in a similar manner to the parallel composition (Section 12.2.3), that is, the actions of the composing systems are merged into a single synchronous composition.

If in the asynchronous approach our focus is on the input state, which determines the way a certain variable is updated, in synchronous design we will focus on deciding which variables are simultaneously updated. Following this, we define the manner in which they are updated, remembering that the resulting actions must be always enabled. In order to ensure this requirement, it is sufficient to specify the output considering all the possible input situations.

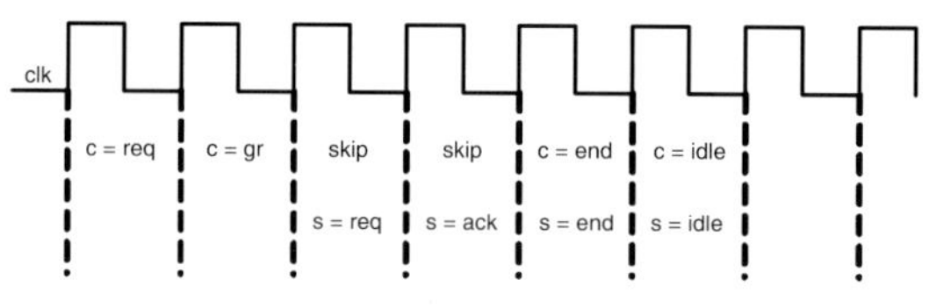

Figure 12.1. Single data processing, single clock period.

Communication model. All the characteristics of communication presented in previous sections stand valid for a synchronous model, too. However, in the following, we analyze just a very simple communication scheme involving a master, an arbiter and a slave (Fig. 12.2), all sharing the same clock signal. We only model the actions that update the communication variables. Thus, between master and arbiter we have a channel of the type $MastChan : com_{2,2}(req, end, gr, idle)$. Between the master and the slave, communication is performed along a channel of the type $SlChan : com_{2,2}(req, end, ack, idle)$. We assume that the operation performed by the slave on the input data is executed in one clock cycle. Then, a communication cycle is represented in Fig. 12.1, where *skip* stands for "no operation", or the preservation of the previous value (for instance: $c := c'.(c' = c)$).

```
sys Arbiter(c : MastChan) ::
|[ actions
    C : c := c'.(c = req ⇒ c' = gr)_
                ∧ (c = end ⇒ c' = idle)_
                ∧ (c ∉ {idle, req} ⇒ c' = c))_
  init c := idle
  exec   do C od
]|
```

```
sys Master(c : MastChan; s : SlChan) ::
|[ actions
    C : c := c'.(c = idle ⇒ c' = req)_
                ∧ ((c = gr) ∧ (s = ack) ⇒ c' = end)_
                ∧ (c ∉ {idle, gr} ⇒ c' = c))
    S : s := s'.((c = gr ∧ s = idle ⇒ s' = req)_
                ∧ (c = gr ∧ s = ack ⇒ s' = end)_
                ∧ (c ≠ gr ⇒ s' = s))
  init c, s := idle
  exec   do C ∇ S od
]|
```

```
sys Slave(s : SlChan) ::
|[ actions
    S : s := s'.((s = req ⇒ s' = ack)_
                ∧ (s = end ⇒ s' = idle)_
                ∧ (s ∉ {req, end} ⇒ s' = s))
  init s := idle
  exec   do S od
]|
```

Figure 12.2. Elements of a bus-based action system.

```
sys Bus_Sys() ::
|[ var c : MastChan; s : SlChan
  actions
    Master.C : c := c'.(c = idle ⇒ c' = req) ∧ ((c = gr) ∧ (s = ack) ⇒ c' = end)_
                    ∧ (c ∉ {idle, gr} ⇒ c' = c))
    Master.S : s := s'.((c = gr ∧ s = idle ⇒ s' = req) ∧ (c = gr ∧ s = ack ⇒ s' = end)_
                    ∧ (c ≠ gr ⇒ s' = s))
    Slave.S : s := s'.((s = req ⇒ s' = ack) ∧ (s = end ⇒ s' = idle)_
                    ∧ (s ∉ {req, end} ⇒ s' = s))
    Arbiter.C : c := c'.(c = req ⇒ c' = gr) ∧ (c = end ⇒ c' = idle)_
                    ∧ (c ∉ {idle, req} ⇒ c' = c))
  init c, s := idle
  exec   do Master.C ∇ Master.S ∇ Slave.S ∇ Arbiter.C od
]|
```

Figure 12.3. The system $\mathcal{B}us\_Sys$.

The synchronous composition of the three systems in Fig. 12.2 results in the system $\mathcal{B}us\_Sys \mathrel{\hat{=}} \mathcal{M}aster \nabla \mathcal{S}lave \nabla \mathcal{A}rbiter$. The flattened description of $\mathcal{B}us\_Sys$ is given in Fig. 12.3.

Within the context of the $\mathcal{B}us\_Sys$ system, we can rewrite:

$$Master.S \ \nabla \ Slave.S \leq Bus.S, \ \ Master.C \ \nabla \ Arbiter.C \leq Bus.C,$$

where the new actions and the corresponding new system are:

sys $\mathcal{B}us\_Sys^1()$ ::
$|[$ **var** $c : MastChan; s : SlChan$
actions
$Bus.C: \; c := c'.(c = idle \Rightarrow c' = req) \wedge (c = gr \wedge s = ack \Rightarrow c' = end)\_$
$\wedge (c = req \Rightarrow c' = gr) \wedge (c = end \Rightarrow c' = idle)$
$Bus.S: \; s := s'.((c = gr \wedge s = idle \Rightarrow s' = req) \wedge (c = gr \wedge s = ack \Rightarrow s' = end)\_$
$\wedge (s = req \Rightarrow s' = ack) \wedge (s = end \Rightarrow s' = idle) \wedge \; (s \notin \{req, end\} \Rightarrow s' = s))$
init $c, s := idle$
exec **do** $Bus.C \; \nabla \; Bus.S$ **od**
$]|$

Expansion to multiple masters. Naturally, a bus-based system does not include a single master and a single slave. By performing a data refinement step together with the introduction of a new local variable for the system $\mathcal{B}us\_Sys$, we can obtain a new representation that models the existence of several (N_M) masters in the system.

The data refinement step is performed on the communication variable c, which is expanded to a vector, $c[1..N_M] : MastChan$. The connection between the original c and the vector $c[1..N_M]$ is expressed by the abstraction relation:

$$R \mathrel{\widehat{=}} (c = idle \equiv \forall j \in \{1, \ldots, N_M\}.c[j] = idle) \wedge \; (c = req \equiv \exists j \in \{1, \ldots, N_M\}.c[j] = req)$$
$$\wedge \; (c = gr \equiv \exists j \in \{1, \ldots, N_M\}.c[j] = gr \wedge \forall k \in \{1, \ldots, N_M\} \backslash \{j\}.c[k] \neq gr)$$
$$\wedge \; (c = end \equiv \exists j \in \{1, \ldots, N_M\}.c[j] = end \wedge \forall k \in \{1, \ldots, N_M\} \backslash \{j\}.c[k] \neq end)$$

In the same step, as described in [3], we also introduce a new variable, GrM, that keeps track of the current master owning the bus. It is assigned every time the arbiter will grant a new master, and it is set to 0 after the current master ended its allowed transactions and releases the control of the bus. This means that there is no master granted control over the bus.

Using R and carefully placing references to GrM we can write

$$Bus.C \;\; \leq_R \;\; [\,[\!]\; 0 \leq i \leq N_M - 1 : GrM = 0 \vee GrM = i \rightarrow Bus.C(i) \; \nabla \; Bus.G(i)],$$
$$Bus.S \;\; \leq_R \;\; Bus^1.S,$$
$$\mathcal{B}us\_Sys^1 \;\; \sqsubseteq \;\; \mathcal{B}us\_Sys^2,$$

where

sys $\mathcal{B}us\_Sys^2()[N_M : natural]$ ::
$|[$ **var** $c[1..N_M] : MastChan; s : SlChan; GrM : 0..N_M$
actions
$Bus.C(i): \; c[i] := c'.(c[i] = idle \Rightarrow c' = req) \wedge (c[i] = gr \wedge s = ack \Rightarrow c' = end)\_$
$\wedge (c[i] = req \Rightarrow c' = gr) \wedge (c[i] = end \Rightarrow c' = idle)$
$Bus.G(i): \; GrM := g.((c[i] = gr \Rightarrow g = i) \wedge (c[i] = end \Rightarrow g = 0)\_$
$\wedge (c[i] \notin \{gr, end\} \Rightarrow g = GrM))$
$Bus^1.S: \; s := s'.((c[GrM] = gr \wedge s = idle \Rightarrow s' = req)\_$
$\wedge (c[GrM] = gr \wedge s = ack \Rightarrow s' = end)\_$
$\wedge (s = req \Rightarrow s' = ack) \wedge (s = end \Rightarrow s' = idle) \wedge \; (s \notin \{req, end\} \Rightarrow s' = s))$
init $c, s := idle$
exec
do $[\,[\!]\; 0 \leq i \leq N_M - 1 : GrM = 0 \vee GrM = i \rightarrow Bus.C(i) \; \nabla \; Bus.G(i)] \; \nabla \; Bus^1.S$ **od**
$]|$

The communication scheme corresponding to the system $\mathcal{B}us\_Sys^2$ is illustrated in Fig. 12.4.

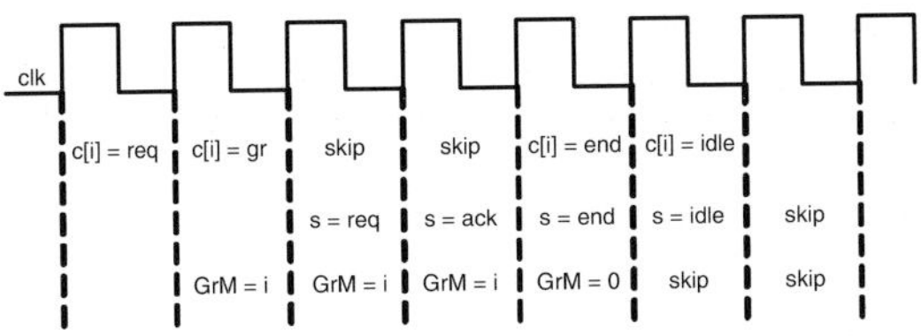

Figure 12.4. Single data processing, single clock period.

Observe that we can extract the new arbiter from the above system $\mathcal{B}us\_Sys^2$. We obtain:

sys $\mathcal{A}rbiter^1(c[1..N_M] : MastChan; GrM : 0..N_M)[N_M : natural]$ ::
|[**actions**
$Bus.C(i) :\ c[i] := c'.(c[i] = idle \Rightarrow c' = req) \wedge (c[i] = gr \wedge s = ack \Rightarrow c' = end)\_$
$\wedge\ (c[i] = req \Rightarrow c' = gr) \wedge (c[i] = end \Rightarrow c' = idle)$
$Bus.G(i) :\ GrM := g.((c[i] = gr \Rightarrow g = i) \wedge (c[i] = end \Rightarrow g = 0)\_$
$\wedge\ (c[i] \notin \{gr, end\} \Rightarrow g = GrM))$
init $c := idle$
exec **do** $[\,\|\ 0 \le i \le N_M - 1 : Bus.C(i)\ \nabla\ Bus.G(i)]$ **od**
]|

4. Refinement of Communication Channels

As a correctness-preserving step towards an implementable system model, each abstract communication channel is refined into a more concrete form. This means that the communication variables of the type $com_{m,n}$ and the other channel variables are turned into a collection of concrete boolean variables or signals by the means of *data refinement* (Section 12.2.4). In order to carry out such a transformation, an appropriate application-specific *abstraction relation* R has to be specified for each channel. It defines how the abstract channel variables are replaced with the concrete ones in the resulting system. As an example, implementation of a communication variable $c : com_{m,n}$ requires at most m request signals $req_0, \ldots, req_{m-1} : bool$ and n acknowledge signals $ack_0, \ldots, ack_{n-1} : bool$ which are related to the values of the original variable c by an abstraction relation of the form

$$R \,\widehat{=}\, \bigwedge_{i=0}^{m-1} Rr_i \ \wedge\ \bigwedge_{j=0}^{n-1} Ra_j$$

with

$$\begin{array}{lcl} Rr_i & \widehat{=} & (c = req_i) \equiv Fr_i(req_0, \ldots, req_{m-1}, ack_0, \ldots, ack_{n-1}) \\ Ra_j & \widehat{=} & (c = ack_j) \equiv Fa_j(req_0, \ldots, req_{m-1}, ack_0, \ldots, ack_{n-1}) \end{array}$$

where Fr_i and Fa_j are application-specific functions on the boolean handshake variables of the refined channel. For instance, if c is of the simple type *com*, we could have:

$$Rr \;\widehat{=}\; (c = req) \;\equiv\; (req \wedge \neg ack) \quad , \quad Ra \;\widehat{=}\; (c = ack) \;\equiv\; (\neg req \vee ack)$$

indicating that the request value of c corresponds to the unique state with req high and ack low, while the acknowledge value corresponds to the three other possible value combinations of the handshake variables req and ack.

In general, a channel transformation is a non-trivial system-level operation which involves all modules attached to the channel in question and includes adding new actions to these modules. It can be proven correct using a condition of the form $R \wedge I$, where R is the defined abstraction relation, and I represents other invariace properties and constraints, specified for the system, which are to be preserved by the refinement step.

5. Example: Specification of a System Bus

In this section, we consider a more concrete on-chip bus specification and its refinement. The bus is inspired by AMBA AHB [11], an existing open bus standard targeted for SoC design. However, not all features of the AMBA AHB are included in the presented specification. While the AMBA bus is synchronous, the bus presented here is asynchronous, based on self-timed signaling and aimed for globally-asynchronous locally-synchronous design, where bus masters and slaves are locally clocked entities. Because of the level of abstraction and the asynchronous structure, some of the original AMBA bus signals are not explicitly modeled. On the other hand, asynchronous communication requires some new signals not present in the synchronous AMBA AHB.

The description we show below is not the starting point in designing such a system. If we begin from an initial specification similar to the abstract bus representation given in Section 12.3.1, by incremental refinement steps we introduce the AMBA-like control variables and protocol. These steps are not shown; instead we describe how by applying a data-refinement procedure we reach an even more concrete level of representation, where one channel is implemented with boolean signals.

5.1 Specification

The Action Systems model of the target bus is given below as the module $\mathcal{B}us$ supporting N_M masters and N_S slaves. It is composed of two subsystems: $\mathcal{A}rbiter$ and $\mathcal{S}witch$, also specified below, which are instantiated as the component instances *Arb* and *Sw* within $\mathcal{B}us$. The $\mathcal{A}rbiter$ module takes care of deciding which requesting master may access the bus in a given moment. The $\mathcal{S}witch$ module, in turn, deals with passing signals sent by the granted master to the addressed slave, and vice versa. At the master side, $\mathcal{B}us$ has the ports $m[i]\langle\, Lck \,\rangle$, $i \in \{0, \ldots, N_M - 1\}$, through which the N_M masters request access and signal the completion of a bus transaction or a data burst, and the port $sm\langle\, a, adm, dm, Wm, Rspm \,\rangle$, shared by all the masters, through which the granted master sends and receives data and control information to and from the selected slave. Correspondingly, at the slave side, $\mathcal{B}us$ has the port $s\langle\, ads, ds, Ws, Rsps, Mst \,\rangle$ through which the selected slave receives and sends data and control information from and to the granted master. In addition, the slaves share the auxiliary interface variable *Splt* of $\mathcal{B}us$, via which the slaves can enable those bus requests at the ports $m[i]$, sent by the masters, that have been temporarily blocked (masked).

```
sys 𝓑us (m[N_M] : mchanT; Lck : bool; sm : com; a : 0..N_S − 1; adm : natural;
        dm : int; Wm : bool; Rspm : respT; s : com_{N_S,1}; ads : natural;
        ds : int; Ws : bool; Rsps : respT; Mst : 0..N_M − 1; Splt[N_M] : bool)
      [N_M, N_S, N_G : natural; G[N_G] : set of 0..N_M − 1] ::
|[ subsys  Arb :  𝓐rbiter (m, Lck, Rsps, Mst, Split) [N_M, N_S, N_G, G[N_G]]
           Sw :  𝓢witch (sm, a, adm, dm, Wm, Rspm, s, ads, ds, Ws, Rsps) [N_S]
   init m := idle;  sm, s := ack;  Splt := false
   exec  Arb ∥ Sw
]|
```

```
sys 𝓐rbiter (m[N_M] : mchanT; Lck : bool; Rsps : respT;
            Mst : 0..N_M − 1; Splt[N_M] : bool)
          [N_M, N_G : natural; G[N_G] : set of 0..N_M − 1] ::
|[ var mask[N_M], disable[N_M], reserve : bool
   expressions Enable(i, j) : (∀k . (k ∈ G[j]) ∧ (k ≠ i) . ¬disable[k])
   actions
     Grant(i, j) :
       (m[i] ∈ {rq, rql}) ∧ (¬mask[i] ∨ Splt[i]) ∧ Enable(i, j) ∧ ¬reserve → ₋
         Lck, Mst, mask[i] := (m[i] = rql), i, false; ₋
         reserve, disable[i] := true, false;  m[i] := gr
     ChkResp(i) : m[i] ∈ {done, end} → (Split(i) ▯ Retry(i) ▯ OkErr(i));  reserve := false
     Split(i) : Rsps = split → mask[i] := true;  Req(i)
     Retry(i) : Rsps = retry → disable[i] := true;  Req(i)
     OkErr(i) : Rsps ∈ {okay, error} → ₋
         (m[i] = done → (disable[i] := true;  Req(i)) ▯ m[i] = end → m[i] := idle)
     Req(i) : ¬Lck → m[i] := rq ▯ Lck → m[i] := rql
   init mask, disable, reserve, Splt := false;  m := idle
   exec
     do [// 0 ≤ j ≤ N_G − 1 :  [ ▯ i ∈ G[j] :  (Grant(i, j) ▯ ChkResp(i))]] od
]|
```

```
sys 𝓢witch (sm : com; a : 0..N_S − 1; adm, ads : natural; dm, ds : int;
            Wm, Ws : bool; Rspm, Rsps : respT; s : com_{N_S,1};) [N_S : natural] ::
|[ actions
     ReqW : sm = req ∧ Wm → ds, ads, Ws := dm, adm, Wm;  s := req_a
     AckW : s = ack → Rspm := Rsps;  sm := ack
     ReqR : sm = req ∧ ¬Wm → ads, Ws := adm, Wm;  s := req_a
     AckR : s = ack → dm, Rspm := ds, Rsps;  sm := ack
   init sm, s := ack
   exec do (ReqW ; AckW) ▯ (ReqR ; AckR) od
]|
```

The mentioned interface variables are defined as follows:

• $m[N_M] : mchanT$. The communication variables between the masters and $\mathcal{A}rbiter$. The type $mchanT$ is defined by **type** $mchanT$: $com_{4,2}(rq, rql, done, end, gr, idle)$, where the meaning of the 6 values is the following. rq: *bus request*, set initially by the master and restored by $\mathcal{A}rbiter$ after receiving the $done$ value assigned by the master; rql: *request for a locked transfer* in which data is transferred between a master and a slave as a continuous burst, set initially by the master and restored by $\mathcal{A}rbiter$ after receiving the $done$ value assigned by the master; $done$: *bus transaction completed*, set by the granted master; end: *data burst completed*, set by the granted master; gr: *bus grant*, set by $\mathcal{A}rbiter$ as the response to the rq or rql request; $idle$: *communication initialized*, initial state and set by $\mathcal{A}rbiter$ as the response to the end value.

• $sm : com$; $s : com_{N_S,1}$. The communication variables between the masters and $\mathcal{S}witch$ (sm), and between $\mathcal{S}witch$ and the slaves (s). The value req of sm is set by the granted master and the value ack by the $\mathcal{S}witch$ as the response to the acknowledgement on s. The value req_a of s, where a is the address of the selected slave, is set by $\mathcal{S}witch$ as the response to the request on sm, and the value ack of s is assigned by the selected slave.

• $Lck : bool$. The transfer mode indicator which is $true$ only if the granted master requested a locked transfer, set by $\mathcal{A}rbiter$.

• $Mst : 0..N_M - 1$. The identification number of the master currently accessing the bus, set by $\mathcal{A}rbiter$ and read by the slaves.

• $Wm : bool$. The transfer direction indicator which is $true$ in the case of a write operation (from a master to a slave) and $false$ in the case of a read operation (from a slave to a master), set by the granted master.

• $a : 0..N_S - 1$. The slave address, set by the granted master.

• $adm, ads : natural$. The address for accessing a memory location in the address space of the selected slave. adm is set by the granted master, ads is a copy of adm assigned by $\mathcal{S}witch$.

• $dm, ds : int$. Data to be transferred during a bus transaction. In a write operation, dm is assigned by the granted master and ds by $\mathcal{S}witch$ (dm is copied to ds). In a read operation, ds is set by the selected slave and dm by $\mathcal{S}witch$ (ds is copied to dm).

• $Rsps, Rspm : respT$. The slave response variables. $Rsps$ is set by the selected slave, and $\mathcal{S}witch$ copies $Rsps$ to $Rspm$ which is read by the granted master. The type $respT$ is defined by **type** $respT$: $(okay, error, retry, split)$, where the meaning of the 4 values is the following. $okay$: *bus transaction completed successfully*, operation proceeds normally; $error$: *bus transaction failed*, data is skipped and the granted master decides whether to ignore the failure or to abort the current transfer burst; *retry*: *slave busy – retry requested*, data is skipped and the transaction is asked to be tried again as soon as possible. *split*: *slave busy – split transfer requested*, data is skipped and the transaction cannot be resumed before the slave has explicitly allowed the involved master to proceed.

• $Splt[N_M]$: *bool*. Split transfer control array assigned by the slaves and read by *Arbiter*. In the case of a *split* response from a slave, the request of the involved master i is masked within *Arbiter*. By setting $Splt[i]$ to *true*, the slave overrides the mask in question allowing the master i to participate in arbitration again.

Arbiter. The subsystem *Arbiter* has two main actions for each master i in its **do-od** loop: $Grant(i, j)$, where j is the number of the priority group the master i belongs to, and $ChkResp(i)$ which is composed of a set of subactions also specified in the **actions** clause. *Arbiter* supports N_G priority groups given as generic parameters $G[j]$ where $j \in \{0, \ldots, N_G - 1\}$. Each group is a set of master identification numbers $i \in \{0, \ldots, N_M - 1\}$. The masters mentioned in the set $G[0]$ have the highest priority, while the masters mentioned in the set $G[N_G - 1]$ form the lowest priority group. We assume here that $N_G \geq 1$. The priority control is explicitly modeled in *Arbiter* by the quantified prioritized composition // in the **do-od** loop. The idea is that the action $Grant(i, j)$ can be enabled only if there are no higher priority requests ppresent, i.e., if none of the actions $Grant(i', j')$, where $j' < j$, is enabled. The inner quantified construct is a nondeterministic choice between the compositions $Grant(i, j) \parallel ChkResp(i)$ for the masters i belonging to the same priority group $G[j]$, modeling arbitration between the requests sent by these masters.

The action $Grant(i, j)$ is responsible of granting the bus to the requesting master i from the priority group $G[j]$ by assigning the value *gr* to the communication variable $m[i]$. When executed, it also sets the interface variables *Lck* and *Mst* to the appropriate states and the local boolean control variable *reserve* to *true* disabling temporarily all N_M *Grant* actions. The action $ChkResp(i)$, in turn, is executed when the granted master i has completed a bus transaction and set $m[i]$ to either *done* or *end*. The value *end* ("burst completed") is actually possible only when the slave response *Rsps* is either *okay* or *error*. Then $ChkResp(i)$ sets $m[i]$ to the initial value *idle*. Otherwise, i.e, if $m[i] = done$ ("transaction completed"), $ChkResp(i)$ acts according to the value of the slave response signal *Rsps*, sets $m[i]$ back to the previous request state *rq* or *rql*, and initializes the variable *reserve* back to *false* allowing the appropriate *Grant* actions to compete for bus access again. If $Rsps = split$, the request of the granted master i is masked by setting the local array element $mask[i]$ to *true*. This disables the action $Grant(i, j)$ until the involved slave sets $Splt[i]$ to *true*, thus overriding the mask. If $Rsps \in \{retry, okay, error\}$, the *Grant* actions of the other masters belonging to the same priority group $G[j]$ as the master i are disabled by setting the local array element $disable[i]$ to *true* (the expression $Enable(i', j)$ becomes *false* for $i' \neq i$). Hence, the action $Grant(i, j)$ is guaranteed to be re-selected for execution, if there are no higher priority requests present. Both $mask[i]$ and $disable[i]$ are initialized to *false* on the next execution of $Grant(i, j)$.

The *Arbiter* module and the attached masters satisfy the following essential invariant I which states that only one of the communication variables $m[i]$, $i \in \{0, \ldots N_M - 1\}$, can have the value *gr*, *done*, or *end* in any given moment, i.e., only one master can be accessing the bus at a time:

$$I \;\widehat{=}\; (\forall i, j.\, (0 \leq i, j < N_M) \wedge (j \neq i).m[i] \in \{gr, done, end\} \Rightarrow m[j] \in \{rq, rql, idle\})$$

Switch. The subsystem *Switch* has a quite simple structure. In the case of a write operation ($Wm = true$), the action sequence $ReqW \;;\; AckW$ takes care of transferring data and control information from the granted master to the selected slave a, and the slave response message back to the master. In the case of a read operation ($Wm = false$), the action sequence $ReqR \;;\; AckR$ transfers control information from the granted master to the selected slave, and data along with the slave response message from the slave back to the master.

5.2 Channel Refinement

As a refinement example, consider the communication variables $m[i]$ of the subsystem *Arbiter*. Since all these N_M variables of the type *mchanT* are refined similarly, we can here focus on only one of them and denote it simply by m. The variable m has 6 possible values: *rq*, *rql*, *done*, *end*, *gr*, and *idle*, each corresponding to a certain phase of a communication cycle on m. In order to develop the *Arbiter* specification into a more concrete form, closer to the circuit level, we implement m using 4 separate *boolean* handshake variables or signals, namely *rq*, *rql*, *done*, and *gr*. They are related to the original communication variable m by the following abstraction relations, needed for proving the correctness of the transformation:

$$\begin{aligned}
R_1 &\;\widehat{=}\; (m = rq) \equiv (rq \wedge \neg rql \wedge \neg gr) \\
R_2 &\;\widehat{=}\; (m = rql) \equiv (\neg rq \wedge rql \wedge \neg gr) \\
R_3 &\;\widehat{=}\; (m = done) \equiv ((rq \neq rql) \wedge done \wedge gr) \\
R_4 &\;\widehat{=}\; (m = end) \equiv (\neg rq \wedge \neg rql \wedge \neg done \wedge gr) \\
R_5 &\;\widehat{=}\; (m = gr) \equiv ((rq \neq rql) \wedge \neg done \wedge gr) \\
R_6 &\;\widehat{=}\; (m = idle) \equiv (\neg rq \wedge \neg rql \wedge \neg done \wedge \neg gr)
\end{aligned}$$

Hence, the refined communication channel between a master and *Arbiter* is considered idle, when all of its handshake variables are *false* (R_6). The channel is put to a request state when the master initially sets either *rq* or *rql* to *true*, or when *Arbiter* sets the grant signal *gr* to *false* while the master is keeping the request *rq* or *rql* high (R_1, R_2). *Arbiter* puts the channel to the grant state by setting the signal *gr* to *true* as the response to the enabled request signal *rq* or *rql* when the signal *done* is low (R_5). The done state indicates that the master has set the signal *done* to *true* while keeping the request *rq* or *rql* enabled, and the grant signal *gr* is still high (R_3). The end state is entered, when the master finally initializes the asserted request signal *rq* or *rql* back to *false* while the grant signal *gr* is high (R_4). As the response, *Arbiter* puts the channel to the initial idle state by setting *gr* to *false* (R_6). The signaling protocol on the refined channel is depicted in Figure 12.5.

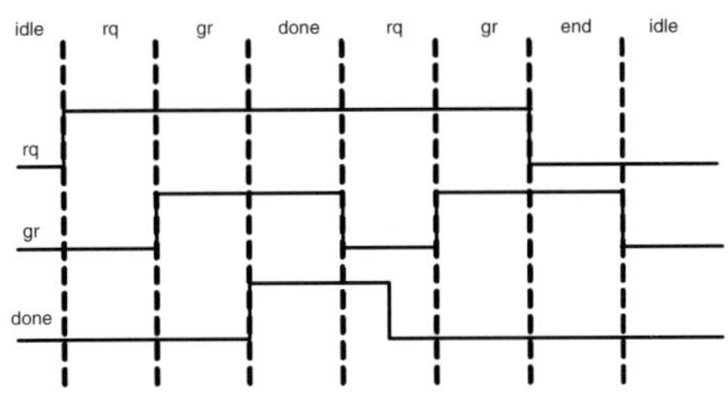

Figure 12.5. Signaling protocol.

The transformation changes the action *Grant*(i, j) of *Arbiter* into

$$\begin{array}{l}((rq[i] \vee rql[i]) \wedge \neg done[i]) \wedge (\neg mask[i] \vee Splt[i]) \wedge Enable(i,j) \wedge \neg reserve \rightarrow \\ \quad Lck, Mst, mask[i] := rql[i], i, false; \\ \quad reserve, disable[i] := true, false;\ gr[i] := true\end{array}$$

and the action *ChkResp*(i) into

$$(done[i] \vee (\neg rq[i] \wedge \neg rql[i])) \wedge gr[i] \rightarrow (Split'(i) \ [\!]\ Retry'(i) \ [\!]\ OkErr'(i));\ reserve := false,$$

where the primed subactions differ from the original ones in the included action *Req*(i) which is replaced in the refined actions with the simple assignment $gr[i] := false$. The subaction *OkErr*(i) is changed even more, so that the refined action *OkErr'*(i) is given as

$$Rsps \in \{okay, error\} \rightarrow \ (done[i] \rightarrow disable[i], gr[i] := true, false \ [\!]\ \neg done[i] \rightarrow gr[i] := false)$$

6. Conclusions

In this chapter we applied Action Systems formalism to the specification and refinement of communication channels between digital system components, within the larger concept of platform-based design. Issues such as abstraction level – reflected in the representation of the communication channels, and asynchronous as well as synchronous communication schemes were addressed. The refinement techniques were used to correctly transform the initial abstract specification into a more concrete form, closer to implementation level.

While the communication platform was analyzed in more detail, the actual computation units were only viewed as communication partners. Hence, their individual internal behavior can be analyzed and specified later, as a separate refinement process that preserves the external behavior established earlier.

References

[1] K. Keutzer, S. Malik, R. Newton, J. Rabaey, and A. L. Sangiovanni-Vincentelli. *System Level Design: Orthogonalization of Concerns and Platform–Based Design.* IEEE Trans. on CAD, 2000.

[2] J. Plosila. *Self-Timed Circuit Design - The Action Systems Approach.* Ph.D. Thesis, University of Turku, Finland, 1999.

[3] T. Seceleanu. *Systematic Design of Synchronous Digital Circuits.* Ph.D. Thesis, Åbo Akademi, Turku, Finland, 2001.

[4] R. J. R. Back, R. Kurki-Suonio. *Decentralization of Process Nets with Centralized Control.* Proc. of the 2nd ACM SIGACT–SIGOPS Symp. on Principles of Distributed Computing, p. 131–142, 1983.

[5] R.-J. R. Back, J. von Wright. *Trace Refinement of Action Systems.* In B. Jonsson and J. Parrow, ed., CONCUR'94: Concurrency Theory, 5th International Conference, LNCS, vol. 836. Springer-Verlag, 1994.

[6] E. Sekerinski, K. Sere. *A Theory of Prioritizing Composition.* The Computer Journal, VOL. 39, No 8, pp. 701-712. The British Computer Society. Oxford University Press.

[7] E. W. Dijkstra. *A Discipline of Programming.* Prentice-Hall International, 1976.

[8] R. J. R. Back, J. von Wright. *Refinement calculus: A Systematic Introduction.* Springer-Verlag. April 1998.

[9] R. J. R. Back. *A Calculus of Refinements for Program Derivations.* Acta Informatica, 26(6), p. 593-624, 1988.

[10] A. Peeters. Single-Rail Handshake Circuits. PhD Thesis, Eindhoven University of Technology, The Netherlands, 1996.

[11] ARM Limited. AMBA Specification (Rev 2.0), 1999.

Chapter 13

NETWORK-CENTRIC SYSTEM-LEVEL MODEL FOR MULTIPROCESSOR SOC SIMULATION

Jan Madsen
Shankar Mahadevan
Kashif Virk
Informatics and Mathematical Modeling, Technical University of Denmark
Richard Petersens Plads, building 322, DK2800 Kgs. Lyngby, Denmark
{jan,sm,virk}@imm.dtu.dk

1. Introduction

The primary goal of system-level modelling is to formulate a model within which a broad class of designs can be developed and explored. Ultimately, this allows designers to efficiently implement instances of systems within a single modelling style as supported by common design tools and methodologies. Figure 1 illustrates a possible System-on-Chip (SoC) design methodology, where the application, represented as a set of communicating tasks, is mapped onto a heterogeneous multiprocessor platform. Such a hardware platform can be developed either as part of the design process (core-based design), or configured from an existing reconfigurable platform (platform-based design). Both design scenarios allow for the implementation of parts of an application as dedicated processors (ASICs). On the other hand, as an increasing portion of applications is implemented in software which, in turn, is growing larger and more complex, dedicated (real-time) operating systems will have to be introduced as an interface layer between the application software and the hardware platform [4].

Therefore, the application has to be partitioned into sets of tasks which can then be mapped onto software or hardware, or a combination of both, while optimizing a number of non-functional design metrics, such as, performance, power consumption, resource utilization, reusability, and flexibility. This process of system-level design is known as the hardware/software codesign [20]. The codesign process depends not only

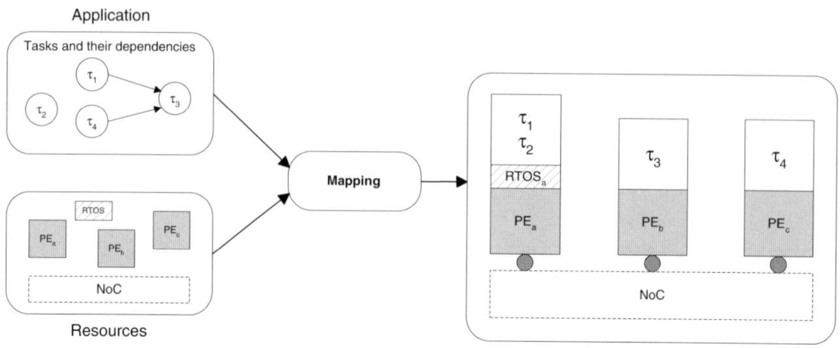

Figure 13.1. System-level SoC design methodology

on the type and the number of processors on a platform but on the topology and protocol of the inter-processor network as well [14].

It is important to realize that a Network-on-Chip (NoC) is a subset of a SoC. The architectural choices provide the set of the processing elements (PE's) required to service the tasks. The granularity (coarse or fine) and the placement (for example global vs. distributed memory) of the PE's in a SoC considerably impact its NoC design and the services demanded off it. On the other hand, the choice of a NoC solution may impact the selection and/or placement of the PE's. The general idea is to make the many available PE's to work together to achieve the desired result.

At the system-level, the details of the PE's and the NoC need to be abstracted in a way that allows for an accurate modelling of the global performance of the system, including the interrelationships among the diverse processors, software processes and physical interfaces and interconnections. To support the designer of single-chip based embedded systems, which includes multi-processor platforms running dedicated RTOS's as well as the effects of on-chip interconnect network, a modelling/simulation framework is required to support the analysis of:

- Network performance under different traffic and load conditions.
- Consequences of different mappings of tasks to processors (software or hardware).
- Effects of RTOS selection, including scheduling, synchronization and resource allocation policies.

An on-chip network model can provide provisions for run-time inspection and observation of the communication. Using this approach, implementations of the most promising network alternatives can be prototyped and characterized in terms of performance and overhead. Taking communication into account during hardware/software mapping is essential in order to obtain optimized solutions as emphasized in [14]. Also [13] and [21], show the importance of evaluating the communication media and how the choice of communication clearly impacts the overall architecture of a SoC.

In this chapter, we present a modelling framework which supports design space exploration at the system-level for mapping an application onto the architecture platform while giving a central role to the effects of the NoC. The rest of the chapter is organized as follows: In Section 2, we discuss various issues related to NoC modelling such as the requirements for modelling general network structures, the interface between the processing elements and the network, and the possible usages of a NoC model. Section 3 gives an overview of our abstract SoC model with emphasis on modelling the NoC, while Section 4 gives a detailed description of the implementation of the NoC model. In Section 5, we present a simple example to illustrate the capabilities of our NoC model. Finally, Section 6 gives a summary and concluding remarks.

2. Issues in NoC Modelling

Architecturally, an on-chip network is defined by its topology and the protocol running on it. The topology concerns the geometry of the communication links while the protocol dictates how these links are utilized. Many combinations of topology and protocol exist for an efficient communication of one or more predominant traffic patterns. For example, in [15], packet-switched NoC concepts have been applied to a 2-D mesh network topology whereas in [11], such concepts have been applied to a butterfly fat tree topology. While there are several mature methodologies for modelling and evaluating the PE architectures, there is relatively little research done to port the on-chip communication to the system-level. In [24], however, attempts have been made to fill this gap by proposing a NoC modelling methodology based upon ideas borrowed from the object-oriented design domain and implementing those ideas in Ptolemy II.

The performance of a network is closely connected to its architecture. Network performance is measured in quantitative terms, such as latency, bandwidth, power consumption, and area usage, as well as, in qualitative terms, such as network configurability (static or dynamic), quality of service (QoS), etc. Predictability of performance is necessary for NoC designers to take early decisions based on the NoC performance before actual implementation. Numerous studies have been done for deadlock-, livelock- and congestion-avoidance, error-correction, connection setup/tear-down, etc. to provide a certain predictable network behavior [7]. Even lower-level engineering techniques like low-swing drivers, signal encoding, etc., have been proposed to overcome network communication uncertainties [3, 5, 12]. Many of these network aspects are custom-tuned to fit the requirements of the application running on top of it.

Throughout this chapter, we use latency as the primary metric to ascertain the performance of a NoC. Network latency is defined as the time taken to move data from the source PE to the destination PE. It includes the message processing overhead, link delay, and the data processing delay at the intermediate nodes. Network latency is a function of the network topology (which determines the number of nodes and links comprising a network) and the communication protocol (which determines the processing requirements for routing and flow control). If two communicating tasks are allocated to two different processing elements, data will have to be transferred over a communication medium and the message transfer time will depend on the message size and the state of the network.

The state of an on-chip network at any instant is given by the number of actively transmitting PE's and the messages within its nodes and links. The state of a network dictates which resources of the network are currently in use and which ones can be available for future use. This provides a measure of the network services available to the system, which would affect its performance. We define *network services* as the system-level characterization of the network resource allocation and scheduling activities. For a given topology-protocol combination, the effect of changes in network services changes the resources available for a given communication event, thus, affecting its latency.

2.1 Network Aspects

Since most of the future embedded applications are likely to be real-time applications running on multiprocessor SoC's, the fundamental properties required of future NoC's to provide these services are: multi-hop, concurrency, and sharing. Although different on-chip networks manifest different subsets of the above-mentioned properties, a network should contain all of them in order to be a successful NoC.

- *Multi-hop* implies segmented communication in which communication events (majority of messages) pass through intermediate nodes while traversing from the source to the destination.

- *Concurrency* implies multiple simultaneous communications. It represents the ability of the network to successfully carry out more than one communication at the same time.
- *Sharing* implies quasi-simultaneous resource usage. It, inherently, allows many communication events to occupy some or all of the resources in an interleaved fashion.

Though defined separately, these properties are closely related to each other through the underlying topology and protocol implementations. If no direct path exists between two communicating PE's, then multi-hop is required. In multi-hop networks, links are connected via nodes. This, inherently, introduces sub-divisioning (segmentation) of the network. Many communication events can, thus, occur in different segments of the network allowing concurrency. Concurrency, in essence, allows co-existence of different communication events. Sharing requires that links be connected to multiple source and destination pairs at the same time. Sharing, essentially, allows the creation of multiple communication events in the network. Sharing can be either spatial (resource sharing) or temporal (time sharing).

Example 1: Consider three sample communication network topologies (see Figure 13.2): (a) fully-interconnected or point-to-point, (b) bus, and (c) mesh. When modelling a fully-interconnected network, illustrated in Figure 13.2(a), at the system level, the inter-task communication can be assumed to be negligible. This type of network has no multi-hop or sharing capability but it does allow concurrency. The reuse and scalability potential of such a network is limited. In the case of a bus network, as illustrated in Figure 13.2(b), the inter-task communication cannot be neglected. Therefore, the task graph of an application can be extended by the insertion of message tasks (τ_m's) which represent the transfer of data between tasks. Such a network is shared but does not allow multi-hop or concurrency. Since only one message transfer can take place at a time, the bus quickly becomes a critical resource.

As the number of processors is increased, a careful selection of task and communication scheduling policies is required when mapping tasks onto processors and deciding upon the system implementation either in software or in hardware. The fully-interconnected and the bus-based network models can provide the best-case and the worst-case performance limits, respectively, while carrying out an initial estimate of the communication requirements of an application. However, a network-on-chip solution requires more sophisticated modelling and design techniques in order to handle muti-hop communication where the message transfer

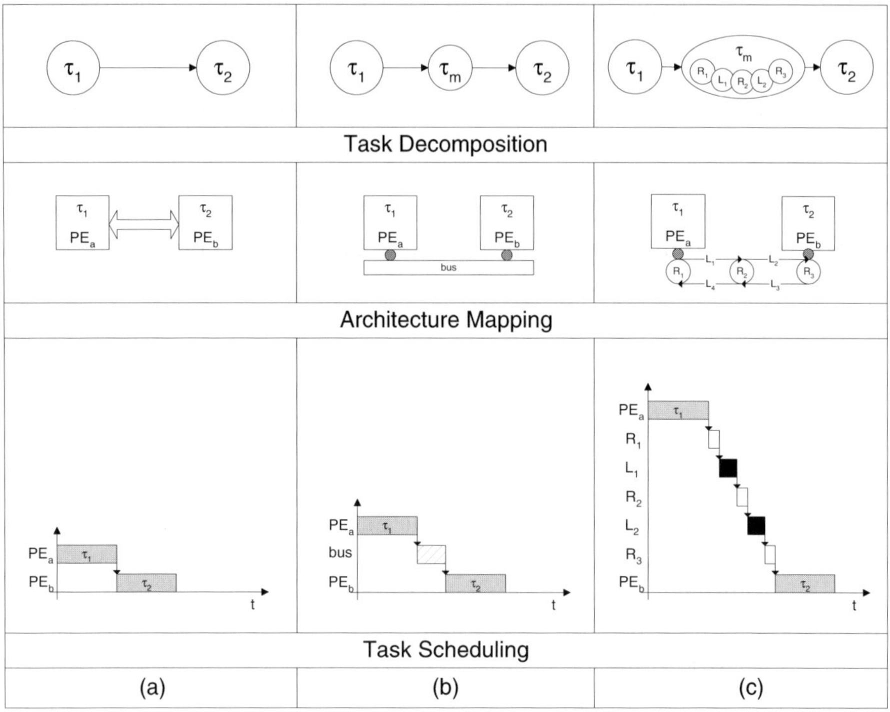

Figure 13.2. Communication modelling.

time may depend upon the traffic and the actual routing path through the network. In Figure 13.2(c), communication in a bi-directional mesh is shown. It has all the requisite network properties listed above and, therefore, it allows successful inter-task communication. □

2.2 Network Boundary Issues

For modelling purposes, it is important to define a precise boundary between the functionality of the NoC and the PE's. For example, in Figure 13.2(b), if a PE is responsible for performing communication tasks as well, then there is an overlap between the tasks running on the PE's and the NoC. Similarly, in Figure 13.2(c), it is possible to design the nodes (R_1, R_2, and R_3) to buffer messages before transmission. On the other hand, referring back to Figure 13.2(b), if a provision exists for the PE's to "dump" their messages for communication to some temporary location without actually performing the communication tasks, it effectively decouples computation from communication. This temporary location is called a network interface (NI).

The OSI [1] layered communication model is a convenient way to handle the complexity inherent in the inter-task communication. The network layer interfaces with the medium-access control layer and provides network access and message transmission services to the transport layer. When requested to send a message by a local task, the source transport layer places the message in the output buffer. From there, each outgoing message is delivered to the network under the control of the source network layer. The input and output buffers are jointly maintained by the transport layer and the network layer protocols. The transport layer interfaces with the tasks and provides them with the message transport services. After the message has traversed the network, the destination network layer places the message in the input buffer and notifies the destination transport layer. The destination transport layer then moves the message to the address space of the destination task and notifies the task of the arrival of the message.

The activities of sending a message can be represented by a chain of tasks. The source and the destination tasks are the predecessor and the successor of this chain of tasks, respectively. At the beginning and the end of the chain of tasks are the source and the destination transport protocol processing tasks. In between them, each task that accesses the network or transmits the message becomes ready for execution after its immediate predecessor completes, possibly with some delay introduced by the execution synchronization protocol used. In short, for modelling purposes, the various options for implementing these layers determine the NoC boundary [2]. In our model, these options boil down to two types of transitions in the task model:

- *Overlapped model:* In this model, a PE not only initiates communication but also contributes to set it up. Thus, the computation capability of a PE is utilized to carry out the communication processing functions like message encapsulation, header creation, encoding, etc. In the OSI context, the PE implements all the layers upto the physical layer to handle communication. As a result, an inter-task overlap occurs between the task generating a communication request on a PE and the communication task complying to such a request.

- *Triggered model:* In this model, a PE only triggers a communication event but the communication is actually handled by the NI. This frees up the computation capability of the PE for other tasks scheduled on it. An NI takes data from a PE, encapsulates it, and ensures its successful transmission through the NoC. In the OSI context, the PE implements all the layers upto the transport layer, while the NI implements the rest of the layers below it, including the transport layer, to handle communication. As a result, there is no overlap of PE tasks demanding communication services with the communication task performing communication.

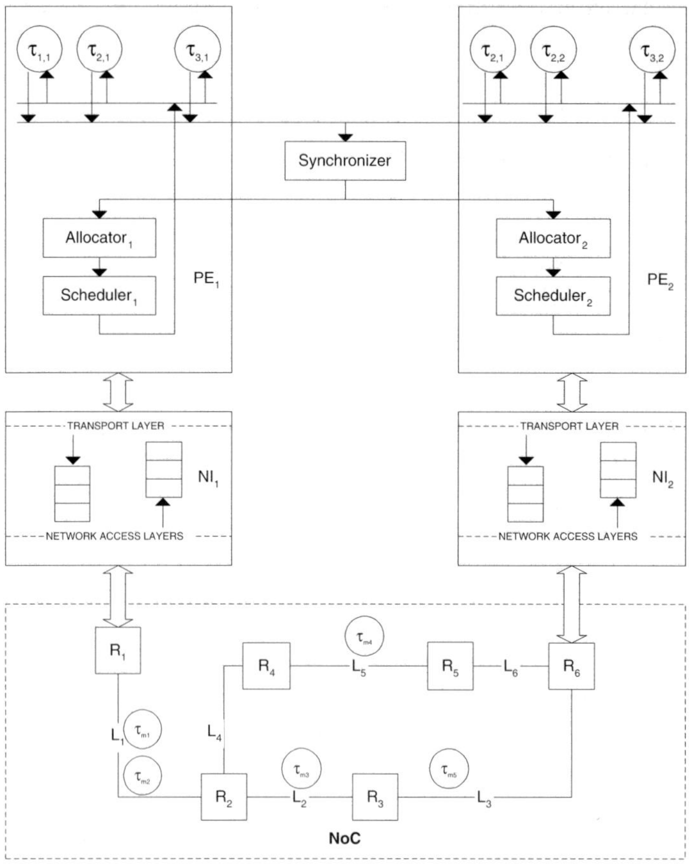

Figure 13.3. Network Interface.

From the above discussion, it is obvious that the choice of a NI can have global consequences on system-level scheduling (especially for large message sizes). When deciding the network boundary, it is important to break down the inter-task communication process into a sub-process that initiates communication and a sub-process that performs communication. Figure 13.3 illustrates this concept. For the overlapped model, the NI will be part of PE while for the triggered model, the NI is a separate entity as shown in the figure. The PE and the NoC models shown in this figure will be discussed later in the Section 4 of this chapter. The implementation of these sub-processes, within the perimeter of either a PE silicon or the NoC silicon, only impacts power and area considerations but not task scheduling. An NI introduces an additional complexity of resource management to the scheduling problem. Its implementation can range from just a set of wires, in a simple case, to a dedicated network processing capability with memory in a more complex case.

2.3 NoC Usage

Application design for embedded systems is a special challenge because embedded systems do not simply perform computations; they also interact with their environment. Thus the embedded systems must interface with and react to real processes. To achieve this goal, system designers must juggle real-time constraints, concurrency, and heterogeneity. The future SoC designers are faced with two major design challenges:

- *Platform Design:* Finding good solution templates for the architecture platform under the constraints and characteristics set by the semiconductor technology on one side, and the application domain in question on the other side.

- *Platform-based Design:* Given a platform architecture, how should it be configured (or instantiated) and how should the application, in terms of a description of multiple, concurrent processes, be mapped onto the platform while optimizing a number of design metrics, such as, performance, power consumption, memory utilization, and size, reusability, and flexibility.

Platform-based design is an efficient way to design complex system-on-chip products. It follows a meet-in-the-middle approach, starting with a functional system specification and a predesigned SoC platform. Performance estimation models can help analyze different mappings between the functional modules of the application and the platform components. During these iterations, designers can try different platform customizations and functional optimizations.

Example 2: As the number of components in architecture platforms increases, the type of on-chip interconnection and communication schemes for processing elements, memories, and peripherals becomes important.

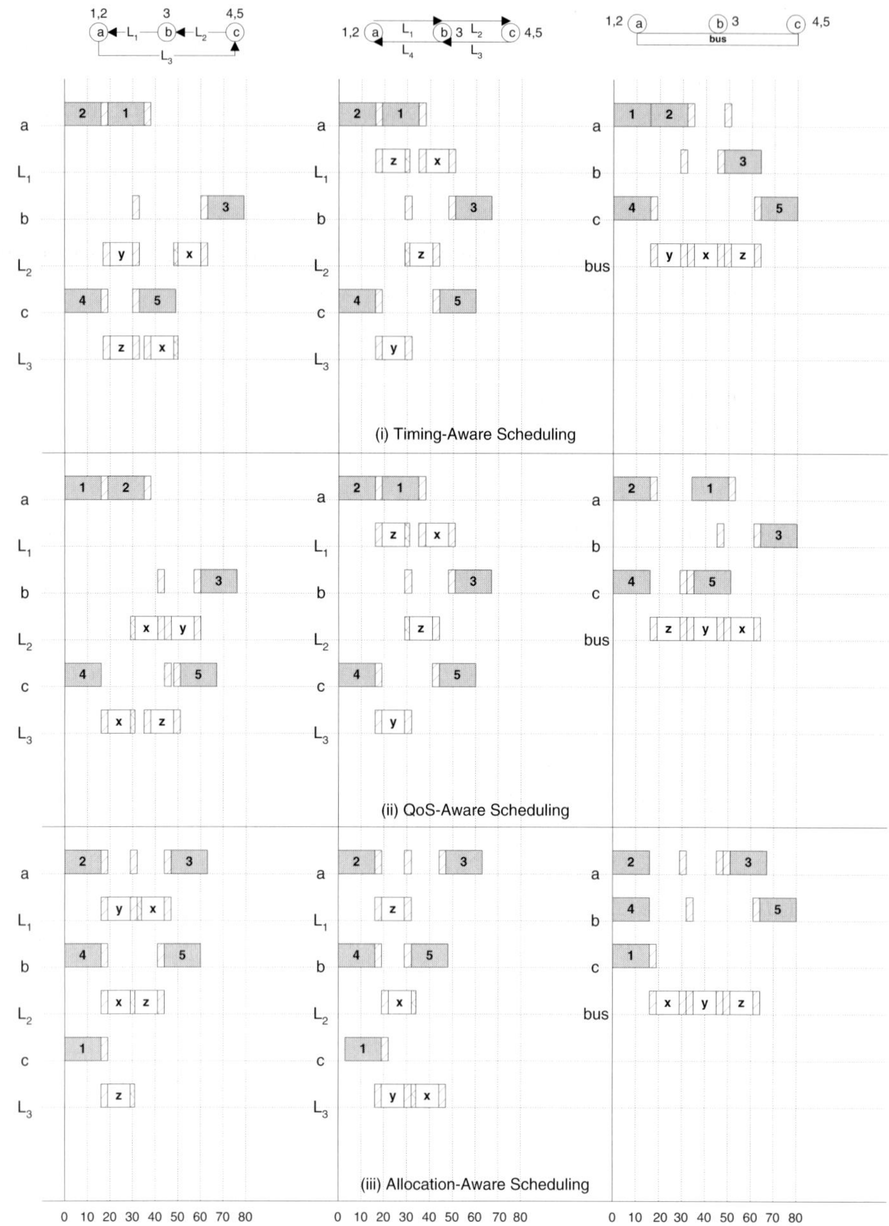

Figure 13.4. System-level modelling illustration ([18] ©2003 IEEE)

Figure 13.4 illustrates a simple example explaining how NoC modelling can facilitate design-space exploration. In this example, we consider a system that can use three sample network topologies (1-D torus, 1-D mesh, and bus), each interconnecting three processing elements $\{PE_a, PE_b, PE_c\}$ executing five tasks $\{\tau_1, \tau_2, \tau_3, \tau_4, \tau_5\}$. The initial mappings are: $\{\tau_1, \tau_2\} \mapsto PE_a$, $\{\tau_3\} \mapsto PE_b$, $\{\tau_4, \tau_5\} \mapsto PE_c$ (where, $\mapsto$ means 'maps to'), and having dependencies: $\{\tau_1, \tau_4\} \prec \tau_3$, $\tau_2 \prec \tau_5$ (where, $\prec$ means 'precedes'). Each task dependency is manifested by the insertion of a communication task between the inter-dependent task pairs. Thus $\tau_1 \prec \tau_{mx} \prec \tau_3$, $\tau_4 \prec \tau_{my} \prec \tau_3$, and $\tau_2 \prec \tau_{mz} \prec \tau_5$ (where τ_{mx}, τ_{my}, and τ_{mz} represent message tasks).

For simplicity, all the tasks have been assigned the same period (T = 100), execution time (BCET[1] = WCET[2] = 15), and deadline (d = 100). The time is measured in absolute time units. The tasks are mapped in such a way that none of them misses its deadline. For the purpose of this example, the overlapped NoC model is used and the system performance criterion is defined to be the time when all the tasks finish their execution, i.e., the earlier all the tasks finish their execution, the better is the system performance. The results of this example are analyzed below. □

2.3.1 Timing-Aware Scheduling. In the first row of Figure 13.4, Timing-Aware Scheduling scheme is depicted. In this scheme, the tasks executing on a system using the bus or the torus topologies finish in 80 time units whereas if the same system is using the mesh topology, the tasks take 65 time units to finish. The link contention is resolved randomly. As a bus comprises a single link, so the NoC resource allocator/arbiter does not have much freedom in its allocation. Therefore, the communication tasks are scheduled sequentially. On the other hand, the torus and the mesh networks have multiple ways to allocate their resources and schedule message tasks. The full potential of the torus network to handle concurrent communication is not exposed very well in this example but for the mesh network, it can be seen that its performance is better than the other two networks by over 10 time units.

From the point of view of link usage, the torus and the mesh networks only use two out of three links and three out of four links respectively. So a possible network optimization can occur for these networks. On the other hand, if the tasks, τ_1 and τ_4, are scheduled concurrently on a

[1]BCET is the Best-Case Execution Time
[2]WCET is the Worst-Case Execution Time

torus network, there is a contention on link L_1. So the network has to be optimized accordingly to meet the timing requirements.

2.3.2 QoS-Aware Scheduling. In the second row of Figure 13.4, QoS-Aware Scheduling scheme is shown. In this scheme, the traffic originating from PE_a is assigned a higher priority and, therefore, it is allocated the contentious link whenever a link contention arises. In the case of a mesh network, there is no link contention, so it has no effect on system performance. However, in the case of a torus network, such a message scheduling scheme results in a performance gain of 5 time units. For a complex network with lots of nodes and links, such a performance gain can be significant (both for the torus and the mesh topologies). The bus architecture, on the other hand, would result in a communication bottleneck.

2.3.3 Task Allocation-Aware Scheduling. In the third row of Figure 13.4, we illustrate the effects of altering the allocation of tasks to the processing elements while selecting different networks. The new task allocation is: $\{\tau_2, \tau_3\} \mapsto PE_a$, $\{\tau_4, \tau_5\} \mapsto PE_b$, and $\{\tau_1\} \mapsto PE_c$ (the task dependencies are kept the same). Compared to the bus, the system performance advantage is significant with the segmented (the torus and the mesh) networks. The reasons for the poor system performance with the bus are the same as described above. From the point of view of link utilization[3], it is now higher with the torus and the mesh networks. Most of the links, though not all, are now used simultaneously without any contention.

2.4 Discussion

As mentioned earlier, the timing-, QoS-, and allocation-aware scheduling analysis of Figure 13.4 is based entirely on the finish-deadline of the task mapped to the allocated resource, where the resource is either a PE or a network resource, such as a node or a link. Additional quantifications such as memory, area and power are also possible to incorporate into the model. For example, if the power consumed per communicated-bit is assumed equal, then the comparison of the power profile of the allocation-aware 1-D mesh with the timing-aware bus will show a power spike within the time duration of 20 to 40 time units for the 1-D mesh network while exhibiting a stable power profile for the bus. Even within

[3]Link Utilization is defined as the aggregation of the number of links occupied in one unit of time.

the 1-D torus network, the communication power profile down the column (in Figure 13.4) shows an aggregation which may be disadvantageous although link utilization has improved.

The main aim of this example exercise is to show how various options for performing design trade offs like resource types, resource allocation, task/message scheduling, etc. can be explored via the proposed NoC framework.

3. Framework for NoC Modelling

For the purpose of abstracting a system-level model, an embedded, real-time application can be represented as a collection of multiple, concurrent execution threads that are modelled as a set of dependent *tasks* under certain precedence and resource constraints which have to be executed on a number of programmable processors under the control of one or more RTOS(s). A system-level model can, thus, comprise three types of basic components: tasks, RTOS services, and communication network, where the communication network is meant to provide communication services between the other system components.

The RTOS services can be further decomposed into independent modules that represent different basic RTOS services like task *scheduling*, resource *allocation*, and execution *synchronization*, where a scheduler models a real-time scheduling algorithm; a synchronizer models the dependencies among the tasks and, hence, both the intra- and inter-processor communications; and an allocator models the mechanism of resource sharing among the tasks. A modeling framework provides the mechanism by which the various components comprising a model interact [16]. It is a set of constraints on the components and their composition semantics and, therefore, it defines a model of computation which governs the interaction of components [8]. Using a modeling framework, a system-level model can be composed from the basic components in such a way that the nature of services provided by any of the components can be altered in a simple and straightforward manner independent of the other components. In our discussion of the system-level modelling framework so far, we have not incorporated the effects of a NoC but we are going to consider that aspect now.

In order to communicate, the tasks executing on different processing elements generate *messages* and submit them to the *communication network* for transmission. The real-time, inter-processing element *traffic* consists of messages that are continuously generated by their sources and delivered to their respective destinations. Such traffic includes *periodic* and *sporadic* messages that require some degree of guarantee for on-time delivery. In addition, there are also *aperiodic* messages. Aperiodic messages have soft timing constraints and expect the system to deliver them on a best-effort basis.

Periodic Messages are generated and consumed by periodic tasks, and their characteristics are similar to the characteristics of their respective source tasks. Therefore, the transmission of a periodic message can be represented by a periodic message task. By a similar argument, the transmission of an aperiodic message can be represented by an aperiodic task. Although an aperiodic message task, like an aperiodic task, does not have a relative deadline, it is still desirable to keep the average delay suffered by aperiodic message tasks to be as small as possible. Sporadic message tasks have widely varying lengths and/or inter-arrival times. In general, sporadic messages represent burst communication and a sporadic message can be characterized in the same way as a sporadic task [17].

For the purpose of efficient transmission through the communication network, messages are fragmented into smaller-sized segments. The unit of data transmission at the network level is called a *packet.* Therefore, a message can be considered as a set of packets, where the packet size is bounded. Packet transmission is non-preemtive. Thus, a communication network can be modelled as a *communication processor* on which *message transmission tasks* are scheduled nonpreemptively on a fixed-priority basis. In this way, the effect of the inter-processing element communication is modelled automatically by the response times of the message transmission tasks on the network [17].

Modelling an on-chip communication network as a communication processor can reflect the demands on the network services. As a communication event within a network is modeled as a message task (τ_m) executing on the communication processor, therefore, when one PE intends to communicate with another PE, a τ_m is fired on the communication processor. Each τ_m represents communication between a set of two fixed, predetermined PE's only. Since a NoC supports concurrent communication, τ_m's need to be synchronized, allocated resources and scheduled accordingly. This reflects the property of the underlying NoC implementation, where the *NoC Allocator* reflects the topology and the *NoC Scheduler* reflects the protocol. Additional flow-control aspects, such as deadlock-avoidance, session-maintenance, acknowledge-based completion, etc. can also be implemented. Though the handling of those aspects either by the NoC Allocator or the NoC Scheduler depends upon the specific NoC architecture.

A resource database which is unique to each NoC implementation, contains information about all its resources. In a segmented network, these resources are laid-out as two-dimensional interconnects and comprise nodes (routers) and links. The algorithm for NoC allocation and scheduling map an τ_m onto the available network resources. The main focus of our discussion here is the networks which allow parallel communication, such as the segmented networks.

3.1 NoC Allocator

In a system-level NoC model, the role of the NoC Allocator is to translate the path requirements of the τ_m in terms of the resource requirements such as link bandwidth, storage buffers, etc. It has to minimize conflicts over the network resources. The links and the nodes in the communication path can be set aside dynamically (i.e., for the requested time-slot) in the resource database. If the resource reservation process is successful, the τ_m has to be queued for scheduling. When an τ_m releases a resource after usage, the resource is free to be assigned to another τ_m. However, if there is a contention over a resource, then resource arbitration has to occur. The NoC allocation patterns for two sample networks are shown in Table 13.1. The resource arbitration can be based on the underlying network implementation and will be discussed further shortly.

Table 13.1. A sample reservation for two sample networks ([18] ©2003 IEEE).

Message Task	Path	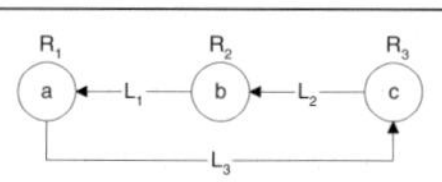			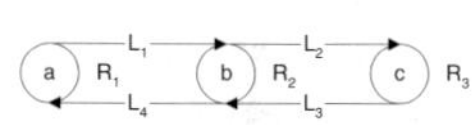	
		1-D Torus			1-D Mesh	
		Resource Allocation	Scheduling Needs		Resource Allocation	Scheduling Needs
			Small Message Size	Large Message Size		Small or Large Message Size
τ_{mx}	a→b	L_1	Immediate	Preemptive	L_1	Immediate
τ_{my}	c→b	L_3, R_1, L_1	Immediate	Immediate	L_3	Immediate

3.2 NoC Scheduler

Message Scheduling constitutes a basic function of any distributed real-time system. The scheduling of real-time messages aims to allocate the medium shared between several nodes in such a way that the time constraints of the messages are respected. As outlined above, not all of the messages generated in a distributed real-time application are critical from the point of view of time. Thus, according to the time constraints associated with the messages, the following message scheduling strategies can be applied:

- *Guaranteed Strategy*: According to this strategy, a message is always guaranteed to be delivered within its deadline requirements. This strategy is generally reserved for messages with critical timing constraints.
- *Stochastic Strategy*: In stochastic message scheduling strategy, the time constraints of messages are met in a best-effort fashion at a pre-computed probability. This strategy is used for messages with soft timing constraints.

In a distributed real-time system, the above strategies can cohabit, to be able to meet various communication requirements, according to the constraints and the nature of the communicating tasks.

As messages have similar constraints as tasks (mainly deadlines), the scheduling of real-time messages uses techniques similar to those used in the scheduling of tasks but with a difference. Whereas, tasks, in general, can accept preemption without corrupting the consistency of the end result, the transmission of a message does not admit preemption. If the transmission of a message starts, all the bits of the message must be transmitted, otherwise, the transmission fails [6].

The NoC Scheduler has to execute the τ_m according to the particular network service requirements. It has to minimize resource occupancy while making sure that the τ_m's are delivered within the specified timing bounds. In a network, resource occupation is dictated by the message size. The concept is better illustrated using the example in Table 13.1, where scheduling needs for the same two sample networks are shown. For a mesh there is no conflict. The τ_m's get the required resources scheduled "immediately". But in the case of the torus, it may experience resource allocation conflict for link L_1. Here, in the event of a small message size, where τ_{mx} is finished before τ_{my} asks for L_1, there is no scheduling problem. The resources can be "immediately" assigned to the τ_m's. But in the case of a large message size were τ_{mx} is still running when τ_{my} asks for the link L_1, resource contention occurs. Thus the resource L_1 is required to be scheduled "preemptively". Preemptively, here, implies the degree of contention resolution.

Example 3: Let us consider the resource conflict from the network-designer's and the system-designer's view point. At the network-level, where the resource conflict may be seen as a network problem, the network designer may over-design link L_1 by providing excess bandwidth or introduce processing overhead, such as TDM-based message interleaving. These techniques would restore fair servicing for both τ_m's, reducing the degree of contention. However, at the system-level, it may be possible to reschedule the communication event between the PE's (either τ_{mx} or τ_{my}). This opens the possibility of alternate path allocations for the τ_m's or simply stall one of the traffics until the other has finished. System designers may even realize that "large messages" (to the extent where L_1 is contentious) never occur within the given system. This could save potential scheduling/computation overhead in terms of hardware real-estate, power, etc. at router, R_1, and on link, L_1, as was envisioned by the network designer. Thus, when seen from the system-level, a trade-off between the NoC resource allocation and the NoC scheduling would not only complement better self utilization, but might give other useful insights for design improvements. □

In the following section, we present an implementation of the NoC model in SystemC [10, 22]. This implementation can be viewed as an extension of the implementation an abstract RTOS model described above.

4. NoC Model Implementation

The above-mentioned ideas about forming an abstract system-level NoC model can be validated by implementation in a system-level modelling language such as SystemC. We will briefly describe here the SystemC implementation of the NoC model described above. Further details regarding the implementation can be found in [9, 18, 19].

For the purpose of implementation, tasks are considered to be an abstract representation of the application and, therefore, have been characterized through a set of parameters, such as the worst- and the best-case execution times, context-switching overhead, deadline, period (if it is a periodic task), offset, resource requirements, and precedence relations. A task is modeled as a finite state machine (FSM) that sends the messages: `ready` and `finished`, to the scheduler which, in turn, sends one of the three commands to a task: `run`, `preempt`, and `resume`. In between the schedulers and the tasks, we have the synchronizer and the allocator acting as "logical command filters". By this scheme, each component can handle its relevant data independently of the other. For example, a task can determine when it is ready to run and when it has finished. The scheduler behaves in a reactive manner; scheduling tasks according to the indications received from them. Thus, as many tasks and schedulers can be added as desired. The same is the case with the synchronizer and the allocator models. They can hold the information regarding their services, i.e., which tasks depend on each other or, for the case of the allocator, what resources are needed by a given task.

The NoC model has exactly the same structure as the abstract RTOS model [9] but with some modifications. The message routing scheme implemented in the NoC model is that of fixed routing but the framework has provisions for implementing other routing schemes. The effects of the network interface or the task overlap due to message processing in the PE, as discussed in Section 2, have not been implemented for simplicity but the model can be extended to incorporate any of those effects.

4.1 Message Task

The message task (τ_m) has the same finite state machine (FSM) structure as the task model in the abstract RTOS model with some modifications to take out preemption and introduce resource requirements. The τ_m implementation accepts a number of arguments for its characterization.

- *Message Task ID*: enables the synchronizer and the NoC Scheduler to identify the τ_m sending the message.
- *NoC Scheduler ID*: is meant for the τ_m's to recognize their scheduler for exchanging various control messages.
- *Best Case Transmission Time (BCTT)*: is the lower bound on the transmission latency of a τ_m through the NoC.
- *Worst Case Transmission Time (WCTT)*: is the upper bound on the transmission latency of a τ_m through the NoC.
- *Offset*: is the setup time for a τ_m.
- *Resource ID*: is the ID tag for a resource (link, router, etc.) required by a τ_m.
- *Critical Section Length (CSL)*: the time duration for holding a resource.

The implementation of an τ_m can be viewed as an FSM that manages various counters after sending indications to the NoC Scheduler and the NoC Allocator and upon receiving commands from the NoC Scheduler.

4.2 NoC Allocator

The NoC Allocator manages its resource database upon receiving `request` and `release` indications from the τ_m's. The resources are

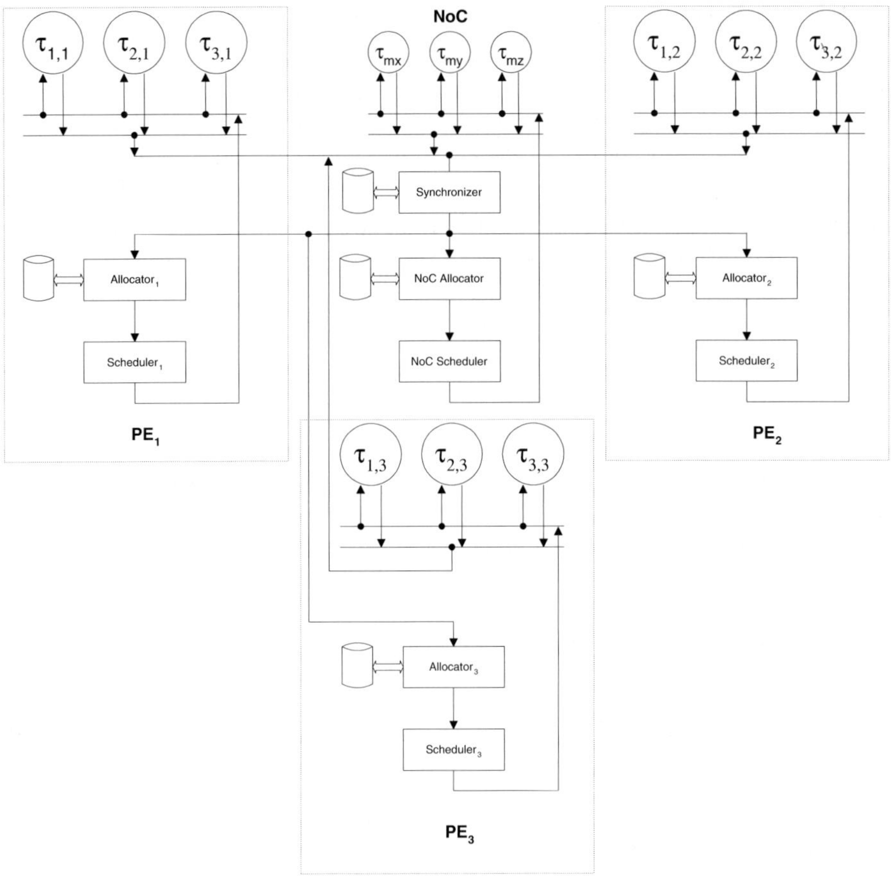

Figure 13.5. The system-level usage of the NoC model with the RTOS model ([18] ©2003 IEEE).

allocated to the τ_m's dynamically and they are released by the τ_m's immediately after usage. This makes resource management very flexible allowing sharing and concurrency. In this implementation, the resources are served by the NoC Allocator on a 'first-come-first' basis but other allocation policies can be implemented as well. Whenever a requested resource is available, the NoC Allocator sends a `grant` indication to the NoC Scheduler and whenever a requested resource is occupied, there is a resource contention and the NoC Allocator sends a `refuse` indication to the NoC Scheduler for appropriate action.

4.3 NoC Scheduler

The NoC Scheduler receives the `ready` and `finished` indications from the τ_m's through the Synchronizer and the `grant` and `refuse` indications from the NoC Allocator. It then issues the `run` and `buffer` commands to the τ_m's. Whenever a task running on a PE, is finished and needs to communicate with a task running on another PE, it sends a `finished` indication to the synchronizer which maintains a task dependency database and passes the `ready` indication for the corresponding τ_m to the NoC Scheduler which issues the `run` command to that τ_m.

Whenever there is a resource contention, the NoC Allocator issues a `refuse` indication to the NoC Scheduler which then either terminates the execution of the requesting τ_m (equivalent to a message dropping) or blocks the τ_m from execution (equivalent to message buffering) till the requested resource becomes available again which is indicated by the `grant` indication sent by the NoC Allocator to the NoC Scheduler. The message dropping or buffering decision is taken by the NoC Scheduler according to its underlying network implementation.

5. Simulation Results

The results of our SystemC implementation of the NoC model from Figure 13.5 are presented in Figure 13.7 and Figure 13.8. The sample SoC-NoC setup is as shown in Figure 13.6. Here, the application is assumed to be decomposed into four tasks. Three PE's are selected to execute these tasks. The task mapping is: $\{\tau_1\} \mapsto PE_a$, $\{\tau_4\} \mapsto PE_b$, and $\{\tau_2, \tau_3\} \mapsto PE_c$. τ_2 has a higher priority than τ_3, so it can preempt τ_3 on PE_c. In this example, we look at a simple case where all the tasks are modeled identically with a period of 25 time units (except for the τ_2 with a period of 24 time units due to the priority-assignment scheme in the Rate-Monotonic Scheduling), execution time (both BCET and WCET) of 10 time units, and finish deadline of 22 time units.

The communications between the tasks are modelled as τ_m's (as described in Section 4) which run on a torus network processor using store-and-forward routing protocol (with infinite buffer at the source and the destination nodes). The message task paths and dependencies are: τ_{mx}, from PE_a to PE_c using L_1, R_2 and L_2, and τ_{mz}, from PE_c to PE_b using L_3, R_1 and L_1. Thus, the link L_1 experiences possible contention. In our test SoC-NoC setup, these resources are tagged by an ID which is given in brackets (in Figure 13.6). We present two cases of interest.

In Figure 13.7(a), modelling of two concurrent communications is shown. As mentioned earlier, there is a link contention between τ_{mx} and τ_{mz} for L_1. It is resolved by scheduling L_1 at different times among

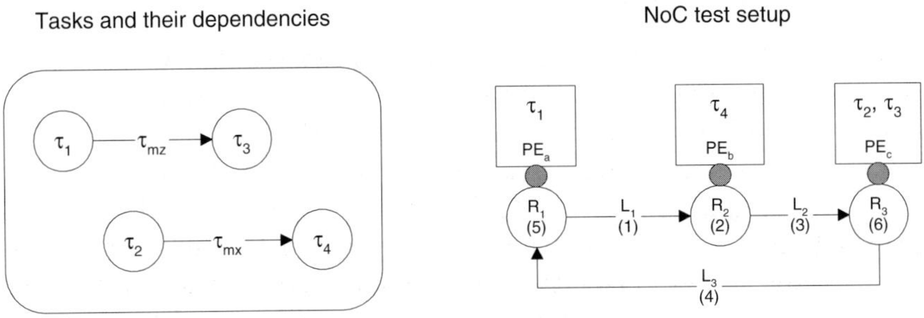

Figure 13.6. System simulation model ([18] ©2003 IEEE).

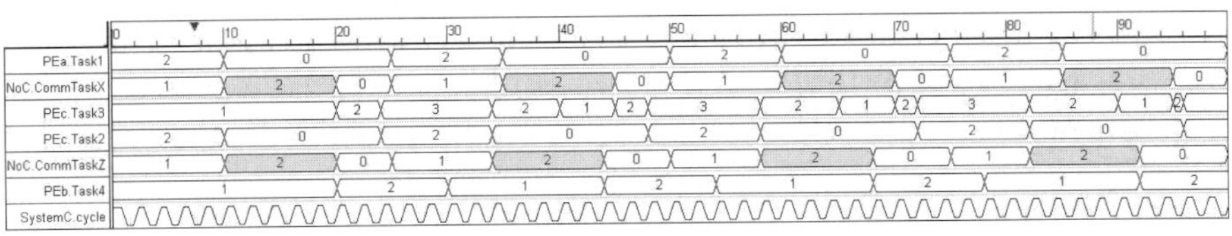

(a)

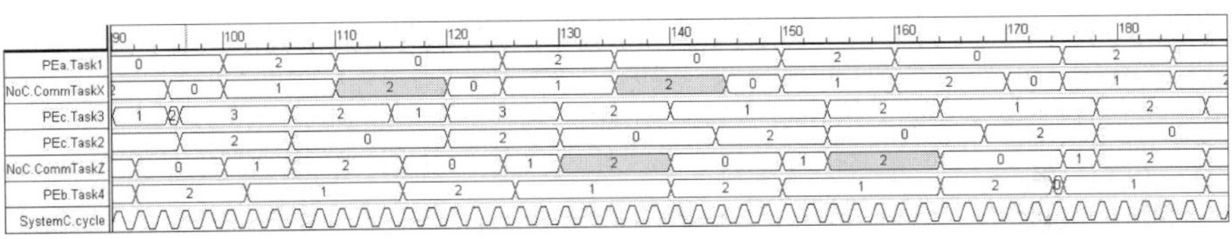

(b)

Figure 13.7. Simulation results for communication events (from 0 to 190 time units). State enumeration: 0=inactive, 1=ready, 2=running, 3=preempted ([18] ©2003 IEEE).

the τ_m's within the time-slot of 10 to 20 time units (and subsequent time slots). L_1 is used from 11 to 14 time units in τ_{mx} and from 17 to 20 time units in τ_{mz}. Figure 13.8 shows the log file of resource occupancy (Resource# 1, that is, the link L_1). The accompanying plots on the right provide a graphical representation (Note that 1 time unit is lost in network setup during simulation). Thus, our model clearly supports concurrent communication as observed in segmented networks.

```
0  Initializations
10 CommTask X released by the synchronizer
10 CommTask Z released by the synchronizer
11 task x (request resource# 1)-> allocator
11 NoC_allocator (granted)->NoC_scheduler
11 task z (request resource# 4)-> allocator
11 NoC_allocator (granted)-> NoC_scheduler
14 task x (release resource# 1)-> allocator
14 task x (request resource# 2)-> allocator
14 NoC_allocator (granted)-> NoC_scheduler
14 task z (release resource# 4)-> allocator
14 task z (request resource# 5)-> allocator
14 NoC_allocator (granted)-> NoC_scheduler
17 task x (release resource# 2)-> allocator
17 task x (request resource# 3)-> allocator
17 NoC_allocator (granted)-> NoC_scheduler
17 task z (release resource# 5)-> allocator
17 synchronizer (release)-> allocator
17 task z (request resource# 1)-> allocator
17 NoC_allocator (granted)-> NoC_scheduler
20 task x (release resource# 3)-> allocator
20 task x (finished)-> scheduler 2
20 synchronizer (finished)-> allocator
20 NoC_allocator (finished)-> NoC_scheduler
20 task z (release resource# 1)-> allocator
20 task z (finished)-> scheduler 2
20 synchronizer (finished)-> allocator
20 NoC_allocator (finished)-> NoC_scheduler

and so on...
```

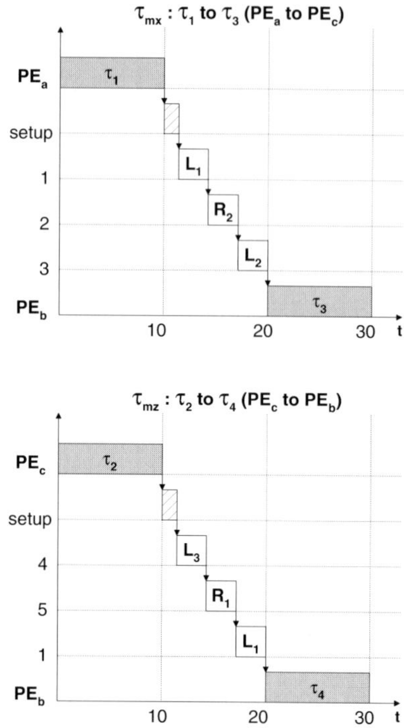

Figure 13.8. NoC allocation and scheduling for the first communication cycle along with the simulation log ([18] ©2003 IEEE).

Figure 13.7(b) shows the interplay of process modelling and interconnect activity. Consider the signal near the time period of 95 time units. Here, it is clear that τ_3 starts accepting the communication message and is then preempted by τ_2 on PE_c because of its higher priority. Once τ_2 is finished, τ_3 resumes and completes in time before deadline. Now consider the next execution of τ_3. Both τ_2 and τ_3 are in contention. The τ_3 does not even start instead, τ_2 starts on the PE_c. τ_3 here is not able to accept the message communicated to it by τ_1. This brings us to an interesting role of the NoC. In this simulation, we have enabled the routers to be able to buffer messages. Thus τ_{mx} finishes freeing up its resources although τ_2 has yet to begin. The τ_3, when finished, is, thus, able to initiate τ_{mz}, which is when τ_2 resumes.

Consider the case where the same torus network processor is running the wormhole routing (plots not provided). Then, in the preemption case, the τ_{mx} stalls, holding the link L_1. As τ_2 has already preempted τ_3 on PE_c, when it is complete, it would preempt τ_{mz}. But this would not be possible as the L_1 required here is busy in τ_{mx}, thus stalling τ_{mz}. This causes deadlock in the system. As seen earlier, we can resolve it either by introducing buffering in the routers or we have the freedom to choose an alternate network implementation or scheduling strategy. Thus, this example clearly demonstrates the global performance evaluation for co-design when both SoC and NoC are jointly modelled.

6. Summary

In this chapter, we have presented an abstract modelling framework based on SystemC which supports the modelling of multiprocessor-based RTOS's and their interconnection through a NoC. The aim is to provide the system designers of single-chip, real-time embedded systems with a simple modelling and simulation framework in which he/she can experiment with different task mappings, RTOS policies and NoC architectures in order to study the consequences of local decisions on the global system behavior and performance.

Our system model is based on a modular decomposition of the RTOS and NoC services, where information is shared through a sequence of events. This simple approach allows the behavior of each of the services (i.e., synchronization, allocation and scheduling) to be implemented independent of the others, and hence, allowing different model configurations with different services. A potential limitation of this approach is that information between the services is shared through a one-directional event flow. We have, however, successfully implemented the priority inheritance protocol, in which scheduling and resource allocation have to be tightly coupled [23]. Our system model assumes that tasks are allocated during compilation, i.e. we do not consider task migration where tasks are moved between the PEs during execution.

From a NoC point of view, the main limitation is that our framework supports only a single NoC instance. Hierarchical NoC's may be handled by appropriate implementations of the three services provided by the NoC. Any behavior that fulfills the event-based decomposition of services, can be implemented in our system model, i.e., it is possible to handle advanced routing protocols such as adaptive routing and virtual channels. In order to handle multiple NoC instances and hence, NoC's composed of different NoC types, we will have to introduce network bridges that share the same pool of message tasks. This however, should be rather straight forward as a network bridge can be viewed as a PE with very limited services; buffering and protocol conversion.

We have demonstrated the potential of our model by simulating and analyzing a small multiprocessor system connected through different

NoC architectures, showing multi-hop, concurrency and sharing. We have shown how the simulation model can be used for design space exploration.

Acknowledgements

The work described in this chapter has been funded by SoC-MOBINET (IST-2000-30094), an EU-funded project on System-on-Chip.

References

[1] L. Benini and G. D. Micheli. Network on Chips: A New SoC Paradigm. *IEEE Computer*, 35(1):70–78, January 2002.

[2] P. Bhojwani and R. Mahapatra. Interfacing Cores with On-chip Packet-Switched Networks. In *IEEE Proceedings on VLSI Design*, pages 382–387, January 2003.

[3] W. Brainbridge and S. Furber. Delay Insensitive System-on-Chip Interconnect using 1-of-4 Data Encoding. In *International Symposium on Asynchronous Circuits and Systems (ASYNC)*, pages 118–126, 2001.

[4] A. S. Cassidy, J. M. Paul, and D. E. Thomas. Layered, Multi-Threaded, High-Level Performance Design. In *Design Automation and Test in Europe, DATE*, pages 954–959, March 2003.

[5] J. Cong. An Interconnect-Centric Design Flow for Nanometer Technologies. In *International Symposium on VLSI Technology, Systems, and Applications*, pages 54–57, 1999.

[6] F. Cottet, J. Delacroix, C. Kaiser, and Z. Mammeri. *Scheduling in Real-Time Systems.* John-Wiley & Sons, 2002.

[7] D. E. Culler, J. P. Singh, and A. Gupta. *Parallel Computer Architecture: A Hardware/Software Approach.* Morgan-Kaufmann, 1998. 1st edition.

[8] S. Edwards, L. Lavagno, E. A. Lee, and A. Sangiovanni-Vincentelli. Design of Embedded Systems: Formal Models, Validation, and Synthesis. *Proceedings of the IEEE*, 85(3):366–390, March 1997.

[9] M. J. Gonzalez and J. Madsen. Abstract RTOS Modeling in SystemC. In *Proceedings of the 20th IEEE NORCHIP Conference*, pages 43–49, November 2002.

[10] T. Grotker, G. M. S. Liao, and S. Swan. *System Design with SystemC.* Kluwer Academic Publishers, New York, 2002.

[11] P. Guerrier and A. Greiner. A Generic Architecture for On-Chip Packet-Switched Interconnections. In *Design Automation and Test in Europe, DATE*, pages 250–256, March 2000.

[12] R. Ho and K. W. Mai. The Future of Wires. *Proceedings of the IEEE*, 89(4):490–504, April 2001.

[13] J-M. Daveau, T. B. Ismail, and A. A. Jerraya. Synthesis of System-Level Communication by an Allocation-Based Approach. In *Proceedings of the 8th International Symposium on System Synthesis (ISSS)*, pages 150–155, September 1995.

[14] P. V. Knudsen and J. Madsen. Integrating Communication Protocol Selection with Hardware/Software Codesign. *IEEE Transactions on Computer-Aided Design of Integrated Circuits and Systems*, 18(8):1077–1095, 1999.

[15] S. Kumar, A. Jantsch, J-P. Soininen, M. Forsell, M. Millberg, J. Oberg, K. Tiensyrja, and A. Hemani. A Network-on-Chip Architecture and Design Methodology. In *IEEE Computer Society Annual Symposium on VLSI*, pages 117–124, April 2002.

[16] E. A. Lee. What's Ahead for Embedded Software? *IEEE Computer*, 33(9):18–26, September 2000.

[17] J. W. S. Liu. *Real-Time Systems*. Prentice-Hall, 2000.

[18] J. Madsen, S. Mahadevan, K. Virk, and M. Gonzalez. Network-on-Chip Modeling for System-Level Multiprocessor Simulation. In *The 24th IEEE International Real-Time Systems Symposium*, December 2003.

[19] J. Madsen, K. Virk, and M. Gonzalez. Abstract RTOS Modelling for Multiprocessor System-on-Chip. In *International Symposium on System-on-Chip*, November 2003.

[20] G. D. Micheli, R. Ernst, and W. Wolf. *Readings in Hardware/Software Co-Design*. Morgan-Kaufmann, 2001. 1st edition.

[21] V. J. Mooney and D. M. Blough. A Hardware-Software Real-Time Operating System Framework for SoC's. *IEEE Design & Test of Computers*, 19(6):44–51, Nov/Dec 2002.

[22] SystemC Workgroup. http://www.systemc.org.

[23] K. Virk and J. Madsen. Resource Allocation Model for Modelling Abstract RTOS on Multiprocessor System-on-Chip. In *The 21st Norchip Conference*, November 2003.

[24] X. Zhu and S. Malik. A Hierarchical Modeling Framework for On-Chip Communication Architectures. In *International Conference on Computer-Aided Design (ICCAD)*, pages 663–670, 2002.

Chapter 14

SOCKET-BASED DESIGN USING DECOUPLED INTERCONNECTS

Drew Wingard
Sonics, Inc., Mountain View, CA, USA
wingard@sonicsinc.com

Abstract: Advances in deep sub-micron semiconductor process technologies offer the SoC designer the promise of more functionality than can be realized using existing tightly coupled architectures and EDA tools. As the ability to integrate IP cores increases linearly, the complexity of interconnecting these cores also increases, often geometrically. The overhead in design, integration, and verification of inter-core communications quickly becomes greater than the design savings in using pre-verified functional cores. The solution requires a matching of interconnect implementations to computational blocks. Socket-based design is a design methodology that can greatly reduce the time and effort expended on design and verification of complex SoCs. We will discuss the challenges of tightly coupled design, explain why decoupled interconnect design is essential, define socket-based design; explore the OCP socket specification and give examples of both processor-centric and I/O-centric SoC design using OCP-based sockets and decoupled interconnect cores.

Key words: socket-based design, IP reuse, smart interconnect IP, OCP socket standards, SoC design methodology, interconnect cores

1. INTRODUCTION

Today SoC (System-on-Chip) devices of greater than 100 million-gate complexity can be developed using 90-nanometer semiconductor process technologies. But existing design architectures, EDA tools and IP core-based design techniques limit the design potential of SoCs to much lower complexity levels, particularly for manageable design teams and timelines. Designing multifunction, system-level SoCs at this complexity level requires new design approaches and a radical shift in thinking. The chip design and EDA industry has historically focused on the development of functional blocks rather than the development of interconnects. But today, a shift must occur as interconnects become the gating factor in developing successful SoC devices. Interconnect architectures must now match the sophistication of functional core architectures if the industry is to realize the potential of deep submicron silicon process technology, a point noted by Mead and Conway in 1980 [1].

Historically, electronic system design has focused on functional design. When systems consisted of processor, memory and I/O devices on printed circuit boards, the interconnect was the passive connection between active chip-level functional blocks. Abstractions such as buses and control signals were used to group chip-to-chip communications into logical clusters. With improved process technology, integration of these devices became feasible and design methodology maintained the passive interconnect model from board-level design. At the level of one or two processing cores surrounded by a handful of I/O blocks and a single memory controller, design tools and designers are capable of managing the complexity of the total system on chip. However, as the integration of many heterogeneous processing units with tens to hundreds of I/O blocks becomes possible, the task of managing the entire design and the complexity of the interconnect schemes are overwhelming the design team and their traditional EDA tools. The focus of much of the industry remains only on the functional aspect of design. As this chapter will explain, the interconnect side of design must now evolve from passive and tightly coupled to active and decoupled in order to support the integration of large numbers of functional cores in a SoC device.

2. EVOLUTION OF DESIGN ABSTRACTIONS

Traditional approaches to managing system design complexity rely upon the creation of abstractions. Abstractions serve to both minimize the number of elements that are managed directly and to simplify the interactions between the elements. The refinement of electronic system design approaches that has led to SoC design has made significant use of design abstractions, particularly with respect to functional design. As shown in Figure 1, the level of functional abstraction has progressed from gate to block to core, and will be migrating to tile-based abstractions.

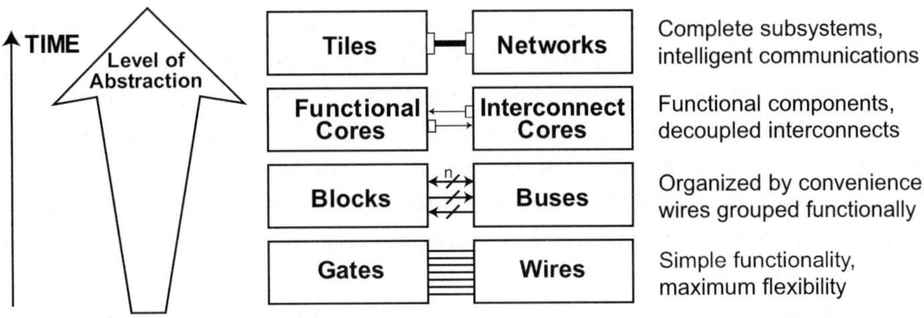

Figure 14-1. Evolving levels of design abstraction

Gates present the simplest functionality and have the least limitations in how wires are used to connect them. Tiles present the highest functionality and leverage a high-level network protocol to efficiently structure and optimize inter-core communications.

In between, blocks are designed using HDL rather than gates and have boundaries based on function and convenience that enable the placement of multiple, identical copies throughout a design, simplifying design and verification. Buses are used to group wires with similar functions together and provide some abstraction of communication through vectors of signals and simple protocols. Most interfaces between blocks have been evolved in ad-hoc fashion rather than having been designed for optimum system integration. Since the functional blocks are the key design element, communication between them appears free. At this level, integration is tightly coupled by definition and block functionality drives the design.

At the next level of abstraction, we would expect a reasonably sized functional core to solve a significant portion of the design problem transparently from the rest of the system. With multiple design teams developing cores and integrating them together, the design methodology must support independent and concurrent design. This abstraction must therefore embrace both intelligent functional partitioning and communications abstraction. Embedding communications logic into these functional cores defeats this purpose. The communications logic should be accumulated into a separate interconnect core. This active interconnect core manages the communication between the functional cores to meet the system level requirements. The result is that each functional core is decoupled from the system-level communications responsibilities.

The tile and network abstraction is above that of functional and interconnect cores. Tiles are defined [3] as complete functional subsystems that can be connected using on-chip networks. Such a tile will execute a set of functions with little or no dependence upon the rest of the device. These self-sufficient operating characteristics may typically be implemented in an embedded processor in the tile. A tile normally manages its own I/O locally. It communicates with other tiles via messages composed of data structures rather than through transactions composed of data words. Since tiles normally contain sufficient local memory for latency-critical traffic, on-chip networks offer and attractive interconnect for routing tile-to-tile messages. Tiles are either application specific, such as a GSM modem or a GPS receiver, or generic, such as a host processor for application software or a reconfigurable computing fabric. An example of a generic tile without a processor would be a shared memory subsystem that provides storage services to other tiles.

Elements at each level of abstraction are composed of elements from the lower abstraction. For instance, an interconnect core is composed of blocks and buses. The result of each abstraction is reduction in complexity by minimizing the number of objects, and their interdependencies.

2.1 Functional Cores and Buses – Mismatched Abstractions

Today, it is typical for the SoC architect to begin by defining the system-level functionality, partitioning the design according to known guidelines that define roughly what can be implemented in a single-chip design (area, power consumption, timing characteristics), and then passing the design on to the functional core designer, the interconnect specialist, and the physical layer designers to implement the final silicon device. Ideally, this approach should enable the rapid development of complex ICs using pre-verified functional cores and tiles. As the number of pre-verified cores increases over time, the process of developing systems should become simpler and faster by reusing these proven functional cores. Ideally, this should lead to "SoC design by feature combination." [2]

The reality is that as more functional cores are added to a design, the system completion time increases dramatically. The flaw lies in the increasing complexity of the inter-block interconnections and the lack of system-level modeling and design tools. It has been estimated that 30 to 70 percent of the design effort of a complex SoC design is now in designing and verifying the interconnect [4].

A useful way of understanding how this occurs is to examine the underlying abstractions used to implement an SoC and how they are mismatched in a typical design.

The reason to move to a higher level of abstraction is to isolate the designer from underlying implementation complexity so each abstracted element can be used independently. Thus a designer integrating functional cores should be isolated from the intricacies of blocks and buses. However, today's SoC designers typically attempt to integrate functional cores using passive buses. This requires the core designer to embed awareness of low level communications schemes within the functional core. Such an approach forces all functional cores to be "bus aware." Embedding communications logic into the functional core results in a fragile design element that is not independent of the intricacies of other cores or the system-level communications scheme. The embedded low level bus functionality forces the functional core to interact at its level, essentially turning the core into a highly complex, but still low level block element. This complexity is evident to the SoC integrator, who must then work at too low a level. The functional core abstraction has failed.

Because of this mismatch, there are numerous problems that occur at the core integration stage as shown in Figure 2. At each stage of integration and verification, designers discover system-level interactions that cause reengineering of the functional cores. At first integration, signaling and protocol mismatches are frequently detected due to poor specification of inter-core communications. When revalidating a core's functionality in the context of the integrated SoC, functional defects are discovered as transaction ordering and timing is discovered to be different that what was anticipated. The logic synthesis and timing analysis processes often uncover unexpected timing arcs that cross functional core boundaries. Later the placement and routing process, where the actual wire lengths between the functional cores are first discovered, uncovers further timing challenges. Finally, if the schedule permits, system-level performance verification, which attempts to verify that the intended application will perform properly on the SoC, is likely to uncover new problems. Each time these functional cores are re-engineered, their verification test benches, synthesis scripts, and test plans must be updated to match the revised core definition.

To achieve the benefits of concurrent SoC design, functional cores must become decoupled from each other. This will avoid the necessity of redesigning functional cores as the system-level communications are refined into the interconnect core.

Because of the problems of long design and verification cycles, the current SoC design approaches are no longer able to fully take advantage of the available on-chip gates. Some observers point to a current lack of front-end system-level design tools and the inability of existing front-end design tools to account for back-end physical layer realities at the beginning of the design as the cause of a growing design productivity gap [5]. This gap is defined as the difference between available on-chip gates and the actual gates used. Better tools may improve, but not solve this problem.

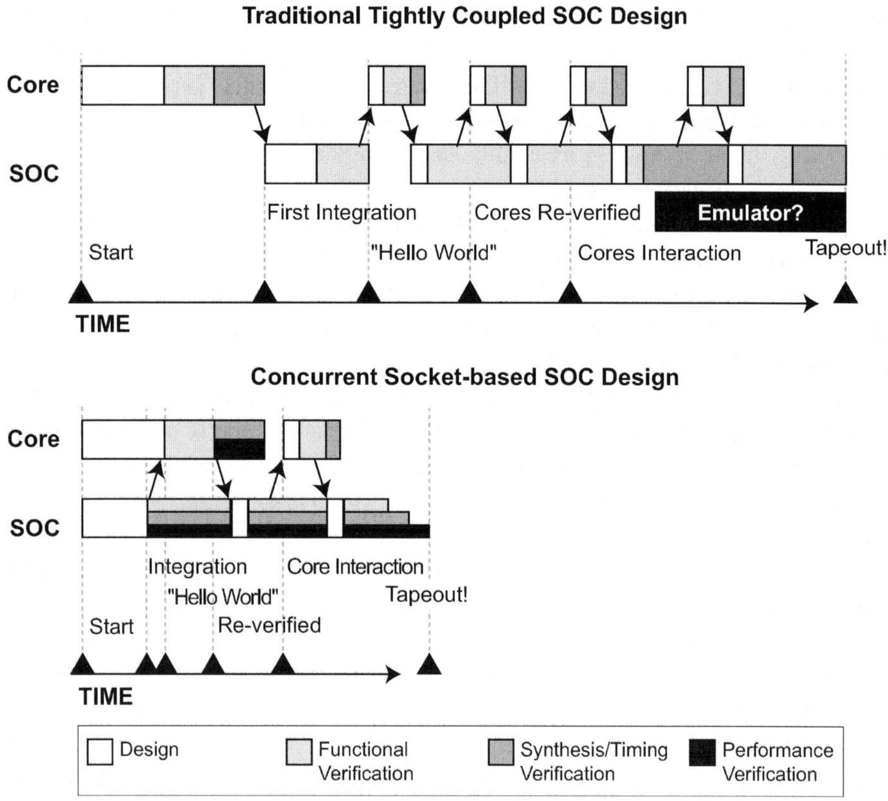

Figure 14-2. Comparison between tightly coupled and concurrent socket-based SoC design

2.2 Leveraging Functional Core and Interconnect Core Abstractions

The problems of long development times and the inability to realize on-chip gate potential can be resolved by eliminating the mismatch of functional cores with buses.

The most obvious solution is decoupling the computational tasks of the functional cores from inter-core communication tasks. The thesis of this chapter is that the practical approach to decoupling is to isolate the functional cores using a standard interface socket and to optimize inter-core communications using an active interconnect core. By applying these decoupled design techniques, a concurrent design methodology can be achieved using existing design tools and styles.

A concurrent design methodology can be implemented in the following manner. First, the design team can begin to separate the functional core from the intricacies of the inter-core communications task by adopting a standard functional core interface socket. Second, by shifting from a "passive wire" model of inter-core communications to an "active interconnect core," the designer can develop new interconnect cores that are independent of core functionality and that can be independently optimized for both local and system-wide communications. This decoupled approach will speed the development of complex SoCs, yield more highly optimized functional and interconnect cores, and greatly increase the potential for core reusability. This approach enables the benefits of NoC (Network On Chip) based platforms using the "design by feature combination" technique put forth by Jantsch and Tenhunen in 2003 [2].

A potential candidate for a standard functional core interface is the OCP (Open Core Protocol) interconnect specification. This flexible, scalable and well-defined socket is processor-agnostic and capable of supporting a wide range of interconnect schemes. Most importantly, it is enjoying growing and widespread support throughout the industry [6].

The shift in modeling from the passive wire to the active interconnect core has already begun as designers are realizing that interconnect cores can be the architectural underpinnings of modern multifunction and reusable SoC designs [7]. A decoupled interconnect-based architecture can deliver scalable latency and bandwidth where needed while enabling separate optimization of its component functional and interconnect cores. The rest of this chapter outlines the challenges of traditional tightly coupled design, describes the requirements for a standard socket, defines the goals of a decoupled interconnect core, uses the OCP socket standard as illustration of how functional cores can efficiently communicate, and presents two decoupled interconnect options as solutions to application specific requirements found in complex SoCs.

3. LIMITS OF TIGHTLY COUPLED DESIGN

At first glance, tightly-coupled design, where IP blocks or functional cores are tightly coupled to each other, would seem to be the most efficient solution to most design problems. One would expect the smallest silicon area and the highest performance when active IP cores are connected to their associated subsystems with the simplest of passive connections (wires) and no additional overhead. Certainly, this system-level interconnection scheme will offer maximum flexibility, as there are no limits on how the interconnection between subsystems takes place.

But these advantages come with some important limitations even in the simplest systems. The expertise gained in accomplishing a given interconnect task, e.g. interfacing one processor to a given bus structure, can be best leveraged in future designs only through shared experience between members of the design team. Put more generally, design team members must work closely together in order to achieve optimum results.

The essential fact limiting tightly couple design is that communication is no longer free – wires are no longer elastic, zero delay elements connecting cores or blocks. The increasing relative wire delays at the distances associated with connecting far-flung cores across the chip plays a major role in the performance of the system. The implication of wire delays approaching or passing cycle times is that the transformation of the functional block diagram into the physical floor plan has profound performance implications. Managing this transformation requires active interconnect cores that isolate functional cores from inter-core communication issues and enable accurate up-front modeling of communications.

In addition to the difficulties in resolving low level issues such as signal timing issues at inter-core boundaries, there are the challenges of managing a large number of inter-dependencies as the design grows in size and scale. Inter-dependencies are much harder to manage when design teams grow to multiple groups in multiple locations, and changes are more difficult to implement. Worse still, it becomes increasingly difficult to understand the complete set of inter-dependency constraints. Whenever one core is inter-dependent upon another, the dependency must be managed at the system level. Design complexity using tightly-coupled design techniques overwhelm the designer's mental capacity to understand the total design interactions.

As tightly coupled designs scale upward, it becomes increasingly difficult to create levels of abstraction that promote understanding or enable reuse. This leads to longer design times as poor models of the electrical, logical and protocol behaviors lead to verification issues that require many design iterations. At the SoC design level, there is no formalism, and thus no tools, available to address all of the interactions among tightly coupled cores. Design teams must deploy a better, more decoupled architecture.

As a result, in actual practice, tightly coupled design of systems deploying large numbers of heterogeneous processors is often the least efficient solution in terms of silicon efficiency, design time and cost, flexibility and re-use of IP. Because of the complex interactions and inter-dependencies of the electrical, logical and protocol components, design iterations intended to result in successive design refinement, actually result in successive system and core redesign.

4. TIGHTLY COUPLED VS. DECOUPLED DESIGN

We have seen that tightly-coupled designs are vulnerable to significant redesign due to electrical, logical, and protocol problems detected at each phase of integration and implementation. With such an approach, the redesign occurs inside the functional units, making IP core re-use highly unlikely. This section provides a comparison between tightly-coupled and decoupled design approaches.

Processor-centric on-chip bus protocols such as AMBA (ARM) [8] and CoreConnect (IBM) [9] are modeled on architectures designed for traditional printed circuit board-based systems. They feature predominately single processor-to-memory and processor-to-input/output device transactions. These blocking bus architectures do not support simultaneous, heterogeneous data traffic among multiple processing elements such as those found in modern SoCs. In particular, such architectures do not easily support isochronous data flows often found in networking, multimedia and digital consumer applications. In such applications it is common to find several processing elements (primary system processor, DSP unit, MPEG encoder/decoder, and multiple communication processors) accessing shared memory and I/O resources throughout the chip.

When developing a SoC using bus-centric interconnects, the resulting system is a mixture of processor subsystems linked together using computer buses and a collection of point-to-point links dependent upon the functionality of the on-chip elements as shown in Figure 3. The trouble with such schemes is that they exhibit behaviors (power consumption, signal delay and distortion, cross-talk and noise) that are extremely difficult to predict or model before the final physical layout stage. Since interconnects can be responsible for up to 90 percent of the global signal delay in a submicron design [10], this means that meaningful timing verification cannot begin until late in the design cycle.

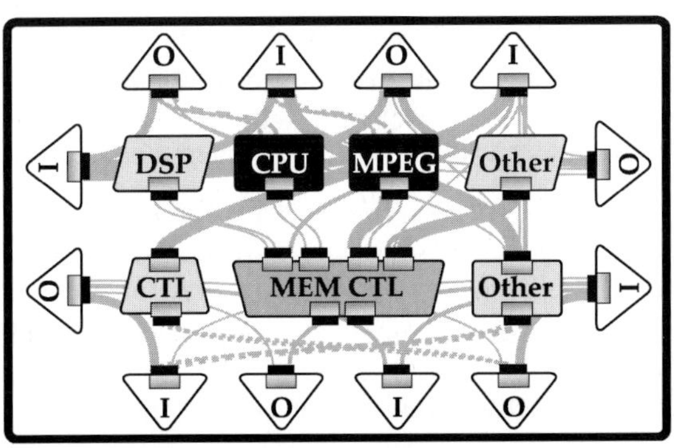

Figure 14-3. Passive interconnect scheme using multiple buses and point-to-point links

At the other end of the design spectrum, a socket-based design approach using active interconnect cores isolates the functional core from the details of the on-chip interconnect architecture as shown in Figure 4. This reduces design time by freeing the core developer to focus on core-related issues rather than system-level interconnect issues and by providing interconnect characterization that can be used to accurately model the entire SoC long before final physical layout begins. In addition, socket-based design results in highly optimized functional cores and tiles that can be easily reused in other systems with no additional redesign.

In the figure, the interconnect cores are represented as multi-drop connections between functional cores. Internal interconnect core topology is unspecified. The core sockets form the connection between the functional and interconnect cores. Tiles are represented as encircled collections of functional and interconnect cores.

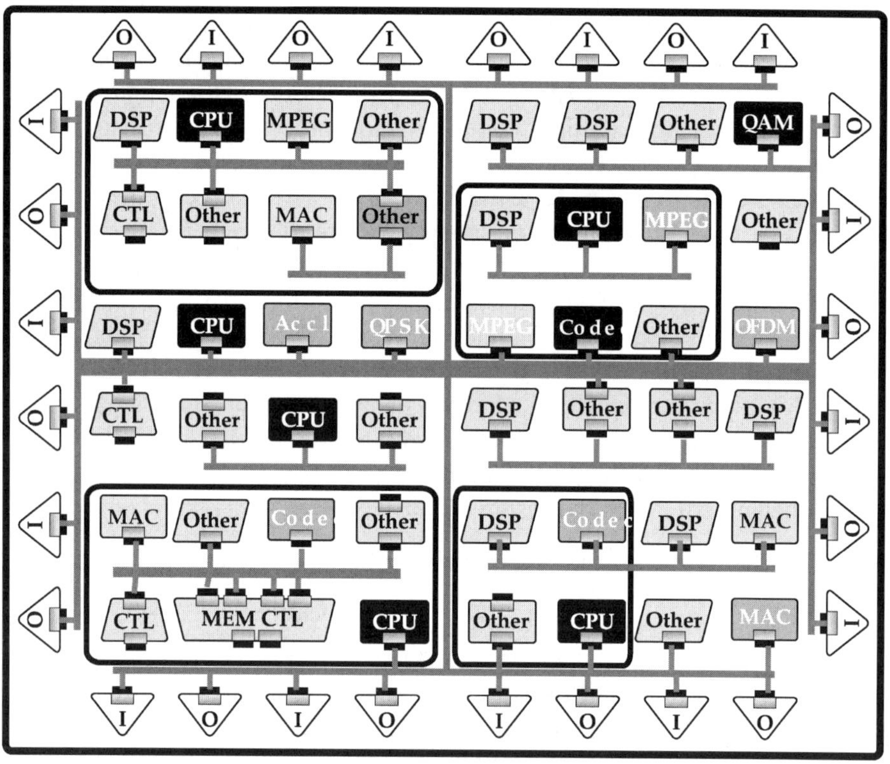

Figure 14-4. Socket-based design concept using decoupled active interconnect cores

Decoupled design methodologies are further needed to respond to the nature of today's electronics business environment. As globalization of the electronics industry has taken hold, there are greater requirements for dispersed design teams working independently. Functional core development must proceed independently from specific device implementation, since the device targets can change during design and the core may be needed for re-use in future designs. Core designers must anticipate the integration and design verification stages to avoid core redesign. Standard interconnect techniques must be deployed that separate core computational functions from inter-core communications and a standard core-to-system interface must be supported.

4.1 Decoupled Interconnect Cores

Only decoupled interconnect cores are able to meet these design constraints. One of the key characteristics of the decoupled interconnect core model is that it enables the selection of independent clock frequencies, data widths and transfer protocols to satisfy both functional core and system-level requirements. To do this, it must be capable of managing total system communications by supplying the required throughput at acceptable latencies.

A decoupled interconnect core can provide an optimum local operating environment for each functional core that meets the core's performance constraints. The functional core designer chooses the appropriate functional and communication architectures for their core, including data path widths, degree of transaction pipelining, and internal buffering. In addition, the designer specifies required bandwidth and latency constraints that the core imposes upon the system as a function of the core's operating mode and/or clock frequency.

While each functional core can now be optimized for its specific functionality, the decoupled interconnect core can be similarly optimized to improve system-level communications between all cores. The interconnect core manages the inter-core communications, satisfying functional core communications requirements while optimizing system performance, power, and area to meet SoC requirements.

System level management includes establishing routing between functional cores, providing an access control mechanism that meet QoS standards in terms of latency and throughput, and generating, routing and monitoring sideband signals. At the silicon floor planning and physical implementation stage, the decoupled interconnect core must be able to span the distances between functional processor core and I/O core placements and to optimized the frequency/latency trade-offs to fit application requirements. It should deliver robust system services such as error management, security/protection, identification, power management and event management, while presenting a set of predictable physical design characteristics that can be easily integrated into existing design tool flows.

In the design examples, we will illustrate two decoupled interconnect IP models that address these topics for processor-intensive applications and for I/O-intensive designs.

5. SOCKET-BASED DESIGN

Socket-based design is a method of using a pre-specified, standard interface to provide electrical, logical and functional connections between cores. Note that the socket need not care about the differences between functional and interconnect cores. The socket is equally applicable for connections between functional core to functional core, functional core to interconnect core and interconnect core to interconnect core. The socket defines the boundary of responsibility between the cores allowing them to be independently designed and verified, with the assurance that they will function correctly when connected.

A complete socket model should provide the designer with a range of abstractions to support various electrical, logical and functional behaviors. At the electrical level, the socket is implemented using wires. At the logical level, the socket uses protocols. At the functional level, the socket delivers transactions. Electrical measures include delay in seconds, capacitive loading in Farads and activity rates in cycles per second. Logical measures include delay in cycles of latency, and activity in peak bandwidth capacity. Functional measures include latency constraints and throughput requirements.

For a socket model to succeed it must have support for and within the existing industry infrastructure. This means that bus functional models, protocol checkers, performance analysis tools, formal property checkers, and multiple language support (HDL, verification language, C++) should be readily available.

To simplify implementation, a socket should be a point-to-point link using only unidirectional, synchronous signaling. This interface model will support the easiest design flow integration, a key consideration of any proposed standard.

To be effective, socket-based design should support the needs of the functional core. This means that communications protocols and timing characteristics must be tuned to optimize for the natural behavior of the functional core. The incorporation of a socket mechanism enables a natural isolation of the functional core and relieves the core from unnatural restrictions in electrical, logical, or functional behavior based on inter-core communications parameters. To enable this isolation, the socket itself must be capable of supporting a wide range of signaling and transfer protocol parameters and be configurable and extensible. Parameterized aspects could include selecting an address width, a data width and the number of interrupts, etc. Configurability could include the ability to enable or disable specific protocol capabilities, tune pipelining, select handshaking/flow control, support threading, select sideband signaling, add command extensions and select burst options. Extensibility could include adding core-specific functionality such as supporting security modes and adding parity or ECC features to transactions.

Socket-based design techniques add value to the SoC design process by allowing the designer to simplify communication issues to local point-to-point transactions at core boundaries, using protocols natural to the core. This approach simplifies the initial design of a core, eliminates artificial performance loss due to bus-centric interface choices, and enables greater independence of the design process. This greater independence facilitates both independent creation and verification of cores as well as much higher reuse of existing core designs. In addition, the socket enables the creation of a wide variety of decoupled interconnect cores using advanced internal protocols that must simply maintain the socket boundary.

6. THE OCP SOCKET

The OCP socket standard simplifies the interconnection of functional and interconnect cores and promotes core reuse. It is an excellent example of an open standard socket in use today. In brief, it is parameterized, configurable and extensible and is well supported by design and verification tools.

Initial efforts to define a standard core interface socket were begun in the mid-90s with the formation of the VSI Alliance (VSIA) [11]. VSIA's ambitious goal was to establish standards that enabled the integration of IP blocks from multiple sources. VSIA defined the Virtual Component Interface (VCI) as its IP core interface. While a valuable first step, this effort did not result in a widely adopted standard.

The non-profit Open Core Protocol International Partnership (OCP-IP) was formed in 2001 to support the first open and complete IP core socket for plug-and-play SoC design. The Open Core Protocol socket is a complete specification capable of supporting a wide range of core communication requirements, bus structures and interconnect technologies.

The OCP socket is bus-independent and handles the three dominant communication types between cores on an SoC – data, control and test. It is supported by a complete set of design, validation, and implementation support tools as well as a SystemC transaction interface and technical support documentation.

The OCP socket is a configurable interface that supports a basic set of data flow signals and an extended set of sideband signals for control and status information. A generic flag bus supports specialized inter-core signaling and a test interface support scan, JTAG and clock controls to simplify SoC debug and test.

The OCP socket can be configured to meet specific core and interconnect requirements. One configuration of OCP can form a simple, low-performance core interface. Another configuration can offer more complex, higher-performance interfacing for advanced processors and memories. The SoC architect can implement whatever system-level interconnect that meets the needs of the SoC, while individual core designers can independently deliver their functional core using OCP.

As defined by the Open Core Protocol Specification [12], an OCP socket uses a master/slave framework with synchronous unidirectional signals sampled by the rising edge of the OCP clock. The OCP socket is fully synchronous with no multi-cycle timing paths and all signals (other than clock and reset) are point-to-point. These constraints simplify timing analysis and physical design.

OCP accomplishes data flow by explicitly separating transfers into a set of phases. In the request phase, the Master provides the Slave with enough information to sequence the transfer. For simple writes, the request phase will accomplish the data transfer. In the response phase, the Slave provides the Master with indication about transfer completion. For reads, the response phase will include the data transfer. These phases may be pipelined with respect to each other, so the Master may attempt to sequence multiple requests before the Slave has responded. A rich set of handshaking and other flow control signals allows each side to pace the transfers and the pipeline depth.

The basic OCP signal set as shown in Figure 5 includes Master command, address and write data signals (MCmd, MAddr, and MData), Slave handshake, response, and read data signals (SCmdAccept, SResp, and SData), the clock (Clk) signal and the reset (Reset_N) signal. This configuration supports basic address-mapped data flow communication between cores.

If the core needs more communication flexibility, additional OCP signals can be configured for the socket. These include burst controls and sideband flag signals, which are especially useful for supporting interrupts.

To support concurrency and out-of-order processing between transactions, OCP supports the concept of independent threads sharing a single socket. Transactions issuing on different threads have no ordering requirements, and so can be processed out of order. Within a single thread, all OCP transactions must remain ordered. OCP threads are similar in many ways to *virtual channels* [13], with slightly more restrictive end-to-end mapping semantics.

Threads are added by configuring Master and Slave thread identifiers (MThreadID and SThreadID) and optional thread busy (MThreadBusy and SThreadBusy) signals for the socket. The thread busy signals allow the receiving device to assert per-thread flow control, which allows cooperating Masters and Slaves to implement truly non-blocking semantics across an OCP socket. This approach prevents a stalled thread from impacting the progress of other threads, eliminating deadlock scenarios while improving OCP's ability to support QoS guarantees.

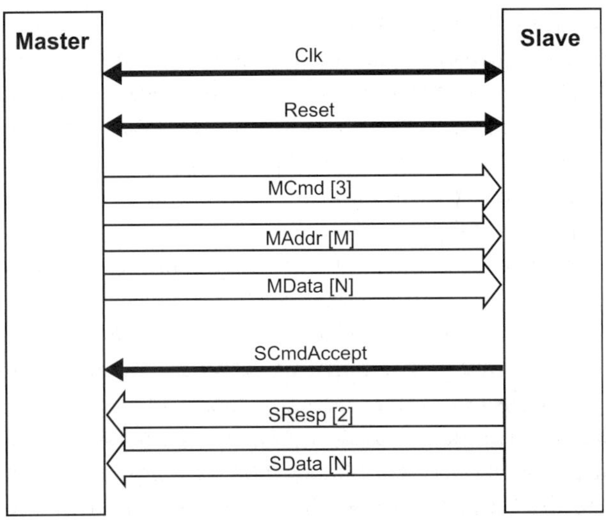

Figure 14-5. Basic dataflow signals of the OCP socket

A core developer selects a configuration of OCP to match the communications needs of the core, thereby creating a unique socket. For example, a UART designer might choose an OCP socket featuring an 8-bit data word (MData and SData), 3 address bits (MAddr), simple handshaking (via SCmdAccept), and a single interrupt signal (SInterrupt).

In contrast, the designer of an SDRAM controller might choose an OCP socket featuring 128-bit data, 28-bit address, deeply pipelined reads and writes, and 8 independent threads to support effective utilization of the multiple SDRAM banks. By implementing the full set of handshaking and flow control options, the designer can guarantee that the threads will stay independent, rather than blocking each other in either the request or response pipelines.

The Open Core Protocol was defined to be an excellent core interface socket. SoCs designed using OCP-compliant functional and interconnect cores realize the promise of socket-based design.

7. USING DECOUPLED ACTIVE INTERCONNECTS

As noted earlier, the adoption of a standard functional core interface standard such as the decoupled OCP socket is part of the concurrent SoC development solution. The second part is the deployment of interconnect architectures based on active interconnect cores.

The recognition of the need to decouple the computing tasks of the functional core from the inter-core communication tasks of the system is the basis of "communications-based design" [14]. A decoupled active interconnect methodology can be the foundation for this approach.

Decoupling frees the active interconnect core to be independently optimized for the targeted SoC, resulting in better overall results in much less time. The resulting SoC designs are more scalable. The decoupled active interconnect core supports increasing logical connectivity and increasing total communications bandwidth by increasing the total amount of communication resource, without requiring changes to any of the functional cores.

From the perspective of the active interconnect core, the functional cores appear to be simply communications sources and sinks. Each functional core may have different communications characteristics and requirements, but all may be described by appropriate annotation of the associated sockets. Such annotation is frequently simple to provide based on the aggregate characteristics of the algorithm implemented by the functional core, and can therefore be accurately modeled in advance of the existence of a new functional core. For instance, an MPEG2 video decoder could be modeled based primarily on the expected and worst-case read and write rates associated with compressed and decompressed MPEG macro-blocks and frames of decompressed video. By examining the composite actual or estimated characteristics of the various functional cores, the SoC architect can readily determine the overall throughput, latency, and quality of service (QoS) requirements for the active interconnect core. These performance requirements, coupled with the power and area goals for the finished SoC, drive the definition and implementation of the interconnect core.

Since the interconnect core is accomplishing most of the inter-core communications in the SoC, the interconnect core is likely to span the die. As has been mentioned previously, this means that the interconnect core is likely to have internal delays dominated by wiring, rather than active logic. It is therefore essential that the SoC designer be able to predict the impact of wire delays, derived from an actual or estimated SoC floor plan, on the overall performance of the interconnect core. Such concerns are localized to the decoupled interconnect core, since the functional cores are each isolated into local regions where maximum wire delays are based on the physical extent of the functional core, and are therefore both smaller and more predictable than those inside the interconnect core. Long wire delay issues are resolved in the interconnect core, not the functional core.

7.1 Examples of Decoupled Interconnect Cores

Sonics, Inc. has been developing and marketing decoupled interconnect cores since 1997. Sonics has defined the term MicroNetwork as a heterogeneous integrated network that unifies, decouples and manages all communications between on-chip processors, memories, and I/O devices [15]. Sonics has introduced two distinctly different MicroNetworks into the marketplace. This chapter will use examples from these two MicroNetworks to illuminate some key optimizations enabled by decoupled interconnect architectures.

Sonics' SiliconBackplane MicroNetwork was first described in 1998 [16]. SiliconBackplane is designed to service moderate to high bandwidth requirements typical of SoCs that integrate three to ten processing-focused cores together with a similar number of streaming I/O interfaces and a few shared memory pools such as internal or external DRAM or SRAM. From a performance perspective, SiliconBackplane is optimized to service aggregate throughputs in the range of 300-4000 MBytes/sec.

Sonics3220 SMART Interconnect IP (S3220) was first described in 2002 [17]. S3220 is designed to service low to moderate bandwidth requirements typical of SoC subsystems that integrate several dozen peripherals together with links to a few control processors and DMA engines. From a performance perspective, S3220 is optimized to service aggregate throughputs in the range of 50-600 MBytes/sec.

Figure 6 shows a block diagram representing the use of two interconnect cores in an representative multimedia application. The arrows represent unique instances of the OCP socket. Arrows point from master to slave, indicating the direction of request flow. Data flow is normally bidirectional (i.e. read and write) across each socket.

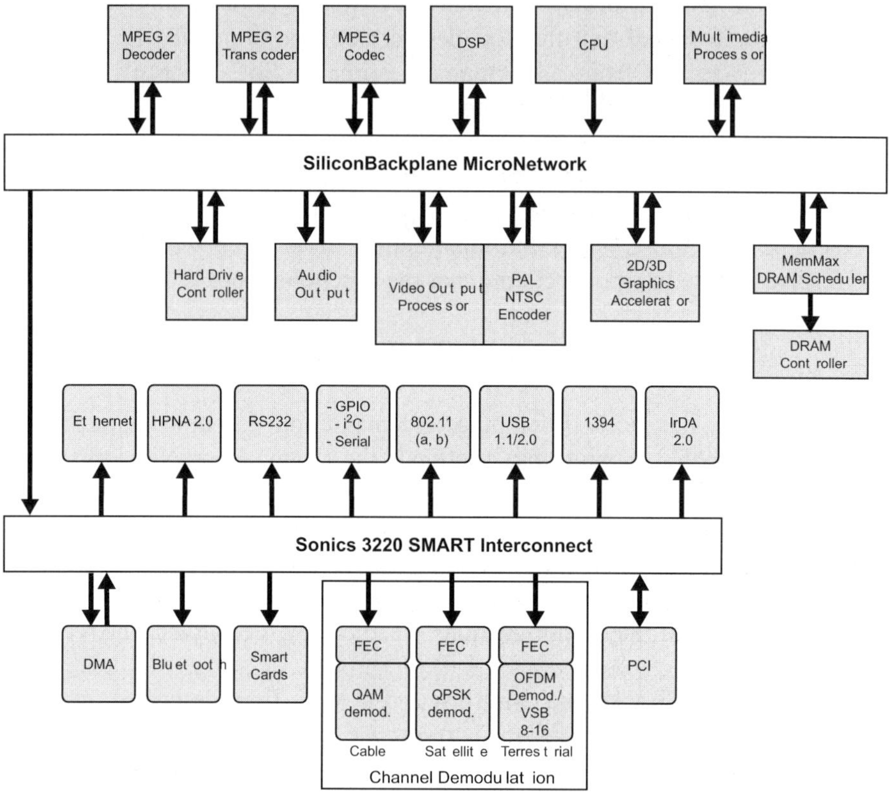

Figure 14-6. SiliconBackplane and S3220 interconnect cores in a digital set-top box

7.2 MicroNetwork Core Decoupling Capabilities

Both MicroNetworks extensively support the OCP socket for interfacing to functional cores. In addition, OCP is used as the MicroNetwork-MicroNetwork interface between separate instances of SiliconBackplane and/or S3220, as shown in Figure 6. Each MicroNetwork supports a rich subset of the total range of allowed OCP configurations [18].

Both MicroNetworks are composed of a set of *agents* that provide the active decoupling logic that bridges the functional core OCP interface(s) to the internal electrical, logical, and protocol environment of the MicroNetwork. Each agent is responsible for creating a local environment in which the attached functional core may operate as designed, isolating it from the different environments present at the other functional cores. This decoupling occurs at a number of layers. Figure 7 shows these layers, in increasing levels of abstraction. The following subsections describe the agent decoupling capabilities in more detail.

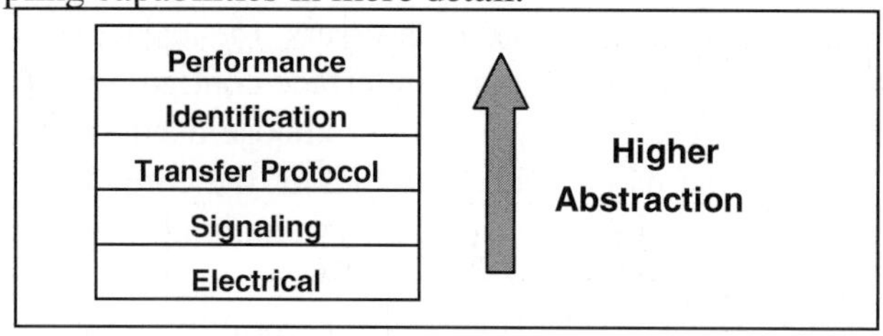

Figure 14-7. Decoupling abstraction layers for active interconnect cores

7.2.1 Electrical Layer Decoupling

The agent provides significant control over the degree of decoupling at the electrical layer. Functional cores may connect using an OCP clock frequency that is identical to the MicroNetwork, or at any integer divisor of the MicroNetwork clock. In addition, the OCP clock may be completely asynchronous to the MicroNetwork clock; in such cases, the agent includes the required synchronization logic to cross the clock boundaries. In all cases, the core sees simple, synchronous timing arcs.

The agents also have very flexible timing. For instance, if the core's outputs arrive late in the interface clock cycle, the agent must sample the signals into registers immediately. However, if the outputs arrive early in the clock cycle, the agent logic associated with the signal may be accomplished in the same cycle as crossing the interface. Thus, agents can cope with different core timings in an optimum fashion, which reduces die area and latency versus the conservative registered input and output approach [19].

The timing described at the electrical layer results from calculations derived from the transistor sizing on outputs, modeled as drive strength, the transistor sizing and internal cell routing associated with gate inputs, modeled as input capacitance, and the wiring lengths associated with connecting outputs to inputs, modeled as wiring capacitance and wiring delay. Accurate early predictions of the wiring induced delays associated with inter-core wiring are increasingly difficult to achieve as the level of integration grows. The MicroNetwork approach to this problem is simple: make these wires as short as possible. The MicroNetwork is specifically engineered to provide predictable mechanisms for spanning distance. As such, the agent should be placed adjacent to, or together with, the attached functional core, resulting in minimum wire length and, therefore, predictable interface timing and rapid timing convergence.

7.2.2 Signaling Layer Decoupling

At the signaling and transfer protocol layers, the agent relies on the OCP socket definition to supply the required interface configuration chosen by the functional core. The core developer should choose a configuration of OCP that closely matches the native characteristics of the core. The agent inherits the chosen configuration and is instantiated containing appropriate logic to handle the signaling and transfer protocol layers.

At the signaling level, the OCP supports several different data word sizes and flexible handshake-based flow control. Bridging different word sizes from the functional core into the MicroNetwork is automated by sequencing (i.e. packing and unpacking) logic built into the agent. Compliant state machine implementations of OCP may accept transfers unconditionally (i.e. independent from the current request), which simplifies timing analysis, or reactively (i.e. dependent on the current request), which offers the greatest design flexibility. Since the OCP uses only unidirectional signaling (i.e. all signals have precisely one driver), interface timing analysis is straightforward.

7.2.3 Transfer Protocol Layer Decoupling

OCP also supports a great degree of flexibility at the transfer protocol level. Compliant implementations offer choices over partial word transfers, burst protocols, request-response decoupling, request pipelining, and concurrency.

The agent implements appropriate request and data storage, plus appropriate sequencing logic, to interface the core-specific OCP configuration to the other agents in the MicroNetwork. Storage requirements are minimal for most agents; only those with proportionally high throughput requirements *and* significant mismatches between the core and MicroNetwork bandwidth require non-minimum storage.

7.2.4 Identification Layer Decoupling

In traditional data networks, source and destination identifiers are integrated in the packet headers, and are visible at many layers of the protocol stack. In contrast, OCP follows the traditional computing model of address-mapped reads and writes, so most identification is localized inside the MicroNetwork.

OCP includes an address field that the initiating core uses to identify the targeted core and intra-core resource. The MicroNetwork includes address-matching logic to implement targeted core selection. This logic is very flexible, supporting positive and negative decoding, variable match width, and optional re-programmability to implement soft address maps. In addition, agents supporting initiating cores splice on source identifiers to core requests, allowing the identifiers to be fully specified at the system level, rather than forcing them to be compiled into the cores. The MicroNetwork uses source identifiers to accomplish thread mapping, selective connectivity, and response routing.

Only intra-core resource identification (i.e. partial address) is typically passed to the target core. Exceptions include bus bridges, where full addresses are typically required, and concurrent multi-bank memory controllers, where source identifiers are useful for establishing transaction priorities.

7.2.5 Performance Layer Decoupling

The agents work together in the MicroNetwork to implement performance-layer decoupling. Each agent may be configured with FIFO-like storage resources that are based on the performance required from the attached core for the application(s) hosted by the SoC. In addition, the MicroNetwork leverages its internal threading capabilities to time interleave multiple requests and thereby minimize buffering, while providing hardware guarantees of quality-of-service. This helps significantly in latency decoupling by minimizing latency uncertainty.

The cost of performance-layer decoupling varies greatly from agent to agent and application to application. The MicroNetwork-based design approach simplifies the calculation of the required storage, automates its creation, and allows simple re-tuning as refinement proceeds.

7.3 MicroNetwork Communication Optimization

The previous section has shown how the MicroNetwork is composed of a set of agents that provide a wide range of decoupling services to the attached functional cores. The same agents also implement the internal communication protocols of the MicroNetwork to enable agent-agent transactions.

The internal MicroNetwork protocols of SiliconBackplane and S3220 share several common characteristics. Given the dominance of wire delay over gate delay in advanced process technologies – particularly at the MicroNetwork level where the maximum distance between functional cores is often bounded by only the physical dimensions of the SoC die – both MicroNetworks leverage scalable internal pipelining so that the operating frequency of the MicroNetwork may be optimized. For instance, the

SiliconBackplane agents contain a number of optional internal pipelining points and bypass points.

Both SiliconBackplane and S3220 MicroNetworks use shared paths to transmit requests and responses. In this sense, they therefore implement a bus topology. A bus topology is useful because it minimizes routing area with respect to cross-bar and other switch topologies that offer parallel paths between initiators and targets.

Threads are supported internally in a similar fashion to the OCP socket. Threads allow the MicroNetwork to interleave requests and responses from multiple initiators at the granularity of single data words.

Transactions are non-blocking. Individual transfers may never occupy the shared inter-agent routing for more than a single pipelined cycle – regardless of target contention, bandwidth discontinuities, etc. Requests and responses attempt to transfer from a buffer in the sending agent to a buffer in the receiving agent.

The combination of threading and non-blocking mechanisms ensures that transactions between slow functional cores – regardless of their length or latency – never block transactions between higher bandwidth functional cores. In addition, access control (i.e. arbitration) is performed each cycle. This is feasible because threading allows arbitrary choice of which agent gets to send the next request and non-blocking guarantees that there will be a new access control opportunity on every cycle. This results in much higher transfer efficiencies than are normally associated with bus topologies [20] and also offers much tighter guarantees of QoS. For instance, commercial implementations of SiliconBackplane in telecom applications have achieved sustained throughputs above 85% of the ideal (data width * clock frequency) internal peak MicroNetwork bandwidth [21].

There are also several significant differences between the internal protocols of SiliconBackplane and S3220. This supports their different application focuses.

SiliconBackplane is optimized to support total internal bandwidths that are higher than the peak bandwidths of any attached OCP socket. In many applications, this allows SiliconBackplane to deliver cross-bar type bandwidths at the routing cost of a bus, much as a time-domain switch exploits *temporal* concurrency to deliver the same throughput as a space switch (which exploits *spatial* concurrency). In a traditional computer bus, it makes no sense to operate the bus at higher bandwidth than the highest-performance attached blocks. Even if the computer bus could be decoupled from the blocks, the routing savings associated with the bus would be lost in gate area spent on burst FIFO's to accomplish the decoupling. This is because computer buses arbitrate (and therefore block) on burst boundaries, so a slower block would need to store an entire burst into a FIFO at the sending end, then arbitrate for access to the bus, and then move the data in a continuous burst into a FIFO at the receiver. The FIFO's are required because neither sender nor receiver can match the peak bus bandwidth.

SiliconBackplane uses very different internal protocols than a computer bus. The threaded/non-blocking internal fabric allows burst transactions to be interleaved with other traffic at a per-cycle basis. SiliconBackplane's access control scheme uses a combination of bandwidth pre-allocation via TDMA with unallocated and unused TDMA slots being dynamically allocated on a fair best-effort basis to waiting initiators as shown in Figure 8. The SiliconBackplane TDMA scheme is non-uniform so that the bandwidth guarantees may span a wide dynamic range, since the heterogeneous processing cores of a multimedia SoC may have orders of magnitude differences in throughput requirements. The designer implements this by the choice of the number of slots in the time wheel, and the number of slots allocated to each initiating agent. The TDMA scheme, when properly configured, allows transfers associated with *isochronous* or *quasi-isochronous* processing cores to be communicated across SiliconBackplane at the natural production and/or consumption rates of the cores. Such cores require very little buffering in the SiliconBackplane agents (merely enough to pack/unpack enough data to fill a SiliconBackplane data word), regardless of the transaction length/burst size. Thus, SiliconBackplanes protocols enable a bus topology interconnect to effectively use higher, decoupled internal bandwidth without requiring burst-deep FIFO's at ingress and egress.

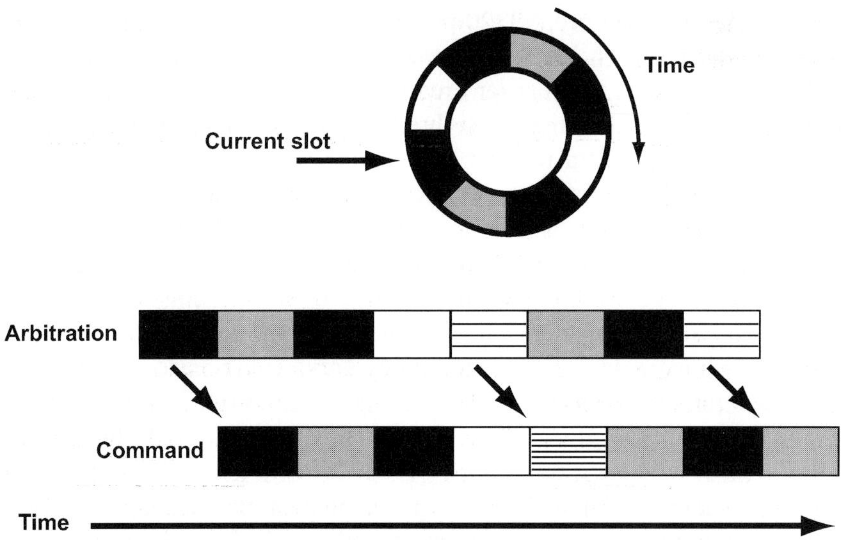

Figure 14-8. SiliconBackplane access control mechanism

Some SoCs (particularly in digital STB and HDTV applications) are characterized by inter-core data flows that mostly terminate in shared off-chip DRAM. In such cases, it is common that SiliconBackplane is configured with peak internal bandwidth equal to the peak DRAM bandwidth. With the overall SoC performance limited by the efficiency of the path to DRAM, it is essential that SiliconBackplane be able to saturate the DRAM channel. To accomplish this, SiliconBackplane is fully pipelined. This means that a single initiator agent to target agent transaction can fully occupy the interconnect, transferring a single SiliconBackplane (and thus DRAM) data word per cycle. However, the more normal case is that the DRAM subsystem is itself multi-threaded at the OCP socket. This allows multiple processing cores to interleave their request streams to the DRAM subsystem, enabling the DRAM controller to re-order DRAM commands so as to maximize the efficiency of the DRAM usage while guaranteeing QoS for real-time traffic. The benefits of such an approach have been described elsewhere [22].

The internal protocols of S3220 are optimized to efficiently connect peripherals spread around the pad ring of the SoC. Since the attached functional cores are typically numerous and small, optimization of die area per agent is crucial. S3220 implements pipelined internal protocols to ensure scalability in operating frequency while spanning large distances. However, full pipelining (as practiced in SiliconBackplane) can cost registers deep enough to cover the pipeline delay to be integrated into each agent, which would take too much area. Peripheral functional cores are rarely able to sustain full throughput, and are even less likely to require full throughput at the system level. Therefore, S3220 protocols are optimized to support a peak issue rate between each initiating thread and its target of one transfer every two cycles. This cuts the register area associated with covering the pipeline delay in half. Note that the multi-threaded, non-blocking nature of S3220 allows it to sustain a data transfer every cycle, as long as there are multiple initiating threads attempting to communicate with independent targets, as shown in Figure 9.

The figure shows a request 0 propagating from the initiating OCP to the targeted OCP, and the associated response 0 propagating from target to initiator. The internal pipeline cannot issue request 1 until the third cycle (as mentioned above), but the intervening grey cycles are available for independent initiator/target transfers.

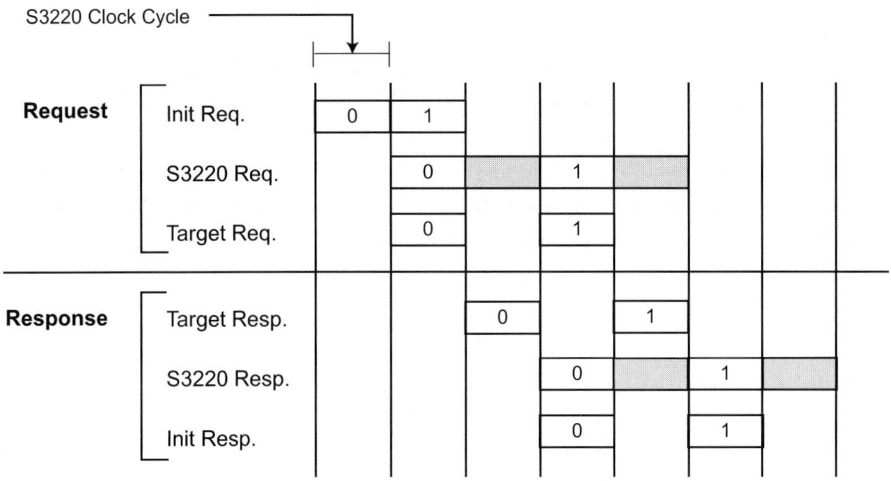

Figure 14-9. S3220 data transfer pipeline

Since the MicroNetwork must span the distance between all of the functional cores, they are constructed using logical structures that offer high performance, physically predictable timing results. Traditional on-chip bus topologies have migrated from distributed tri-state to centralized multiplexor implementations in order to accommodate the needs of ASIC-style design flows. However, the centralized multiplexor is essentially a "star" topology, with each sender driving an input across the SoC to the multiplexor, which generates both uncertain wiring delay (based upon the relative placement of the block and the multiplexor) and high routing congestion around the multiplexor – which is exactly what the designer intends to avoid by choosing a bus topology in the first place!

In contrast, both SiliconBackplane and S3220 implement a distributed multiplexor structured as a tree. As shown in Figure 10, each agent includes a transceiver that implements OR-tree multiplexing on the path towards the root of the tree, and a repeating buffer on the path back from the root. This simple structure has some very attractive physical properties. If the multiplexor tree is connected such that each agent connects directly to its nearest neighbors, then the length of the intermediate wires in the tree is closely related to the agent-agent spacing, which is in turn closely related to functional core-functional core spacing. Since most functional cores are fairly small, the inter-agent wires should be relatively short (1-2mm), and therefore the delay of these wires are unlikely to require additional repeater insertion. A properly-connected distributed multiplexor tree also minimizes routing congestion, since each logical bus wire is represented by only two physical wires at any point in the tree. Perhaps most important, the delay through such a structure is quite predictable as soon as the number of agents and estimated SoC die size are known. This is because the timing behavior is dependent primarily upon the total wire length between the root and leaves of the tree and the depth of the tree. The tree depth is dependent on the number of functional cores. The total wire length is often estimated based on the MicroNetwork spanning the expected die. The evolving floor plan is then used to determine the multiplexor tree ordering, ensuring rapid timing convergence and low routing congestion.

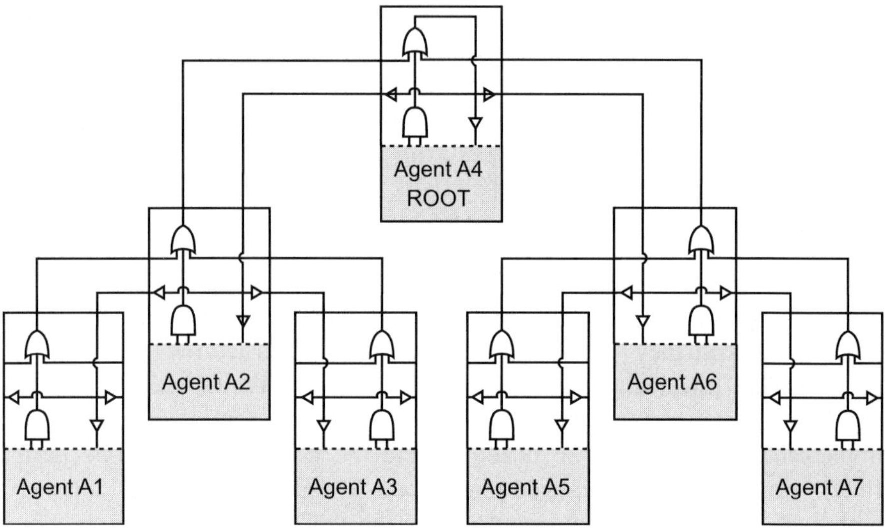

Figure 14-10. Internal MicroNetwork multiplexor tree

S3220 takes the physical layer optimization one step further. S3220 is optimized for low area and very low active and idle power. The designer may segment the S3220 logical bus into several physical branches, each of which contains a distributed multiplexor tree. A pipeline register at the root of the branches ensures that only the branch containing the agent servicing the targeted functional core is activated on a transfer request. This optimization avoids activating the other branches, which are likely to be long and thus have high capacitances that will significantly increase power dissipation.

7.4 Configuration and Generation of the MicroNetworks

To support the design of SoCs and MicroNetworks, Sonics has developed Sonics Studio, a graphical and command line SoC development environment. In developing a MicroNetwork instance, there are a large number of options from which the SoC architect can choose and the Sonics Studio development environment provides a number of automation tools to make this task easier [23]. Each OCP socket can be configured independently and the MicroNetwork can be configured and tuned for a large number of performance and system management characteristics. While it is possible to create a system using just command line tools, it is more efficient to use a graphical interface to perform some of these tasks, with command line based tools and utilities available when needed.

The designer/architect can use the GUI to quickly create a working prototype of the SoC concept as shown in Figure 11. Each interconnect core is shown composed of a set of agents, which terminate one or more OCP interfaces (depicted as arrows). The square blocks represent actual functional cores, or models of cores, depending upon the design phase. The GUI automatically connects objects with compatible interface bundles; the bundle specifies the name and direction for each signal group for each type of interface that connects to the bundle. For example, the OCP interface bundle has different signal directions for Master versus Slave interfaces.

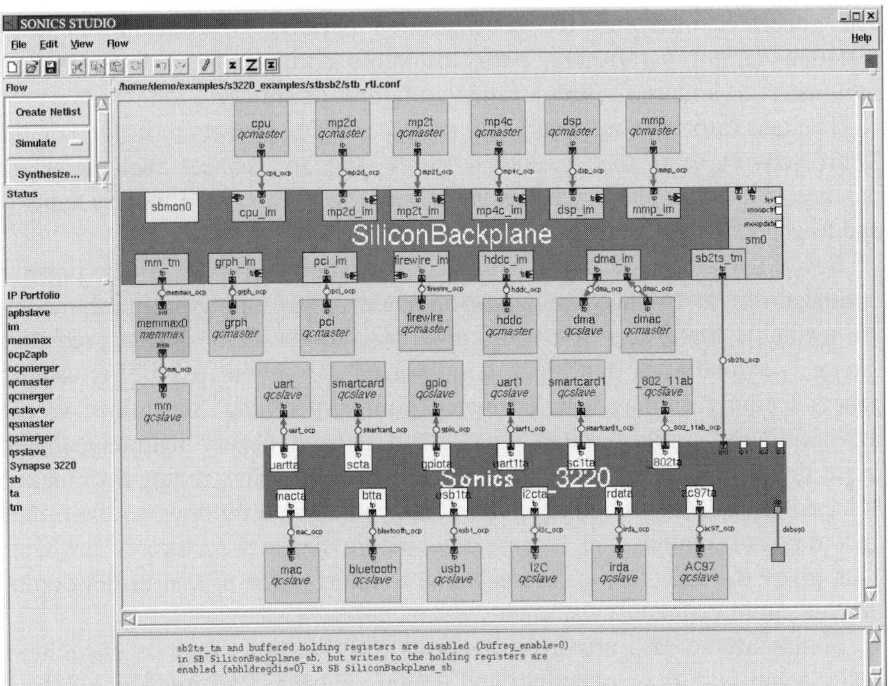

Figure 14-11. Sonics Studio GUI screen shot of behavioral digital STB design

To speed system design, Sonics Studio automatically configures MicroNetwork agents to inherit OCP configuration parameters when attached to functional cores with OCP sockets. The GUI has configuration panes for each MicroNetwork for capturing parameters such as data path width and the characteristics of the arbitration system. For agent level parameters, there are settings for buffer depths, clock frequency ratios among others. These features streamline the MicroNetwork generation and configuration task.

Sonics Studio supports models in C++ or Verilog/VHDL languages. Behavioral models of both OCP masters and slaves are available for describing estimated functional core behavior. These models are configurable to meet the full OCP specification allowing them to be used to accurately model any OCP-compliant functional core at the Bus Functional Model level. Other tools include simulation monitors that capture trace data into ASCII files, disassemblers, protocol checkers, and performance measuring programs. These features facilitate the early data flow modeling that enable rapid performance analysis and optimization.

A finished MicroNetwork may have several hundred configuration settings. The source code for the MicroNetwork is compiled into the Sonics Studio RTL generator with configuration parameters captured from the GUI. The generator interprets the configuration parameters, computes context-sensitive default parameters, performs value checks, and passes the final parameters to a macro processor that configures the final RTL.

The generator also automatically produces HDL net lists to instantiate the MicroNetwork and the functional cores and to connect their interface bundles. This automation makes it possible to create SoC models quickly and to generate new models as the SoC is refined.

Several of the MicroNetwork configuration options affect pipeline optimizations to balance timing convergence versus latency. Timing models are available for agents in a wide range of configurations so that predictive timing is available as the design is optimized. An automated tool is used to generate configurations with estimated boundary timing constraints, to run the configured agent through logic synthesis and library mapping and to parse the static timing report to capture the results. This timing information is based on process technology, cell library and synthesis flow, so the timing and area information can be prepared before the architecture is finalized. This gives the architect accurate physical information to use in developing the SoC architecture.

Sonics Studio also provides interfaces to other tools such as simulation, design synthesis, floor planning, and timing analysis. By providing a central development environment that enables fast generation of MicroNetworks, rapid modifications of performance characteristics, integrated simulation, synthesis and verification tools, Sonics Studio empowers the SoC architect with the tools needed to create complete SoCs using decoupled, active interconnect cores with standard interface sockets.

8. SUMMARY

Effective development of complex SoCs that incorporate many heterogeneous processing elements and tens to hundreds of I/O functions requires a new decoupled active interconnect methodology. Without socket-based decoupling and active interconnect cores, the process of design integration and verification breaks down at the inter-core communications level. Mismatching high-level functional core abstractions with low-level bus-oriented abstractions is the cause of this breakdown. The solution is decoupling the computational tasks of the functional cores from the communications functions Inter-core and system-level communications functions should be placed in interconnect cores. For developing a decoupled interconnect-based system architecture, there are several potential standard solutions available including the OCP socket specification as a standard inter-core interface and the Sonics MicroNetwork family of interconnect cores. Using development tools such as Sonics Studio enables the design of complex SoCs based on sockets and interconnect cores, leading to the realization of "design by functional combination" approach to complex SoC design.

ACKNOWLEDGEMENTS

The theories expressed in this chapter, and the products and protocols described herein, have been defined and refined over the past seven years. Many people have contributed to this work. In particular, the author would like to acknowledge the SMART Interconnect IP development team at Sonics, Inc. for their creativity and dedication in pursuing these goals. In addition, the Working Group members of OCP-IP have helped enhance and focus these theories and practices. Finally, the author would like to thank Chris Bailey for his assistance in preparing this manuscript.

REFERENCES

1 C. Mead and L. Conway. *Introduction to VLSI Systems.* Addison-Wesley Publishing Company, Inc., 1980.

2 A. Jantsch and H. Tenhunen, *Networks on Chip*. Kluwer Academic Publishers, 2003.

3 D. Wingard, Tiles – An Architectural Abstraction for Platform-Based Design, EDA Vision, June 2002, available from www.edavision.com.

4 Mentor Graphics, February 2002.

5 G. Smith, Gartner Dataquest, Design Automation Conference, June 2003 as published in Electronics Engineering Times, June 9, 2003.

6 Open Core Protocol International Partnership, information from www.ocpip.org.

7 S. Kumar, On Packet Switched Networks for On-chip Communication. In A. Jantsch and H. Tenhunen, editors, *Networks on Chip*, Kluwer Academic Publishers, 2003.

8 ARM, Limited, AMBA Specification, Revision 2.0, May 1999, available from www.arm.com.

9 IBM, CoreConnect Bus Architecture, 1999, available from www.chips.ibm.com.

10 A. Doganis and J. Chen, Mentor Graphics Deep Submicron Technical Paper, Interconnect Simple, Accurate and Statistical Models Using On-chip Measurements for Calibration. July 1999.

11 Virtual Socket Interface Alliance. *Virtual Component Interface Standard*, OCB Specification 2, Version 1.0, March 2000., available from www.vsia.org.

12 Open Core Protocol Specification, available from www.ocpip.org.

13 W. Dally. Route Packets, Not Wires: On-Chip Interconnection Networks. In *Proc. of the 38th Design Automation Conference, June 2001.*

14 M. Sgroi, M. Sheets, A. Mihal, K. Keutzer, S. Malik, J. Rabaey, and A. Sangiovanni-Vincentelli. Addressing the system-on-a-chip interconnect woes through communications-based design. In *Proceedings of the 38th Design Automation Conference*, pages 203-211, 1998.

15 D. Wingard, MicroNetworks for Flexible SoC Platforms. In *Proc. of the Sophia Antipolis Forum on MicroElectronics (SAME2000)*, October 2000.

16 D. Wingard and A. Kurosawa. Integration Architecture for System-on-a-Chip Design. In *Proc. of the 1998 Custom Integrated Circuit Conference*, pp. 85-88, May 1998.

17 D. Maliniak, "Interconnect IP Steers SoC Integration into the Fast Lane," *Electronic Design*, November 25, 2002.

18 SiliconBackplane MicroNetwork Overview, available from www.sonicsinc.com.

19 M. Keating and P. Bricaud. *Reuse Methodology Manual for System-on-Chip Designs. Kluwer Academic Publishers*, 1998.

20 H. Chang, L. Cooke, M. Hunt, G. Martin, A. McNelly, and L. Todd. *Surviving the SoC Revolution –A Guide to Platform-Based Design* . Kluwer Academic Publishers, 1999.

21 D. Wingard. Smarter Interconnects for Smarter Chips. Paper presented at Hot Chips Symposium on High Performance Chips, August 2002.

22 W. D. Weber, "Efficient Shared DRAM Subsystems for SoCs," presented at Microprocessor Forum October 2003.

23 D. Wingard. MicroNetwork-based integration of SoCs. In *Proc. of the 38th Design Automation Conference, June 2001.*

IV

APPLICATION CASES

Chapter 15

INTERCONNECT AND MEMORY ORGANIZATION IN SOCS FOR ADVANCED SET-TOP BOXES AND TV

Evolution, Analysis, and Trends

Kees Goossens[†], Om Prakash Gangwal[†], Jens Röver[‡], and A.P. Niranjan[‡]

[†] *Philips Research Laboratories, Eindhoven, The Netherlands*

[‡] *Philips Semiconductors, Sunnyvale, USA*

{ Kees.Goossens,O.P.Gangwal,Jens.Roever,Niranjan.AP } @Philips.com

1. Introduction

In this chapter we show that the organization of the communication and memory infrastructures is critical in today's complex systems-on-chip (SOCs). We further show that resource management in the form of scheduling or arbitration is common to them both. The increasing importance of these issues is illustrated by following the evolution of an advanced set-top box and high-definition digital TV application (ASTB) and its SOC implementations over time.

In Section 2, we introduce the application domain (embedded systems for high-volume consumer electronics), and the application (advanced set-top boxes for high-definition digital and analog TV). The computation kernels and the communication (data rates, latencies) needed for real-time audio and video are demanding. Meeting real-time requirements, while minimizing resources for cost-effectiveness is challenging for such large heterogeneous SOCs.

In Sections 4 to 6, we review two existing and one possible future SOC implementation of the ASTB application, along the following axes, introduced in detail in Section 3. We commence with the application itself, including real-time requirements, the computation kernels, the kinds of traffic flowing between them, and the logical memories used by them. The following axes categorize its implementation; the mapping of computation (types and number of IP blocks) and communication. Then we consider the interconnect organization, communication abstraction (how IP blocks interact with the interconnect), and the memory organization (how the logical memories are implemented). Finally, we review arbitration: how traffic types are supported by the interconnect, and how the system as a whole meets its real-time requirements.

The Viper SOC (Section 4) and its successor Viper2 (Section 5) share some important characteristics (such as a dependence on a single external memory) but have a different philosophy underlying their architecture. The former implements the interconnect grouped by data-rate requirements (leading to low and high-bandwidth interconnects), while the latter groups IP blocks by traffic kind (control traffic, low-latency data traffic, and latency-tolerant data traffic). A possible future implementation (Section 6) builds on Viper2's traffic separation, and additionally integrates multiple on-chip memories and external memories.

In Section 7, we review the evolution of the ASTB application and implementations, and observe some trends.

2. The ASTB Application

We focus on embedded systems for high-volume consumer electronics, in particular, advanced high-quality set-top box and TV systems. The ASTB application comprises audio decoding (e.g. AC3, MP3), video decoding (e.g. MPEG2), video pixel processing (e.g. de-interlacing, noise reduction), and graphics [1, 2]. These functions are combined to provide different products (such as analog, digital, and hybrid TV), as well as different modes of operation within a product. For example, digital TV decodes audio and video before performing further video pixel processing, while analog TV uses noise reduction instead of video decoding, and in hybrid TV both are present.

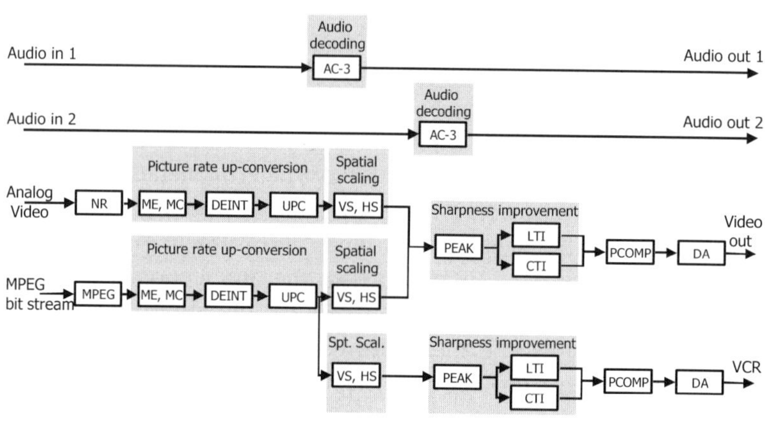

Figure 15.1. Example hybrid (analog and digital) ASTB processing chain.

Figure 15.1 depicts an example hybrid TV application. The audio processing chains for analog and digital remain independent, while the corresponding video-processing chains interact in a convergence to the screen, and a VCR output. We refer to [3, 4] for more information about the particular video functions, such as noise reduction (NR), picture rate up-conversion (motion estimation ME and motion compensation MC, de-interlacing DEINT, up-conversion UPC, etc.), spatial scaling (horizontal scaling HS and vertical scaling VS), sharpness improvement (e.g. peaking, luminance transition improvement LTI), picture composition PCOMP (e.g. mixing and blending of multiple pictures and graphics), and display adaptation DA (color conversion, skin tone correction, blue stretch, green enhancement, etc.). An important characteristic of many of these functions (such as NR, ME, MC) is their *temporal processing* nature, i.e. they use previous, current, and (sometimes) next pictures to generate their output picture. These pictures require significant storage and subsequent retrieval. Table 15.1 lists some typical computation, communication, and memory requirements to implement these functions [5, 6].

Table 15.1. Characterization of ASTB applications. Video pixel processing combines many video-improvement algorithms in a chain. We show typical numbers; however, they depend on the number of (temporal) processing stages.

	computation	in	out	local	memory
audio decoding	100 MOPS	32-640 kbps	5 Mbps	5 Mbps	50 kb
MPEG2 video decoding	4 GOPS	10 Mbps	120 MBps	240 MBps	8 MB
video pixel processing	100 GOPS	360 MBps	360 MBps	360 MBps	4 MB

The audio and video decoding functions conform to standards, such as AC3 and MPEG2. Many implementations complying with a standard are therefore available for these functions [7, 8]. In contrast, video pixel processing functions are often proprietary, and are the differentiating factor for products because they visibly improve the video picture quality [9, 10, 11].

ASTB SOCs have to support various input/output picture sizes, in particular standard (SD) and high definition (HD). Some processing is specific to analog or digital input (e.g. noise reduction is not used for digital input), specific to the display type (e.g. CRT or matrix), and specific to the number of streams processed simultaneously for one screen (e.g. with or without picture-in-picture PIP). Some modes are static, i.e. fixed for a particular product and do not change during its lifetime (e.g. CRT versus matrix). Other modes are dynamic and may change frequently, either triggered by the user (e.g. PIP on/off) or by the environment (e.g. broadcast change from video to film type, or SD to HD for main window).

Audio and video processing have intrinsic real-time requirements: for example, a video field must be displayed at regular intervals (e.g. 1/50th, 1/60th, or 1/100th of a second). Furthermore, audio and video must be synchronized within strict limits (known as "lip synchronization"). Only a limited amount of processing, such as graphics, has less strict timing restrictions.

Application characteristics From the brief overview of the application domain we extract the following characteristics.

First, computational requirements vary substantially (three orders of magnitude, cf. the second column of Table 15.1), from low for audio processing to very high for video pixel processing functions. This leads to a large number of heterogeneous computation elements. For example, Viper2 (Section 5) contains 60 dedicated and weakly-programmable IP blocks, two application-domain-specific VLIW media processors, and one general-purpose RISC processor. (A note on terminology: we divide IP *blocks* in two categories: function-specific *cores*, and programmable *processors*.)

Programmable processors offer the flexibility to implement emerging standards for audio and video, and differentiation of products (when integrating SOCs in products). Low cost, power efficiency, and high computational performance are achieved through the use of function-specific cores.

Video decoding and video pixel processing operate on large amounts of data, necessitating large buffers for communication and temporal data. Multi-tasking processors, especially high-performance VLIW processors, have large amounts of program instructions (code), which also requires storage. For high memory requirements, large external (off-chip) memories are more cost effective than on-chip memories.

The interconnection infrastructure plays a central role in the SOC architecture for a number of reasons. The large amounts of (temporal) video data have to be transported between many IP blocks, via memory or directly. This requires a high-performance interconnect. To support the many static and dynamic modes of operation, data transportation mechanisms must be highly configurable. Moreover, to configure the SOC for a mode of operation, a flexible control interconnect must allow IP blocks to be accessed and programmed [12].

Finally, the combination of cores, processors, and interconnect must guarantee the hard-real time requirements of the application. Resource management (arbitration) of computation (e.g. real-time operating systems), communication (e.g. traffic classes with different performance), storage (by the external-memory controller), and their combinations are essential.

3. System Analysis Overview

Before we discuss individual designs, we define the axes along which the designs are presented. Although the axes are not independent (interconnect and memory organizations are strongly related, for example), for every axis an evolution can be identified for successive designs. For each design we discuss the *application*, its *computational complexity*, and the resulting *traffic types*. These are reflected in the *interconnect organization*, the *communication abstraction* (i.e. how IP blocks interact with the interconnect), the *memory organization*, and finally how all these components are managed or *arbitrated*. We have some general remarks on a few of these points in this section.

3.1 Traffic Types

Traffic types help us to discuss how communication is mapped to interconnect implementations. The heterogeneity of the processing elements (cf. Section 2), results in a variety of traffic types, based on data rate, latency, and jitter characteristics, see Table 15.2. Based on these traffic types, the three designs have different interconnect partitionings, as we shall see in Sections 4 to 6, and summarized in Table 15.4.

Table 15.2. A classication of traffic types.

label	data rate	latency	jitter	example
LRLL	low	low	low	control traffic
HRLL	high	low	low	cache misses
low-jitter HRLT	high	tolerant	low	"hard-real-time" video
jitter-tolerant HRLT	high	tolerant	tolerant	"soft-real-time" video
jitter-tolerant MRLT	medium	tolerant	tolerant	audio & MPEG2 bitstreams
best effort	tolerant	tolerant	tolerant	graphics

Control traffic originates from control tasks that are usually mapped on one or more processors, which must obtain status information from cores and program them. It has a low data rate, but requires low latency (LRLL) to minimize the system response time, e.g. when the application mode changes.

Multi-tasking processors, such as MIPS and especially high-performance VLIW TriMedia processors, do not have sufficient local memory to contain all instructions (code) and data of the multiple tasks. Instruction and data caches are therefore used to automatically swap in and swap out the appropriate instructions and data. This leads to high (instantaneous) data rates, and requires low latency (HRLL).

Dedicated video-processing cores usually operate on and generate streaming (sequential) traffic with high data rates. They are composed in deep chains without critical feedback loops, and their low-latency requirement can therefore be made less critical by using buffers to avoid underflow. The resulting traffic has a high data rate but is latency tolerant (HRLT). Medium-data-rate latency-tolerant traffic (MRLT) is generated, for example, by audio and MPEG2 processing cores.

Jitter (latency variation) can be handled similarly, and we use the distinction between low-jitter and jitter-tolerant HRLT traffic. IP blocks with the latter traffic, such as the memory-based video scaler, have an average data-rate requirement but can be stopped when there is no data, and make up by processing at a higher rate later, or by averaging out data bursts. By contrast, low-jitter HRLT IP blocks do not tolerate variations in data rates, because they cannot make up for any lost processing cycles. Examples are video-processing blocks operating at actual video frequencies, where line and field blanking can not be used as slack.

Some processing, like graphics, operate on best effort traffic, in the sense that it gets by on whatever bandwidth and latency it is given.

3.2 Communication Abstraction

As SOCs increase in complexity, the need for IP re-use and associated standardized SOC design methods has resulted in the notion of a platform [13]. A platform structures and standardizes SOC architectures, by regulating the kind of IP blocks that can be used, how they are combined, and how the system is programmed. Separating the computation (processors and cores) from communication (the interconnect core) has many advantages [14, 15]. In particular, we show how the use of the device-transaction-level communication standard (DTL) [16], part of Philips's Nexperia platform, has been key in allowing the interconnect to evolve over time, while the IP blocks remained unchanged. Communication abstraction has been promoted by the VSI alliance, the OCPIP consortium, and ARM, whose respective VCI [17], OCP [18], AXI [19] protocols are similar to DTL.

The use of communication abstractions, such as DTL, has several advantages. The development of cores is simplified because DTL is tailored to the requirements of IP blocks, rather than the interconnect. Moreover, IP blocks become independent from the interconnect, and hence re-usable. The interconnect and IP blocks are glued together by means of (re-usable) adapters. System-dependent customization is restricted to the adapters, instead being implemented by the IP blocks or the interconnect. Examples include little/big endianness conversions, interconnect-dependent sizing of latency-hiding communication buffers. This enables re-use of both the IP blocks and the interconnect.

3.3 Memory Organization

We distinguish several logical memories for *data use*. These are: *algorithmic* memories (e.g. temporal field memories, field to frame conversion memories, line memories of a scaler), *state* memories (e.g. local variables of a hardware or software task), and *decoupling* memories. Decoupling memories even out differences in data production and consumption rates of IP blocks, e.g. due to field and line blanking, horizontal and vertical data access patterns of video scalers. In the following discussions we omit state memories because, in the designs we discuss, they are always part of the IP block. A second type of logical memories are used to store *instructions* of programmable processors.

Memories that pertain to the communication architecture include pipelining, *packetization*, *latency-hiding*, and *clock-domain-crossing* memories. Pipelining memories are used to increase the operating frequency of interconnects. Packetization memories are required to convert the data of IP blocks to the format used by the interconnect (e.g. 64-bit words). Latency-hiding memories remove or hide latency and jitter introduced by communication networks, memory controllers, RTOS task scheduling, and so on. For example, a 128-byte buffer is used in Viper. A small amount of memory is used to safely cross different clock domains.

All the kinds of logical memories are mapped to (implemented by) physical memories, basically on-chip or off-chip memories. For each of the designs, we show how the different logical memories are mapped to on- or off-chip memories (but we do not further sub-divide in RAM, flash, register files, etc.).

3.4 Arbitration

The ASTB application can be analyzed in terms of its computation and communication, but, additionally, it has real-time constraints for audio and video. To meet real-time requirements computation, communication, and memory resources must be arbitrated, or managed. For example, a simple first-come first-serve bus arbiter cannot distinguish low-latency from latency-tolerant traffic, and cannot offer differential data-rate services. Both are important in meeting real-time requirements. Two key issues in each of the designs are the arbitration of critical resources (such as external-memory bandwidth), and managing the interaction of various arbiters (e.g. those of external memory and on-chip interconnect).

4. Viper

Viper [1] is a highly integrated multimedia SOC targeted at advanced set-top box and digital TV (ASTB) applications. It provides progressive SD, and interlaced SD and HD outputs. Viper's main functions are video decoding (e.g. MPEG2), audio decoding (e.g. AC3), and video pixel processing to improve picture quality and to convert between image formats (e.g. de-interlacing). It simultaneously supports two video streams (e.g. one high definition of 1920x1080 pixels interlaced at 60Hz, and one standard definition of 720x480 pixels interlaced at 60Hz) and three audio streams.

4.1 Computation Mapping

Viper contains two processors (a MIPS-PR3940 and a TriMedia TM3218) and 50 function-specific cores. The cores include video-processing modules such as MPEG2 decoders (with high computation and communication requirements), audio/video input/output modules (e.g. MPEG2 transport-stream parser) and general-purpose peripherals (e.g. UART interface, USB controller). As noted in Section 2 their communication requirements and memory usage vary significantly.

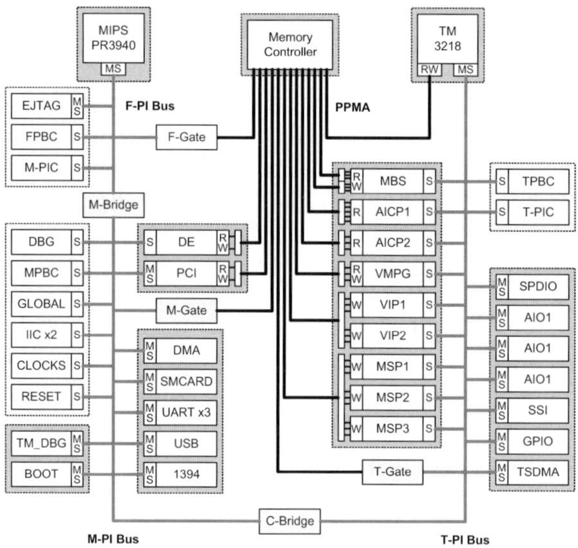

Figure 15.2. Simplified block diagram of Viper.

4.2 Communication Mapping

Memories are required to a) store instructions and data for both processors, b) function as an algorithmic memory (in particular temporal data, and for vertical scaling), and c) decouple IP blocks if their processing rates do not match. On-chip memories cannot be used in many of these cases because they would be too large. Hence it is advantageous to merge all data and instruction memories in one external memory. This results in high data rates for the off-chip memory from streaming cores (HRLT), and from data and instruction accesses of the programmable processors (HRLL). To hide the latency to the background memory the MIPS and TriMedia processors all have data and instruction caches. In Viper, all IP blocks are masters, i.e. they autonomously write to and read from the memory. In Figure 15.2 master and slave ports are labeled M and S, and all read (R) and write (W) ports are masters. This avoids the need for a central DMA agent that has to be programmed by a processor, which could be a bottleneck. Given the particular combination of many masters and one slave, it makes sense to optimize the combination of external-memory controller and interconnect, as we shall see in the next section.

The status and program registers of all cores can be accessed at locations in a global memory map. This control traffic is generated by a few masters (the programmable processors) requiring access to many slaves (the cores), in contrast to the data traffic. Control traffic has low data rates, but requires low latency, to reduce the system response time (e.g. when the application mode changes).

4.3 Interconnect Organization

A single-hop broadcast interconnect, such as a shared bus (multi-master and multi-slave) that connects all IP blocks (including external memory), is not a feasible solution for a SOC like Viper. It cannot fulfill the bandwidth requirements of high-data-rate traffic, and certainly cannot offer the required low latency for control (LRLL) and cache traffic (HRLL). In Viper the traffic is therefore partitioned over several different interconnects, with *traffic separated on the basis of data-rate requirements*. There are two interconnects: one for low-data-rate (LRLL, control traffic) to medium-data-rate traffic (MRLT, audio and encoded MPEG2), and one for high-data-rate traffic, both low latency (HRLL, cache misses) and latency tolerant (HRLT, video). Below, we first discuss each interconnect in turn, and then how they interact.

The low to medium-bandwidth interconnect The idea underpinning this interconnect is that low latency can be ensured for control traffic by isolating it from high-data-rate traffic. Two 32-bit PI busses [20] are used: M-PI (see Figure 15.2) for control traffic from the MIPS and DMA traffic from streaming cores, and T-PI for control traffic of the TriMedia. MRLT DMA traffic is added to the M-PI bus to increase its utilization. The two busses are connected by a bridge (C-BRIDGE, with a master and slave port on both sides), to allow both processors access to all cores for programming.

The high-bandwidth interconnect The high data-rate requirements to access external memory cannot be fulfilled with a tri-state bus, such as the PI bus. Even with large tri-state drivers, it would be too slow because of a high capacitive load due to the large number of IP blocks. Observe, however, that all high-data-rate traffic is destined to the off-chip memory, that is, many masters communicate with a single slave. This leads to the design of a specialized 64-bit point-to-point memory-access interconnect (PPMA) that connects the IP blocks directly to the memory controller, see Figure 15.2. The PPMA allows multiple active high-speed transactions because, essentially, it consists of a set of independent direct (non-pipelined) wires from the IP blocks to the memory controller. The MIPS and memory controller are re-used from earlier designs, and have different interfaces. As a result, the MIPS is connected to the memory controller using a PI bus (F-PI) running at processor speed (the M-PI bus runs more slowly), via a gate (F-GATE, which has only a slave port at the MIPS side). As discussed next section, this arrangement is not ideal because it increases the MIPS's latency when accessing the external memory.

The memory controller can efficiently schedule multiple outstanding transactions, as all client requests are visible. However, in this interconnect design, the very large number of long global (top-level) wires running from the many IP blocks to the single memory controller significantly complicated clock-tree balancing and global timing closure.

IP blocks are connected to the PPMA by means of adapters that packetize (format) the data of the IP blocks to 128-byte data bursts. The packetization memory that is required for this is merged with the latency-hiding memory. The latter hides the access latency an IP block would see before it is served by the memory controller. The use of adapters is essential to implement the communication abstraction, further discussed in Section 4.4.

Combining the two interconnects To allow cores on the PI busses access to the external memory, two additional gates (M-GATE and T-GATE) are used. The F-PI and M-PI MIPS PI busses are connected by the MIPS bridge (M-BRIDGE). The interconnects are summarized in Table 15.4.

In Section 4.6 we discuss the performance and arbitration issues of the chosen interconnect scheme, in particular the role of the gates.

4.4 Communication Abstraction

In this design, there are two different interconnects: a point-to-point PPMA of 64 bits wide, and tri-state PI busses of 32 bits wide. PI busses are used for the control traffic and the MIPS PPMA interface because IP blocks were only available with tri-state PI-bus interfaces. Alternative bus implementations (multiplexed or wired-OR) were therefore not considered although tri-state busses can cause testability and lay-out problems. The F-GATE is another penalty (in terms of additional latency for the MIPS cache misses) because the memory controller does not use the PI bus protocol. These observations motivate the communication-abstraction concept introduced in Section 3.2, i.e. IP blocks use an abstract point-to-point interface and protocol suitable for them, which is converted to different interconnect protocols by means of adapters. In fact, communication abstraction was already used to connect IP blocks to the PPMA (except for those on the F-PI bus), and proved to be very successful in re-using these IP blocks in Viper's successors, even as interconnects evolved.

4.5 Memory Organization

Most logical memories are mapped on the external memory, because they are too large to be kept on chip. This includes algorithmic memories (such as SD/HD fields for temporal processing), decoupling memories (e.g. SD/HD fields), and the instruction memories for the processors. The only exception is the line memories of the memory-based scaler, which are mapped to on-chip local memory. Both processors use instruction and data caches to hide the latency of accessing data and instructions in the external memory.

Latency-hiding memories, which hide the variations in data-access latency (jitter) to the external memory, are kept on chip. This jitter is introduced by the memory controller when it arbitrates between the urgent cache misses of the processors (HRLL), the heavy data rates of streaming traffic (HRLT), and the remaining traffic (e.g. best-effort graphics). These latency-hiding memories are implemented in the adapters (the boxes marked R and/or W in Figure 15.2). They are merged with the packetization memories that are required to convert the IP block's data to the format used. The interconnect, memory controller, and overall system requirements determine this format. None of the interconnects (PI busses, PPMA) are pipelined. The bridges and gates contain only memory to synchronize clock domains. They are therefore *circuit-switched*, that is, they make an end-to-end connection between the master and the slave, occupying all intervening interconnects. While this simplifies transaction handling, it also causes a performance bottleneck, as we shall see below.

4.6 Arbitration

No arbitration takes place in the PPMA. Instead the memory controller considers all outstanding requests from all IP blocks connected to it. The memory controller optimizes the bandwidth to the external memory. The run-time-programmable arbitration scheme uses time-division multiplexing, with two priorities per slot. The higher priority guarantees a maximum latency to low-jitter clients, and the lower priority allows other clients to use the bandwidth left over.

Traffic with similar characteristics is coalesced before entering the PPMA, conceptually adding a first level of round-robin arbitration. This includes both the (multiple) read and write ports of a single core (e.g. VMPG in Figure 15.2), and multiple cores (e.g. VIP1 and VIP2). This reduces the number of top-level wires to the memory controller.

The arbitration of the PI busses proceeds independently, except when addressing a slave behind a bridge or gate. In this case, both busses are locked until the slave has responded (non-pipelined circuit switching). This works well for the inter-PI bus bridges (M- and C-BRIDGE) and F-GATE. However, when accessing the PPMA through the M-GATE and T-GATE a transaction can be stalled for some time, depending on the traffic class of the master (e.g. best effort). During this time the F-PI bus is locked. For example, a DMA access of the USB module on the M-PI bus can in this way stall the MIPS for some duration, if the MIPS tries to program a core at the same time. For this reason, the latency of control traffic can vary considerably: from 10-20 cycles without DMA interference, to 100+ cycles with interference.

Hence, the way in which traffic types are separated (by data rates rather than latency) and mapped (on separate, yet interacting interconnects) causes interacting arbitration schemes (of the PI busses and memory controller). This complicates guaranteeing the real-time behavior required by the application, and forces overdimensioning of parts of the system (in particular the PI bus frequency). These observations suggest improvements for Viper's successors.

5. Viper2

Viper2 is a successor of Viper, and targets mid to high-end analog, digital, and hybrid (both analog and digital) TV, including wide-XGA plasma and LCD displays. Viper2 extends Viper by handling 100Hz interlaced and 60Hz progressive displays. It upgrades the conversion of interlaced to progressive output video from SD to HD. It also includes advanced video improvement algorithms, such as motion-compensated Digital Natural Motion for SD pictures.

5.1 Computation Mapping

Viper2 contains three processors (one MIPS PR4450 and two TriMedia TM3260) and 60 function-specific cores. The cores are similar to those of Viper, but have higher computational and communication requirements due to HD output.

5.2 Communication Mapping

Viper2 has one external memory, like Viper. The second TriMedia adds more HRLL cache-miss traffic, and the larger number of cores, with higher data rates, load the external memory close to its maximum.

5.3 Interconnect Organization

Recall that in Viper, the interconnects are defined by data-rate requirements: PPMA for high-data-rate traffic, and the PI busses for low to medium-data-rate traffic. Their interaction via the gates leads to performance issues on the PI busses (cf. Section 4.6). For Viper2, this problem would have been more acute with the increased number of IP blocks. This is addressed by *partitioning the interconnect on the basis of traffic types* instead. Cache misses (HRLL), streaming data (MRLT and HRLT), and control traffic (LRLL) are separated in three interconnects, respectively. No bridges are required between these inter-

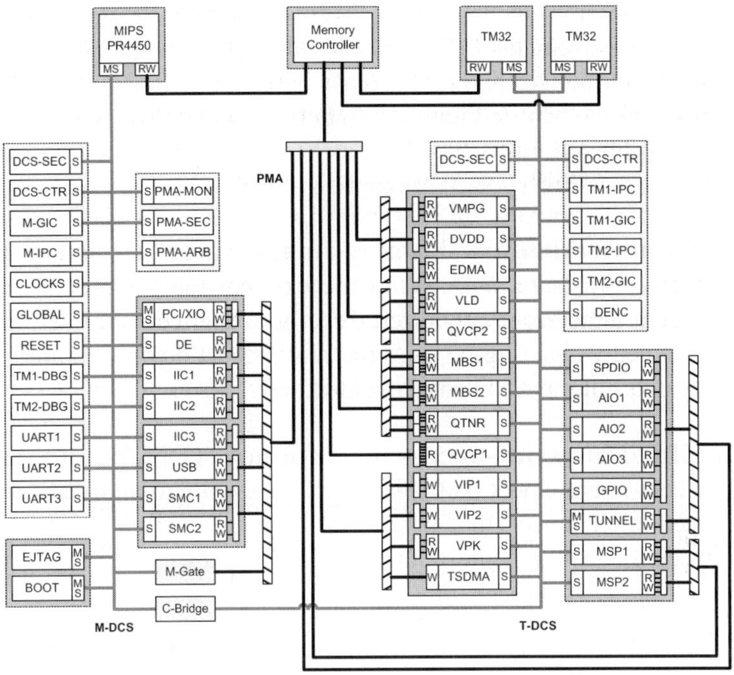

Figure 15.3. Simplified block diagram of Viper2.

connects, and hence there is no interaction between their arbiters. This avoids the traffic interactions of Viper. (Observant readers will notice the M-GATE in Figure 15.3, which, however, is only used during booting and debug.) Below, we describe each of the interconnects, and return to arbitration issues in Section 5.6. The interconnects are summarized in Table 15.4.

The device control and status interconnect (DCS) The DCS interconnect differs from Viper's PI busses in several ways. First, the traffic to be transported over the DCS interconnect is homogeneous (only low-data-rate low-latency traffic), in contrast to the PI busses in Viper. Only single-word transactions are allowed, to ensure low latency. Second, it is a synchronous or asynchronous wired-OR bus, not a tri-state bus, to lower the length and number of wires, and so increase its frequency and improve its testability. Timing closure is alleviated by structured post-lay-out netlist modification. Further, to ensure low latency, the load on the DCS busses is kept to less than 0.5% (one data word per cycle corresponds to 100%). Finally, to hide the significant difference in speed between the processors and the core being accessed, posted writes and reads can be used. A posted read accesses a (possibly out-of-date) copy of a core's registers in the adapter of the core, which is in the fast DCS clock domain instead of the slower core's clock domain. By the combined effects of these improvements the latency of the control interconnect of Viper2 is the same as in Viper, despite the increase in the number of IP blocks.

The bridge (C-BRIDGE in Figure 15.3) between the MIPS (M-DCS) and TriMedia (T-DCS) interconnects functions as before (i.e. both interconnects are locked when addressing a core through the C-BRIDGE).

The high-bandwidth low-latency interconnect The three processors all require low-latency access to the external memory for their cache misses. They are therefore connected directly to the memory controller, using non-pipelined wires. Viper's F-PI bus and F-GATE to connect the MIPS to the memory controller have been eliminated.

The pipelined memory-access interconnect (PMA) The pipelined memory-access interconnect (PMA) is Viper2's medium to high-bandwidth latency-tolerant interconnect. In Viper, the 64-bit point-to-point PPMA has direct wires between the memory controller and all IP blocks with high data rates (except for the MIPS). In Viper2, this is no longer feasible because the Viper2's increased chip area results in longer wires, and the medium-data-rate cores on the PI bus are moved to the PMA. Moreover, the number of medium to high-data-rate IP blocks has increased, as well as their data-rate requirements (to deal with HD instead of SD pictures).

The PMA therefore contains two innovations. Both hinge on the fact that many masters communicate with a single slave, like in Viper. First, the outstanding transactions from multiple cores are presented one at a time to the memory controller, to reduce its complexity, and to decouple the PMA communication arbitration from the memory arbitration (breaking global arbitration into local subarbitrations). The memory controller is now independent of the number of cores, making it more re-usable. We return to the PMA arbitration in Section 5.6.

Second, the point-to-point wires of PPMA of Viper have been replaced by a pipelined multiplexed interconnect to reduce the length and amount of wires, to ease lay-out and timing verification. A tree topology clearly fits well with exposing a single transaction to the memory controller, as transactions of different cores converge towards the top. In Figure 15.3 the cores on the dark shaded background connect to a PMA of 8 nodes (shown as hatched boxes), using 28 ports, but it is clearly scalable to a larger number of masters. Note that the tree topology is motivated by lay-out and timing closure, and that a node in the tree is not necessarily a point of arbitration (further discussed in Section 5.6).

The adapters glueing the cores and PMA together contain combined memories for clock-domain crossings, packetization (to convert data from the cores to 128-byte PMA bursts), and pipelining and latency-hiding memory (to overcome the latency and jitter introduced by the memory controller and PMA).

5.4 Communication Abstraction

In Viper, re-use of legacy IP blocks required the F-PI bus with its associated side effects (cf. Section 4.4). This motivates the creation of IP blocks that are independent of the interconnect, which was already applied to Viper's PPMA interconnect. Many of Viper's IP blocks were re-used in Viper2 without change, proving the value of the methodology. Even the adapters that connected the IP blocks to the PPMA required only minor modifications for the PMA (e.g resizing of the packetization and latency-hiding buffers).

The DTL (device-transaction-level) protocol [16] that is used has several profiles, which are related to the traffic types. The MMIO (memory-mapped IO) profile is chiefly used for control traffic (LRLL). The MMBD (memory-mapped block data) and MMSD (memory-mapped streaming data) profiles are used by IP blocks to communicate via shared on- or off-chip memory. MMBD and MMSD are used predominantly for reading and writing, respectively. Cache misses (HRLL) use MMBD, while streaming (audio MRLT and video HRLT) cores use MMSD or MMBD.

The use of DTL ensures that IP blocks can transparently connect to any of the interconnects (direct IP to IP communication, PMA, DCS interconnect), making it easier to move IP blocks within a design, and to re-use them across designs, with possibly different interconnects.

5.5 Memory Organization

All algorithmic memories (such as SD/HD fields for temporal processing), the decoupling memories, and the instruction code for the processors are mapped to external memory. There are two exceptions. The first, like in Viper, is the line memories of the memory-based video scaler (MBS), which are mapped to on-chip local memory of the MBS. Second, a small local memory is introduced in one place to (significantly) reduce the bandwidth pressure on external memory. Although this introduces another slave, it statically connects only two IP blocks, and hence is not part of the PMA.

All processors use instruction and data caches to hide latency.

The adapters contain latency-hiding memories for cores, to even out variations in data access latency to external memory. This variation is due to the combined effects of the memory controller (like in Viper), and the PMA arbitration (new in Viper2). The adapters contain packetization memories to convert IP-block data to a format suitable for efficient transport over the PMA. The packetization memories and latency-hiding memories are merged for area efficiency.

There are no pipeline memories in the DCS. The PMA is a pipelined circuit-switched interconnect. This means that every node in the tree contains some pipeline stages, to decrease wire lengths, allow higher operating frequencies, and ease timing closure. Circuit switching entails that a transaction that is accepted occupies all nodes on the path from the IP block to the memory controller. To eliminate run-in and run-out of the pipeline, and hence loss of bandwidth, transactions are prefetched as close as possible to the top of the tree, based on their scheduled order.

5.6 Arbitration

The two DCS interconnects are arbitrated independently, using a round-robin scheme (Table 15.3). M-DCS and T-DCS have 6 and 4 masters, and 32 and 38 slaves, respectively. The bridge is circuit switched; it locks both interconnects when addressing a slave at the other side of the bridge.

The memory controller has four inputs, one for each processor and one for PMA. The PMA offers a sequentialized view on the cores, i.e. the multiple active transactions of IP blocks are offered one at a time to the memory controller. The memory controller arbitrates its inputs using a round robin to aim for low latency. To guarantee a maximum latency, burst lengths are limited.

Table 15.3. Arbitration overview.

location	traffic	aim	method
DCS	LRLL	low latency	round robin
memory controller	HRLL	low latency	round robin with cut off
PMA top	low-jitter HRLT	maximum latency	time-division multiplexing
PMA top	jitter-tolerant HRLT	minimum bandwidth	priorities
PMA top	best effort	best effort	round robin
adapters	(all)	coalescing	round robin

Recall from Section 3.1 that jitter-tolerant HRLT cores have an average data-rate requirement but can be delayed when there is no data, whereas low-jitter HRLT cores cannot be delayed. Best-effort cores can operate on whatever bandwidth and jitter they are given.

The PMA uses the fact that many masters contend for a single slave, and arbitrates low-jitter cores using time-division multiplexing (to guarantee a maximum latency), jitter-tolerant cores with priorities (to guarantee a minimum bandwidth), and best-effort cores with a round robin. These arbitration mechanisms are applied (prioritized) in the order listed in Table 15.3. (The TDMA slots are skipped when unused. Note that priorities alone do not guarantee a minimum bandwidth, but their combination with system invariants does.) The arbitration scheme is fully programmable at run time.

The multi-stage arbitration scheme of PMA is motivated by performance requirements, and maps only partially to the physical tree of PMA nodes, which is driven by back-end issues. A node in the physical tree is not necessarily a point of arbitration. In fact, only at the top of the PMA does arbitration take place.

Like in Viper, traffic with similar characteristics is coalesced in the adapters before they enter the PMA, conceptually adding a first level of round-robin arbitration. This includes both the (multiple) read and write ports of a single core (e.g. QVCP1 in Figure 15.3), and multiple cores (e.g. AIO1 to AIO3, SPDIO, and GPIO).

6. An Example Future SOC

In this section we take a leap into the future, and describe a speculative SOC, based on an extrapolation of Viper and Viper2. We present a design that illustrates the trends that we foresee, but intermediate and hybrid solutions are very likely between Viper2 and the future SOC sketched here.

Future applications will contain more advanced video-processing functions, at higher picture resolutions. Examples are motion-compensated high-definition noise reduction and temporal up-conversion, as well as a move to MPEG4 and advanced graphics. We further expect higher and more dynamic data rates.

6.1 Computation Mapping

The number of processors increases to support emerging media-processing applications, and (system-integrator-defined) differentiation of products. The number of function-specific cores also increases, to efficiently implement standard or proprietary application kernels (e.g. PixelPlus and subpixel luminance-transient improvement).

6.2 Communication Mapping

As the number of processors increases, the amount of data and instruction cache misses (HRLL) grows. This is an undesirable trend, because low latency is hard to guarantee for more than a few users of any shared resource, whether it is an external memory or an interconnect. There are several (partial) solutions. First, minimize the use of caches, e.g. reduce multi-tasking to decrease code memory size and increase locality, or improve memory management such as software prefetching in combination with scratch-pad memories. Second, lower the dependency on low latency, e.g. by using hardware multi-threading, or by increasing the emphasis on streaming instead of random-access traffic. Third, use fewer shared resources, e.g. use multiple memories, and interconnects that allow concurrent accesses such as switches and networks.

Control traffic (LRLL) essentially suffers from the same low-latency problems. However, current solutions can be used in the near future because the available head room can accommodate the increase in (low-data-rate) control traffic.

MRLT and HRLT traffic increases because of the growth in the number of function-specific cores, which tend to implement streaming computation. Increasing picture dimensions from standard to high definition, also boosts HRLT traffic (e.g. sixfold for temporal up-conversion).

Thus, future interconnects must address the rise in communication needs, but architectural opportunities to limit the increase in traffic types that are hardest to implement, low latency in particular, should also be exploited.

6.3 Interconnect Organization

We see several trends that affect communication.

First, the number of processors grows, resulting in more masters and LRLL traffic for the control interconnect. Busses like DCS are single-hop broadcast media, for which latency increases for two reasons. The arbiter frequency slows down because global arbitration must take more masters into account, and because the wires from IP blocks to the arbiter become longer. Moreover, without concurrent transactions and with limited operating frequency, only a subset of masters can have low latency.

Second, the increasing number of processors gives rise to more cache-miss (HRLL) traffic, which cannot be supported by only one slave (the external memory). Again, with a single shared resource, not everyone can have low-latency access. Hence multiple memories (slaves) are required. They are probably external because processor instructions are too large to fit on chip. Some kind of switch must connect multiple masters (processors) to multiple slaves (memories).

Third, we have seen that the amount of HRLT traffic grows due to larger pictures. Communicating only via a single external memory is no longer feasible, for bandwidth reasons, and we foresee a combination of multiple off-chip and on-chip memories. The former, while not ideal for energy dissipation and pinning, is indispensable because HD (temporal) video data is too large to be kept on chip. The latter reduce the bandwidth pressure on external memories, and lower the latency and power to access data. In Viper2 a similar technique is used once (cf. Section 5.5), but shared on-chip memories will gain in number and importance. In both cases, we see a growing number of slaves, which the PMA interconnect alone cannot address.

Finally, both Viper and Viper2 contain IP blocks (tunnels) to communicate with off-chip components, either in a system-on-package or multi-chip setting.

Similar examples are USB, PCI, and PCI EXPRESS. A further rise in the use of these interfaces (and hence masters and slaves) is likely.

Networks on chip (NOC)

From the preceding discussion we conclude that any future interconnect must deal with many masters and many slaves, with high data rates. Busses cannot fulfill the bandwidth requirements. Switches [12] fare better because they offer concurrent master-slave communications, but are not scalable to the extent we require (50+ masters, and 50+ slaves). The PMA interconnect is optimized for a single slave, and multiple instantiations would be necessary. Multiple switches, or networks on chip (NOC) [21, 22, 23, 24] are scalable, and can solve many of the issues listed here. A NOC consists of a collection of routers (or switches) that transport data in packets. Adapters, now called network interfaces, connect routers to IP blocks and packetize the transactions of the IP blocks. The remainder of this section compares NOCs with the other interconnects.

First, we observe that wires connecting IP blocks are underutilized (as little as 10% [25]). Both PMA and NOCs reduce the number of wires that interconnect IP blocks by sharing them. However, instantiating PMA multiple times to address multiple slaves would increase the number of wires. Therefore, NOCs are better scalable in the number of attached slaves, as illustrated in Figure 15.4.

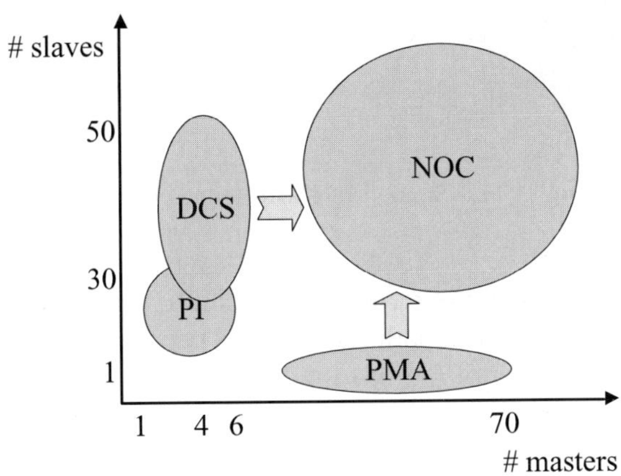

Figure 15.4. Interconnect evolution.

Second, a NOC is scalable in the sense that adding more routers results in more bandwidth. A NOC copes well with many masters and slaves because many transactions can take place concurrently. There are several reasons for this. First, a NOC has distributed arbitration, unlike a bus or switch. This removes a major bottleneck, cf. Section 6.6. Second, in the appropriate topology (such as a mesh or (partial) fat tree, e.g. Figure 15.5) there are many independent paths that can be used simultaneously. Finally, packet switching is commonly used in NOCs, instead of circuit switching, employed in Viper2's PMA. As interconnects increase in size (number of routers), their diameter (distance between master and slave) grows, and reserving wires end to end (from master to slave) for the duration of the transaction becomes inefficient. The set-up and tear-down phases of the circuit take longer, causing congestion (blocking other transactions) [26]. (Fundamentally, multiple interconnects with circuit-switching bridges, such as those in Viper and Viper2, suffer from the same problem, discussed in Section 4.6.) Packet switching reduces these problems, by allowing pipelined transactions (on a single path), possibly at the cost of higher latency. NOCs can therefore offer tremendous bandwidth between many masters and slaves [27].

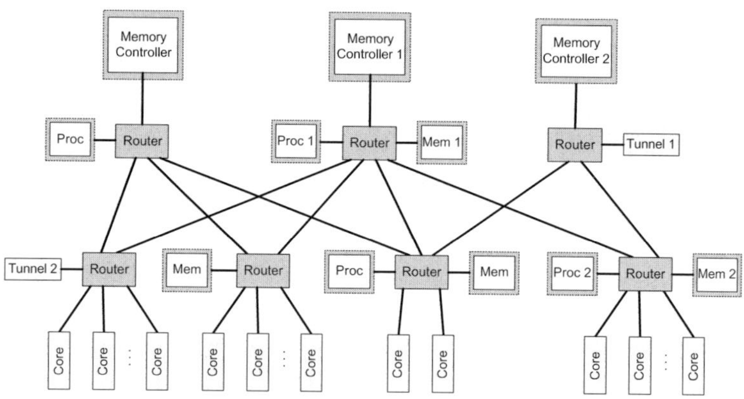

Figure 15.5. Simplified block diagram of future SOC.

Third, while NOCs scale in terms of bandwidth, this is not so clear cut for latency. Packet switching has both positive and negative effects on the latency. High operating frequency of (point-to-point) links and routers, and concurrent transactions reduce the latency, while arbitration per router, and possible congestion may increase it. In any case, placing IP blocks with latency-critical communication close to one another in the NOC topology reduces the number of router hops, and hence the latency (e.g. proc 1 and memory controller 1 in Figure 15.5).

Finally, a major advantage of NOCs is their ability to offer differentiated services [28]. This means that different traffic types can be implemented on a single NOC, and that different traffic types can be multiplexed on a single network port. In Section 6.6 we discuss the strong relation between NOC services and NOC arbitration.

6.4 Communication Abstraction

The re-use of Viper2 IP blocks in a NOC is straightforward due to the use of DTL. The adapters from DTL required little modification to change from Viper's PPMA to Viper2's PMA internal communication protocol. However, moving from Viper2's circuit-switching PMA to a NOC's packet switching is more elaborate (e.g. end-to-end flow control, transaction reordering, distributed memory), but existing IP blocks are unaffected.

A NOC has the ability to offer different services to different connections, on a single network port. For example, a multi-tasking processor can request a connection to shared memory per task, with different properties per connection, such as bandwidth, latency, transaction ordering, and flow control. In fact, this is essential when multiplexing several logical communications over USB or PCI EXPRESS to off-chip components. It also eases the design of real-time multimedia applications [24], like those discussed here. If we want to take advantage of this capability, DTL must be extended to deal with connections (optional for backward compatibility). This tendency can already be observed in the appearance of thread and connection identifiers in OCP and AXI.

6.5 Memory Organization

In Section 6.3 we argued that multiple external memories will be likely in future SOCs to cope with increasing HRLL cache traffic and HRLT traffic (for algorithmic and decoupling memories).

We also foresee multiple on-chip memories for low power, to lower data access latency, and to relieve pin and bandwidth pressures. Viper2's processor and cache memory model can be extended to a memory hierarchy, as is illustrated in Figure 15.5. Proc 2 has its own cache (not shown), but can overflow to memory mem 2, which is relatively close (one hop), or memory mem 1, further away (two hops, but still on chip), or one of the external memories memory controller 1 or memory controller 2 (two hops, but passing through a memory controller). In fact, *all* memories are accessible to any of the IP blocks, but at non-uniform cost (although possibly within a uniform address space, i.e. NUMA). Similarly, all IP blocks can be programmed by any of the processors. A multi-master multi-slave NOC interconnect is therefore useful for both data and control traffic, as suggested by Figure 15.4.

On-chip memories can function as caches or scratch pads (for data and instructions). By keeping inter-IP-block communication on chip the latency and jitter introduced by memory controllers is eliminated, reducing the size of latency-hiding memories (cf. Section 5.5).

Like for Viper2, the network interfaces contain latency-hiding memories for streaming cores, to even out variations in data access latency. If the NOC offers low-latency and/or low-jitter communication, these and on-chip latency-hiding memories can be reduced. The network interfaces also use some memory to packetize the data to the format used by the NOC routers, and to cross clock domains. NOCs have pipelined routers (even in circuit-switched variants), and sometimes pipelined links too, to increase the operating frequency of the network. The ÆTHEREAL NOC from Philips, for example, provides a combined guaranteed-bandwidth-and-latency service with a router pipeline of three words deep [27].

6.6 Network Services and Arbitration

A major advantage of NOCs is their ability to offer differentiated services. This means that different traffic types can be implemented on a single NOC by means of a protocol stack [14], and different traffic types can be multiplexed on a single network port. Different DTL profiles and traffic coalescing, like that of Viper and Viper2, is then taken care of by the network (interface). In particular, the ÆTHEREAL NOC [28] offers guaranteed bandwidth, and best-effort connections, that are useful for the traffic types listed in Table 15.2.

However, sophisticated global arbitration such as PMA's, is more expensive when using the distributed arbitration of NOCs. Time-division multiplexing is relatively cheap, but distributed priority- or rate-based arbitration is not acceptable, in terms of buffering cost of routers [27]. The PMA interconnect can take advantage of its tree topology for its arbitration, but this is harder even in regular NOC topologies such as fat trees and meshes. NOC services are therefore less expressive and flexible than PMA arbitration. Moreover, it is harder for distributed arbitration to be of the same quality as global arbitration (cf. contention and congestion). However, given the abundance of bandwidth [27], this can be addressed by bandwidth overallocation, with best-effort traffic consuming unused capacity.

7. Conclusions

The advanced set-top box and hybrid TV (ASTB) application is demanding in terms of computation (high-definition video pixel processing), memory (temporal video data), and communication (high data rates) requirements. The first results in heterogeneous computation elements (function-specific cores, various processors). Instruction, algorithmic, and decoupling memories are all large and mapped in off-chip memory. The communication infrastructure is critical because it must connect many IP blocks with high data rates, in a flexible manner for product differentiation and run-time mode changes. Finally, the application centers on real-time audio and video, which means that the system resources (memories, interconnect) must be carefully managed (arbitrated).

Viper's interconnect is separated by data-rate requirements, resulting in a low- to medium-bandwidth interconnect (for LRLL and MRLT) consisting of two bridged PI busses, and a high-bandwidth interconnect (PPMA for HRLL and HRLT) of dedicated wiring (Table 15.4). Utilization of both interconnects is high, but the circuit-switched M-GATE between the interconnects causes interference of arbiters for different traffic types, making real-time guarantees more intricate.

Table 15.4. Mapping traffic types to interconnect structures.

	LRLL	MRLT	HRLT	HRLL
Viper	PI	PI	PPMA	PPMA
Viper2	DCS	PMA	PMA	point to point
future	DCS / NOC	NOC	NOC	point to point / NOC

To avoid this, Viper2's interconnect is separated by traffic kind, resulting in three independent interconnects: two bridged DCS interconnects for LRLL control traffic, dedicated wiring for HRLL cache-misses, and the PMA interconnect for MRLT and HRLT audio and video. The PMA uses a sophisticated global arbitration scheme (Table 15.3) that distinguishes low-jitter HRLT, high-jitter HRLT traffic, and best-effort classes, for a high utilization.

Future systems will use multiple on- and off-chip memories, increasing the number of masters and slaves, see Figure 15.4. This motivates a move to multi-hop interconnects, such as networks on chip (NOC). NOCs are scalable in the number of masters and slaves, in bandwidth, and to a lesser extent in latency. NOCs can offer differentiated services and very high bandwidth, but their distributed arbitration favors scheduling simpler than that used in Viper2's PMA.

The challenge for future SOCs for real-time applications is to define advanced memory organizations (e.g. a hierarchy of on- and off-chip memories), and to offer different communication services (different traffic classes) with an interconnect structure that is both scalable and cost efficient, e.g. a NOC.

References

[1] Santanu Dutta, Rune Jensen, and Alf Rieckmann. Viper: A multiprocessor SOC for advanced set-top box and digital TV systems. *IEEE Design and Test of Computers*, pages 21–31, Sept-Oct 2001.

[2] M. Brett, B. Gerstenberg, C. Herberg, G. Shavit, and H. Liondas. Video processing for single chip DVB decode. In *IEEE Transactions on Consumer Electronics*, volume 47, pages 385–393, August 2001.

[3] O. P. Gangwal, J. G. Janssen, S. Rathnam, E. B. Bellers, and M. Duranton. Understanding video pixel processing applications for flexible implementations. In *Proceedings of Euromicro, Digital System Design*, pages 392–401, September 2003.

[4] G. de Haan. *Video Processing for Multimedia Systems.* ISBN: 90-9014015-8, September 2000. Eindhoven.

[5] E. G. T. Jaspers, P. H. N. de With, and J.G.W.M. Janssen. A flexible heterogeneous video processor system for television applications. In *IEEE Transactions on Consumer Electronics*, volume 45, pages 1–12, February 1999.

[6] R. Kramer. Consumer electronics as silicon engine. *International Electron Devices Meeting (IEDM)*, pages 3–7, 1999.

[7] H. Yamauchi, S. Okada, K. Taketa, Y. Mihara, and Y. Harada. Single chip video processor for digital HDTV. *IEEE Communications Magazine*, pages 394–404, August 2001.

[8] Markus Rudack, Michael Redeker, Jörg Hilgenstock, Sören Moch, and Jens Castagne. A large-area integrated multiprocessor system for video applications. In *IEEE Design and Test of Computers*, pages 6–17, January 2002.

[9] G. de Haan. IC for motion-compensated de-interlacing, noise reduction, and picture-rate upconversion. In *IEEE Transactions on Consumer Electronics*, volume 45, pages 617–624, August 1999.

[10] G. de Haan and J. Kettenis. System-on-silicon for high quality display format conversion and video enhancement. In *Proc of ISCE'02*, pages E1–E6, September 2002.

[11] M. Schu, D. Wendel, C. Tuschen, M. Hahn, and U. Langenkamp. System-on-silicon solution for high quality consumer video processing–the next generation. In *IEEE Transactions on Consumer Electronics*, volume 47, pages 412–419, August 2001.

[12] Jeroen A.J. Leijten, Jef L. van Meerbergen, Adwin H. Timmer, and Jochen A.G. Jess. Stream communication between real-time tasks in a high-performance multiprocessor. In *Proceedings of Design, Automation and Test in Europe Conference*, pages 125–131, 1998.

[13] K. Keutzer, S. Malik, A. Richard Newton, Jan M. Rabaey, and A. Sangiovanni-Vincentelli. System-level design: Orthogonalization of concerns and platform-based design. *IEEE Trans. on CAD of Integrated Circuits and Systems*, 19(12):1523–1543, 2000.

[14] M. Sgroi, M. Sheets, A. Mihal, K. Keutzer, S. Malik, J. Rabaey, and A. Sangiovanni-Vincentelli. Addressing the system-on-a-chip intercon-

nect woes through communication-based design. In *Design Automation Conference*, pages 667–672, June 2001.

[15] Drew Wingard. Socket-based design using decoupled interconnects. Chapter 15, this volume.

[16] Peter Klapproth. Architectural concept for IP-re-use. In *VLSI ASP DAC*, December 2002.

[17] VSI Alliance. Virtual component interface standard, 2000.

[18] OCP International Partnership. Open core protocol specification, 2001.

[19] ARM. *AMBA AXI Protocol Specification*, June 2003.

[20] Open Microprocessor Initiative. *OMI/PI-Bus specification*, OMI 324: PI-Bus Rev. 0.3d edition, 1994.

[21] Pierre Guerrier and Alain Greiner. A generic architecture for on-chip packet-switched interconnections. In *Proceedings of Design, Automation and Test in Europe Conference*, pages 250–256, 2000.

[22] Luca Benini and Giovanni De Micheli. Networks on chips: A new SoC paradigm. *IEEE Computer*, 35(1):70–80, 2002.

[23] K. Goossens, J. van Meerbergen, A. Peeters, and P. Wielage. Networks on silicon: Combining best-effort and guaranteed services. In *Proceedings of Design, Automation and Test in Europe Conference*, pages 423–425, March 2002.

[24] Kees Goossens, John Dielissen, Jef van Meerbergen, Peter Poplavko, Andrei Rădulescu, Edwin Rijpkema, Erwin Waterlander, and Paul Wielage. Guaranteeing the quality of services in networks on chip. In Axel Jantsch and Hannu Tenhunen, editors, *Networks on Chip*, chapter 4, pages 61–82. Kluwer, 2003.

[25] William J. Dally and Brian Towles. Route packets, not wires: On-chip interconnection networks. In *Design Automation Conference*, pages 684–689, June 2001.

[26] André DeHon. Robust, high-speed network design for large-scale multiprocessing. A.I. Technical report 1445, Massachusetts Institute of Technology, Artificial Intelligence Laboratory, September 1993.

[27] E. Rijpkema, K. G. W. Goossens, A. Rădulescu, J. Dielissen, J. van Meerbergen, P. Wielage, and E. Waterlander. Trade offs in the design of a router with both guaranteed and best-effort services for networks on chip. In *Proceedings of Design, Automation and Test in Europe Conference*, pages 350–355, March 2003.

[28] Andrei Rădulescu and Kees Goossens. Communication services for networks on silicon. In Shuvra Bhattacharyya, Ed Deprettere, and Juergen Teich, editors, *Domain-Specific Processors: Systems, Architectures, Modeling, and Simulation*, pages 275–299. Marcel Dekker, 2003.

Chapter 16

A BRUNCH FROM THE COFFEE TABLE - CASE STUDY IN NOC PLATFORM DESIGN

Tapani Ahonen*, Seppo Virtanen**, Juha Kylliäinen*, Dragos Truscan***, Tuukka Kasanko*, David Sigüenza-Tortosa*, Tapio Ristimäki*, Jani Paakkulainen**, Tero Nurmi**, Ilkka Saastamoinen*, Hannu Isännäinen*, Johan Lilius***, Jari Nurmi*, and Jouni Isoaho**

**Institute of Digital and Computer Systems, Tampere University of Technology, Finland*

***Department of Information Technology, University of Turku, Finland*

****Department of Computer Science, Åbo Akademi University, Finland*

1. Introduction

As the non-recurring engineering costs of new technologies continue to rise, the importance of mass production grows. Mask costs in the million-dollar range are intolerable for low-volume products that are not targeted for high-end applications. On the other hand, Field Programmable Gate Arrays (FPGAs) are free from such costs, but feature a high unit cost. Beside this, interconnect delay and power consumption penalties in FPGAs may get prohibitively high for some applications. The gap to Application-Specific Integrated Circuits (ASICs) continues to widen in this sense, because wires dominate the performance of FPGAs. Interconnect downscaling increases resistance and sensitivity to electrical and mechanical stress, causing serious timing and reliability problems. These challenges in interconnect processing technology have incurred the fact that wire downscaling is not keeping up with the pace of logic downscaling. At the same time, decreasing wire spacing inevitably strengthens signal coupling effects between adjacent wires, which increases the level of uncertainty in functional correctness. On top of all that, there is considerable leakage current and hence leakage power due to thin isolation layers. The leakage current and interconnect effects together constitute the most important Deep SubMicron (DSM) effects that have to be taken into account already at an early stage of a design process of an ASIC. Obviously, something between ASICs and FPGAs is needed to simplify circuit design while keeping the cost and performance penalties to a minimum. For a more detailed discussion, refer to part 1: Physical and Electrical Issues.

The DSM effects and the growing scale of Network-on-Chip (NoC) designs limit designer productivity by increasing the complexity of verification. NoCs are Systems-on-Chip (SoCs) that are implemented around an on-chip communication network rather than a bus. It is a well-established fact that designer productivity has not grown together with integration level and that there is little to gain by ever increasing the size of design teams. Team sizes between 8 and 15 people have been argued to be the most effective, while several such teams may be working on different aspects or abstraction levels of a single design. The use of pre-designed and verified intellectual property (IP) blocks is being adopted to reduce the aforementioned problems. This increases designer productivity at the architectural level by pushing the required effort more towards system integration. Among other things, system integrators have to focus on interaction issues including communication requirements, compatibility of interfaces, correct co-functionality as well as successful technology migration. Thus a significant design effort is still needed. A natural step forward is to capture this effort for design reuse.

Platform-based design enables better use of new technologies. Reuse of an existing NoC platform effectively lowers system design and manufacturing costs at the expense of a small efficiency overhead due to a more general purpose nature of the underlying hardware (HW). The problem of verification is essentially raised to the application level of abstraction as it has been solved for the lower abstraction levels during platform development. NoC platforms are composed of data acquisition and signal formation elements like Analog to Digital Converters (ADCs) and Digital to Analog Converters (DACs), memories, and processing elements (PEs) such as Digital Signal Processor (DSP) and Reduced Instruction Set Computer (RISC) cores. Equipped with a customized software (SW) environment including a real-time operating system(s) and a compiler(s), one design could effectively apply to a complete application domain. In many cases, however, the interconnection network forms the performance bottleneck of the system and therefore care should be taken in its design. The platform architecture design and customization processes should be automated with parameterizable IP libraries as the basis of such automation. Of the current efforts to automate platform design, the MESCAL (Modern Embedded Systems, Compilers, Architectures and Languages) project [2] is one of the most thorough.

1.1 Aspects of Platform Design

Successful system design has always been based on separation of concerns. This is also true for platform design, where the key issue is to isolate communication and computation from each other. Being the dominating bottleneck, the communication network varies the most when tailored for the needs of the application domain at hand. For communication-centric style of design, computational elements should only be regarded as producers and consumers of communication traffic, whereas the communication network should be made as transparent as possible from the computation point of view. This means, among other things, that the model of computation (MoC) chosen should by no means constrain the implementation of the communication network. It is preferable that the MoC does not imply any communication model at all, even up to the highest protocol layers. For a more detailed discussion, refer to part 3: Design Methodology and Tools.

Networks-on-Chip can be categorized into two main types, circuit and packet switching networks. The packet switching networks are better suited for NoC platform development. In this type of network, varying transaction latency and bandwidth (BW) are the main drawbacks, but compared to the circuit switching networks they are far more flexible and hence easier to abstract by means of protocols. Among other things, higher protocol layers can be utilized to fix a constant latency and bandwidth for desired parts of the network by utilizing virtual circuit switching. Performance is naturally hindered by protocol-related overheads such as bandwidth loss due to packet headers and error correction codes. The inherent need to store packets in the network increases implementation cost. Additional resources are also required for the computational and control tasks like error detection and correction, packet queuing, and routing. On the other hand, the performance of actual circuit switching networks suffer from limitations like rerouting overhead, difficulties in network partitioning, and more probable failure due to lack of alternative paths.

Since even a single communication failure could paralyze the whole system, selection of packet switching over circuit switching could be characterized as an insurance to protect the investments in system design. As established earlier, communication failures will become more frequent with the adoption of smaller technologies unless wire geometry and layout are designed with care and the shielding of the whole chip and individual wires is done appropriately, not to mention thorough manufacturing tests for connectivity. Static communication defects usually originate from manufacturing or incomplete verification whereas dynamic ones are most often caused by electrical shocks such as switching of an adjacent wire and variation in power supply voltage levels. One useful feature of packet switching networks is that they can be designed to tolerate some amount of static and dynamic defects. This increases the manufacturing process yield and thus amortizes implementation costs. In addition, there's less need for redesign and verification efforts because first prototypes are more likely to be functional. For a more detailed discussion on advantages of packet switching networks, refer to chapter 9: From Buses to Networks.

2. Approaching Platform Design

NoC platform design is an elaborate task calling for a more sophisticated design flow compared to ASIC or processor design. In traditional ASIC design, everything is done to satisfy the requirements of a single application. The aim of processor design is efficient hardware resource utilization, usually, setting the framework for application development. Although processors and ASICs are among the building blocks for NoC platforms, the point of view of hardware or a single application is inadequate as a starting point for platform design. Instead, engineers are required to make tradeoffs between varying aspects of multiple systems.

A typical platform design flow, illustrated in figure 16.1, begins with the exploration of the application domain requirements. The supported applications are broken into tasks (SW and HW processes). Their characteristics (execution time, resource utilization, power dissipation etc.) are derived and mutual dependencies specified. Information about appearance of special functions and function sequencies calling for hardware acceleration is also needed. In a system of multiple applications, tasks may be shared. Thus, the supported systems need to be characterized since a system might be different from the sum of individual applications constituting the system.

Exploiting the characteristics of the targeted systems (including applications and tasks) statistical execution models are formed. On basis of these models, the temporal processing power requirements are derived to enable the selection of PEs. It is essential not to set too stringent requirements for the design resulting in ASIC-like optimizations. Instead, careful cost per performance analysis should be carried out for key parameters to enable tradeoffs that minimize the combined cost of characteristics for large sets of tasks while the performance requirements are met.

The target of mapping and scheduling tasks to the execution units should be at minimization of the global communication traffic. Heavily

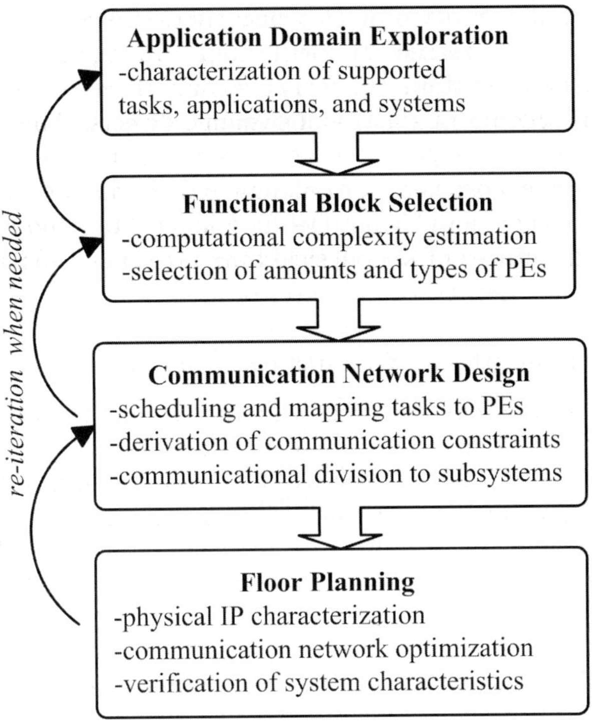

Figure 16.1. NoC hardware platform design flow.

communicating tasks should be mapped to the same PE if possible. After successful mapping and scheduling, inter-PE communication requirements can be derived. Two simple metrics, throughput and latency, are used to characterize communication between modules. For a more precise characterization, also the communication pattern distribution over time should be considered.

The highest priority issue in interconnection design is to keep the paths with most stringent requirements (highest cost) as short as possible down to the layout level of abstraction to ensure low delays and/or power consumption. This is achieved by exploiting hierarchy and communicational partitioning to groups of PEs running tightly coupled tasks. The Globally Asynchronous Locally Synchronous (GALS) [10, 11] design paradigm or a similar kind of functional division to clock islands helps with this partitioning. Inside the partitions the relative placement of PEs is optimized to keep the most congested interconnections short. Short interconnections also reduce the number of required metal layers and hence fabrication costs. With a Network-on-Chip (NoC) communication model the number of links (connections from a network node to another) may be decreased for further optimization as long as performance goals are met. However, reducing connections might compromise system functionality under noisy conditions, since alternative paths are restricted.

Interconnection cost per performance optimization is about optimizing the system-level floor plan. Detailed information about each Intellectual Property (IP) block's physical characteristics on the target technology is needed. Accurate estimates of area, shape of the layout, and locations of communication attachment points enable approximations of interconnect length, required number of repeaters, and metal width. The metal thickness in turn determines the layer used as well as affects the electrical delay constant RC. With the approximations it is possible to estimate the achievable operating frequency at different parts of the network, the minimum communication latencies, and the maximum throughputs. These estimates enable refinement of the communication network. Finally, the performance characteristics of the whole system are approximated. If the design fails to meet the specification at any point of the flow, previous stages are resumed for iteration.

By a quick look at the presented design flow, which is by no means exhaustive, it is obvious that NoC platform design includes a variety of non-trivial tasks that should be automated to some extent. The architectural design process itself is so overwhelming that it should not be complicated further by component design activities. Instead, the design team should be able to rely on well-documented and thoroughly verified IP components throughout the project. Ideally, the metrics for all the components should be known on the target technology. If a component is implemented for the first time using the target technology, a performance model for that technology should be derived on the basis of implementation document on a different technology. These models enable early stage performance estimation [1] for design automation purposes. An idealized design flow goes from collecting and grouping the needed data about target applications and IP-blocks to an abstract graphical entry that is refined to the floor plan level of abstraction, simulated, and synthesized [2].

2.1 Case Study in Platform Design and Application

A simple case study in platform design and application is presented in the subsequent sections of the chapter. The used methodology deviates from the described one where design tasks have not been satisfactorily automated. In section 3, the study concentrates on forming a multimedia processing platform by exploiting previously designed and verified soft and hard IP components. Due to limitations in IP availability and lack of design automation, proper application domain exploration has not been performed. In section 4, the application part of the case study, the formed platform is used as a sub-network that receives, decrypts, and decodes a compressed video stream.

3. A NoC Platform for Networked Multimedia Processing

The multimedia processing platform presented here is built on the PROTEO on-chip communication backplane. This packet switching network has been configured to provide the communication resources and services required by the application domain. System components, which are called network hosts, are connected to the PROTEO network through network nodes. These nodes can be thought of as local protocol processors that perform data routing and buffering. The Virtual Component Interface (VCI) standard [21] has been utilized for connections between the hosts and the nodes. PROTEO also supports the Open Core Protocol (OCP) standard [12], which is a superset of the VCI. An overview of the soft and hard IP components used in the processing platform is given in table 16.1. A soft IP component is such that it can be modified, parameterized, and synthesized to various target technologies [5]. Thus, it has to be delivered to the customer as a synthesizable hardware description, not as a technology-specific netlist like a hard IP component. Additional reuse-facilitating features of soft IPs include good documentation, well-designed verification suites, and synthesis scripts for the most popular tools.

Component	Description	IP type	Interface type
PROTEO	Packet switching network backplane	Soft IP	Any VCI
TACO	Protocol processor platform	Soft IP	BVCI
COFFEE	General purpose RISC/DSP	Soft IP	BVCI
Data cache	Local, high-speed cache unit	Hard IP	COFFEE's I/F
Inst. cache	Local, high-speed memory unit	Hard IP	COFFEE's I/F
RSA	RSA encryption and decryption unit	Soft IP	BVCI
I/O buffer	I/O buffer with FIFO access scheme	Soft IP	PVCI
Memory	Memory unit with FIFO access	Hard IP	PVCI

Table 16.1. Components used in the networked multimedia processing platform.

The TACO protocol processor platform. The TACO[1] platform is used to develop application-specific hardware architecture instances for protocol processing. The TACO design methodology consists of a Transport Triggered Architecture (TTA) based platform, a SystemC framework for simulation, a Matlab model for performance estimation, and a VHDL model for synthesis of modules. The models are integrated to a graphical design tool that automates the model instantiation.

The RSA unit. This special-purpose VHDL component implements the RSA (Rivest Shamir Adleman) [14] algorithm for encryption and decryption of messages. For this platform, the RSA unit is the most expensive resource in terms of implementation cost and performance.

The COFFEE RISC/DSP embedded processor. The COFFEE processor is a customizable general purpose processing element suitable for most applications in either a NoC environment or in a more conventional embedded system. The processor can be configured for diverse applications. Simple interfaces are provided for communication and extension units as well as for local (cache) memories. A suitable platform for an application can be selected by connecting different peripheral modules as needed.

I/O buffer unit. This network entity is a simple high-speed buffer for storing and forwarding a dataflow. In the presence of discontinuities in the flow, it protects the rest of the network by allowing continuous operation provided that the discontinuity is short compared to the buffer length.

3.1 A Flexible Packet Switching Network Solution - The PROTEO NoC

PROTEO is a packet switching network model developed at the Tampere University of Technology, Finland. Its features make it well suited for platform design. These include point-to-point connections, standardized interfaces, parameterizable instantiation from a synthesizable IP library, light behavioral simulation, variability of topologies and protocols, and so on. The concepts behind the development of PROTEO were described in more detail in chapter 9: From Buses to Networks and [15].

[1]Tools for Application-specific hardware/software CO-design, developed in the Turku Centre for Computer Science (TUCS).

3.2 TACO - A Protocol Processor Platform

The challenge with protocol processor design is to find an architecture that is a good compromise between a general purpose processor and a custom, protocol-specific processor (ASIC). Ideally, the architecture should be programmable and optimized for a family of protocols and tasks required in the processing of these protocols. The TACO protocol processor [19, 20] design framework [18] addresses this design problem by providing the tools and methods for helping the designer in specifying, simulating, evaluating and synthesizing programmable protocol processors.

The TACO design flow starts from a high level description or specification of the target application. The application software constrains the processor design work so that the processing requirements of the application determine the hardware architecture of the protocol processor [8, 17]. This is different from most commercial protocol processors available today as they are often multiprocessors with high-performance general purpose processing cores as computing elements.

The TACO protocol processor architecture, as seen in figure 16.2, is based on the TTA model [3, 16]. In TTA based processors, operations are triggered by the programmed data transports. In contrast, traditional processor architectures are programmed by defining operations, which implicitly specify the data transports. A TTA based processor is composed of functional units (FUs) that communicate via an interconnection network. This network of data buses is controlled by a special purpose controller unit. The FUs are connected to the buses through modules called sockets. It was observed in the TACO project that this kind of modularity facilitates both component reuse and hardware design automation.

The functional units of TACO processors have input and output registers. Operations in the FUs are executed every time data is moved to a specific kind of input register, the trigger register. Each FU in a TACO processor has one such register, and performs a specific protocol processing task or operation.

TTAs are in essence single-instruction processors, as instructions only specify data moves between functional units. Thus, the instruction word of a TTA processor consists mostly of source and destination addresses of sockets. These addresses are called socket identifiers (IDs). The socket IDs are transported on ID buses from the interconnection network controller. There are as many ID buses as there are data buses in the interconnection network. Upon finding its socket ID on one of the ID buses, a socket opens the connection between an FU and the correspond-

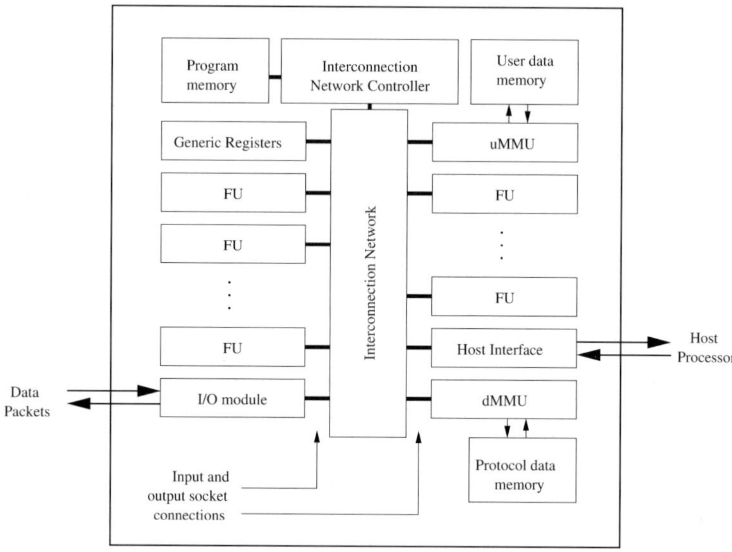

Figure 16.2. A functional view of the TACO architecture.

ing data bus on the interconnection network. The maximum number of instructions (i.e. data transports) that can be carried out in one clock cycle is equivalent to the number of data buses in the interconnection network.

The benefit of TTA-based platforms is their modularity and scalability. Functional units can be added to the architecture or they can be refined and changed as long as they provide the same interface to the sockets connecting them to the interconnection network. The TACO architecture (see figure 16.2) is therefore more of a template for protocol processors, instantiated for a specific protocol or a family of protocols. Module reuse has been a goal for TACO design projects.

3.3 IP for Hardware Encryption and Decryption of 1024-bit RSA Codes

Out of the known message encryption methods, 1024-bit RSA code is one of the most secure. Decryption of such code is based on the following equation:

$$M = C^e (mod\ n) \tag{16.1}$$

where all terms are 1024-bit wide, C represents the encrypted message, e and n together form the key that is required for decryption, and M

represents the decrypted, original message. Message encryption is done using the same equation by merely switching the places of M and C. From equation 16.1, it is obvious that RSA's computational complexity implies costly implementation. Thus in hardware the implementation is either very slow or occupies a considerable silicon area and hence dissipates a lot of power. In the described case study, a block designed to provide high bandwidth was used.

The hardware decryption IP that we use in this platform instance [13], is easily scalable and fast at the expense of high resource utilization. The architecture of the block features 128 16-bit multipliers working in parallel, and a high number of buffer registers. While the architecture is expensive, it provides easy scalability to the desired bandwidth. This combined with the RSA's high level of security, is why hardware encryption and decryption of RSA codes should be considered when basic security schemes are insufficient, for example in a Wireless Local Area Network (WLAN) environment.

3.4 An Open-Source RISC/DSP Machine - The COFFEE Processor

The COFFEE RISC/DSP processor [6] was designed to be a general-purpose processing element, which can be customized to suit most applications in either a NoC environment or in a more conventional embedded system. In practice this means that the COFFEE is not a fixed design, but rather a family of designs. It provides software configurability at run-time to facilitate system integration, while its balanced pipeline ensures good performance. Reusability and configurability were the main directives for the COFFEE processor design.

The basic version of the COFFEE processor provides adequate resources and processing power for many applications in itself, while it can also be enhanced in several ways. These enhancements are facilitated by the modular structure of the core, the well documented interfaces for extension and communication, and a large set of previously designed and verified extension modules. The resources common to all COFFEE cores include a built-in interrupt controller, two timers, and simple memory protection mechanisms. The system designer selects the combination of core modules, memories, and I/O peripherals to result in the desired performance tradeoffs. If none of the possible combinations of previously designed and verified modules yield satisfactory results, custom modifications can be easily made.

A simplified block diagram of the core is depicted in Figure 16.3. As can be observed, the core is a rather straightforward implementation of the Harvard architecture. However, unlike traditional RISCs, the COFFEE core has built-in centrally controlled forwardings to make the core efficient and flexible. Forwarding and hazard detection not only simplify compiler construction but also ease software development.

All arithmetical and logical operations are executed in a single pipeline stage, with the exception of the multiplication operation that extends over three stages. The Arithmetic Logic Unit (ALU) allows variable length bit fields to be extracted from a 32-bit word within one clock cycle. This increases bitstream handling capacity of the core with only a small area overhead. All data address calculations are performed using the ALU. This removes the need for additional arithmetic blocks for address calculations and thus saves chip area and power.

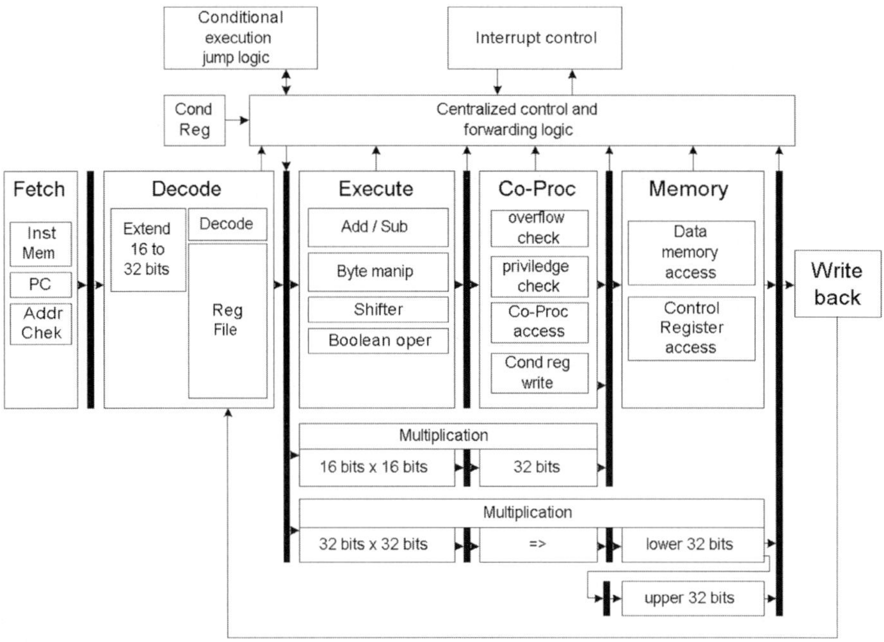

Figure 16.3. Block diagram of the COFFEE processor core.

Since the COFFEE processor is delivered as an RTL level VHDL description, it can be ported to any current technology using industry-standard development environments. Arithmetic operations are coded at boolean level to prevent a synthesis tool from mapping operations to fixed hard implementations, ensuring predictable results. The pipeline is balanced based on relative depths of the logic in each stage. This should ensure equal results between different synthesis tools. Even direct technology mapping without any optimizations should be enough to produce acceptable results.

Because of its relatively simple interface, the COFFEE processor is easy to instantiate. It can be equipped with cache memories and a number of peripheral devices. Peripherals are connected via either direct register interface or any On-Chip Bus (OCB). A series of wrappers for VCI are provided, which allow connections to other so called Virtual Components (VC) [21]. The user is able to map the memory address space freely, since there are no fixed addresses for peripherals or configuration registers. Even the boot address can be defined externally. Address verification is performed in order to detect possible address calculation overflow and to verify data and instruction memory access privileges.

The COFFEE core supports direct connection of up to four co-processors to boost for example floating-point arithmetics or specific signal processing tasks. The co-processor interface has direct access to the pipeline to enable fast operation. The core supports up to twelve interrupt sources with run-time configurable priorities. Additionally an external interrupt controller is supported, so that the number of interrupt sources can extend up to 256.

In its default configuration, the core has two separate register sets, one for application software and the other for the operating system or the privileged software. Besides separate register sets and restricted access to one of them, the core also supports memory protection without any external memory management unit. Access to both data and instruction memory can be controlled independently and the access limits can be dynamically configured at run-time by the privileged software. The operating system defines two limits for both memories and specifies whether the application software can access the memories between the limits or outside of them. The memory protection mechanism requires only a small amount of logic, but enables the construction of a flexible and secure Real Time Operating System (RTOS) for the system.

4. Case Study in Digital Video Reception

In this section, a case study in NoC design, optimization, and evaluation is described. The multimedia platform is configured for use as a subnetwork that receives, decrypts, and decodes a compressed video stream. The subnetwork interfaces to a system through an I/O buffer for a (RF) receiver and a network bridge to the display control part.

In everyday life, this network cluster could be used for applications like confidential mobile videoconferencing. The video stream is assumed to be received with a handheld device like a PocketPC over a WLAN link that uses Internet Protocol version 6 (IPv6). RSA is used for message decryption, and MPEG-2 (Moving Picture Experts Group) standard for decompression to full frame video. MPEG-2 is described here because it has the highest bit rate of the applicable video compression standards and features light computational requirements. The other extreme of possible decoding standards would be MPEG-4 AVC (also known as H.264) that requires more than twice the computing power, but reduces bit rate to approximately a third while preserving the image quality.

4.1 Description of the Case Study Application

The data stream being received is broadcasted from a network server through a WLAN base station using the IPv6 protocol. It is assumed that the video data has been previously encrypted using the RSA encryption algorithm, and compressed using the MPEG-2 standard for video compression. The video compression has been carried out using parameters that best suit current PocketPC-type machines with WLAN support, so that the video bit rate and the video screen size have been set to values that enable reception and playback in a small hand-held device. Key MPEG-2 parameters for this video stream are shown in table 16.2. The received IPv6 data stream is verified and decrypted. After decryption, the video is decompressed to full frame video, and output at the target framerate.

Video BW	Audio BW	Picture Size	Frame Rate	Bits/Pixel
900 kbps CBR (constant)	64 kbps (joint stereo)	320 x 240 pixels	25 fps	24-bit (true-color)

Table 16.2. Key MPEG-2 parameters for the case study.

Current WLANs operate at maximum speeds of less than 100 Mbps. If 1500-octet IPv6 datagrams and a 100 Mbps peak transmission rate are assumed, the target NoC must be able to receive and process more than 8300 IPv6 datagrams per second. This requirement determines the lower bound of processing speed for stream reception. For decryption, the minimum processing speed is defined by the bit rate of the incoming 1 Mbps video stream. The bit rate for the final, uncompressed video stream is 46 Mbps, which sets the lower limit for the video decompression speed.

4.2 Mapping the Application to the Platform

Composition of the studied network cluster is illustrated in Figure 16.4. While the chosen ring topology is simple, the PROTEO communication backplane does not limit the network composition at all. The topology selection is based on the quite monotonic data flow in the application domain that this system is designed to serve. That is, the direction of data flow stays unchanged, and the required bandwidth is virtually constant. Although the data flows in one direction only, bidirectional communication is needed. The return path is required for system control information and urgent service requests like synchronization, forcing a buffer read and so on. Of course, the bit width of the links in the return direction can be chosen to be considerably narrower than in the direction of the data flow.

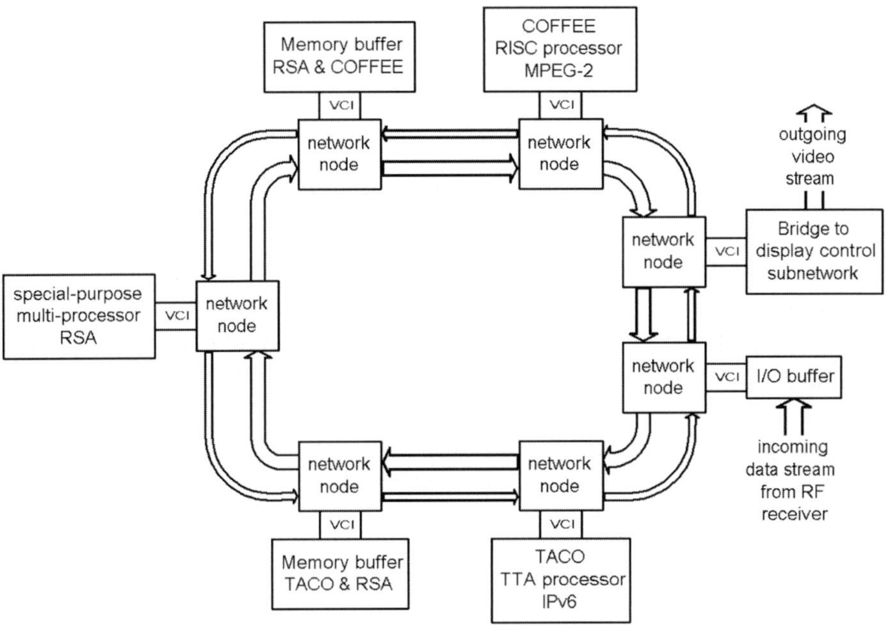

Figure 16.4. Block diagram of the case study platform cluster.

The TACO protocol processor receives IPv6 datagrams from an Input/Output (I/O) buffer unit through the PROTEO on-chip network. Besides removing protocol-related control information from the data, the protocol processor verifies datagram integrity and order. The average amount of control information in IPv6 datagrams is 3-4 %. In this case, there are 44 header octets for each bundle of 1280 to 1500 datagram octets. Finally, the TACO processor writes the encrypted video data into a First-In-First-Out (FIFO) memory module through the PROTEO on-chip network.

A special-purpose processing unit for RSA decryption reads the encrypted video data from the output memory buffer of TACO, decrypts it, and writes it to another FIFO memory module.

The COFFEE RISC/DSP processor reads the decrypted MPEG-2 data stream from the output memory buffer of the RSA unit. The COFFEE processor decompresses the MPEG-2 stream and writes the decompressed data (that is, full frame video) into a PROTEO bridge node. From there the data is transmitted to the display control subnetwork, which is not described here in detail.

4.3 PROTEO Network Implementation

The cost of a PROTEO network implementation is mostly due to the need of packet buffers in the network nodes. For this study, the maximum size of payload per packet has been limited to four cells. This limitation gives the framework for buffer implementation. Since the current version of PROTEO uses registers as buffers, the area occupied by the network is more than double of what could be achieved. Using on-chip RAMs instead of registers is presently being studied to lower the area requirement.

If the presented bidirectional ring topology (Figure 16.4) is implemented with 8-bit links in both directions, it occupies 0.25 square millimeters area. The implementation has a total of 3080 flip-flops of which 784 are pure data buffers. The portion of combinatorial logic in the implementatation is less than one fourth of the required area. If the same topology was implemented with 32-bit links in both directions the needed area would be approximately 0.58 square millimeters. The 8-bit and 32-bit network implementations were verified to operate at 200 MHz. Thus, the network IP does not limit the system bandwidth, but allows narrower links to be considered if the network cost is a primary issue. For this case study, 8-bit links were used in the direction of dataflow as well as in the opposite direction.

Such an off-the-shelf implementation of a general purpose communication topology facilitates solving both communication and computation related issues in upper levels of abstraction, since the interconnection architecture does not notably favour specific implementations. An abstract MoC, for example, is easy to implement, since upper layer communication mechanisms can be chosen quite freely without significant performance variation due to the interconnection architecture.

4.4 TACO IPv6 Client Implementation

The TACO processor serves as an input processing unit that receives IPv6 datagrams, verifies their correctness and addressing, and extracts the upper layer payload to be passed to the next processing unit through the PROTEO network.

IPv6 [4] is the latest version of the Internet protocol introduced mainly to overcome the address restrictions of IPv4. It features 128-bit addresses (compared with 32-bit addresses of IPv4), and an improved addressing hierarchy. Additionally, the structure of the IPv6 datagrams has been simplified by introducing an extensible datagram format consisting of a header and a number of optional extension headers.

For this case study we assume that the IPv6 client receives IPv6 datagrams only. As presented in Figure 16.5 we also assume that a datagram is composed of an IPv6 header, an upper layer header and upper layer payload. The IPv6 header has a fixed size (40 octets) and is composed of a number of fields, e.g. Version (6 for IPv6), Payload Length, Next Header type (the type of upper-layer protocol), Hop Limit (time-to-live of the datagram), and address fields for the source and destination of the datagram.

For the upper layer protocol we use our proprietary transport protocol that provides mechanisms for error checking (checksum) and packet re-ordering (sequence numbers). The upper layer message structure is composed of three fields: the Sequence Number, Checksum and Payload fields. The Sequence number allows the IPv6 Client to deliver the incoming data in the specified order, while the Checksum field ensures the integrity of the datagram.

Upon receiving a datagram, the IPv6 client verifies that the Version field is set to value 6 (IPv6), that the Next Header value indicates our proprietary upper layer protocol type, and that the datagram length is smaller than the maximum datagram size allowed on the connection. In addition, the datagram length should be larger than 44 octets, ensuring

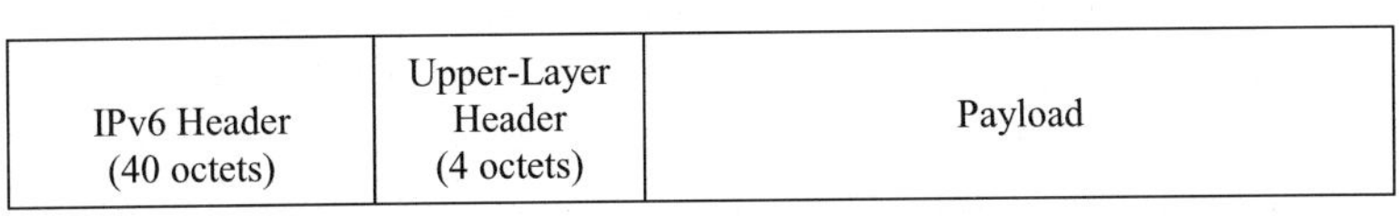

Figure 16.5. IPv6 Datagram structure

that the datagram carries upper-layer payload. Incoming datagrams are also checked for addressing to the client's interface address and origination from the correct address. The IPv6 client also verifies for the integrity of the received datagrams by computing the checksum and then comparing it with the value in the Checksum field.

The TACO IPv6 client core used in this NoC case study is composed of FUs originally created for an IPv6 router project [9, 20]. Figure 16.6 shows a functional view of the TACO IPv6 client processor instantiated for the presented NoC platform. The functional units needed to constitute an IPv6 client are also shown in the figure. In the general case, a TACO processor can have more than one bus, as well as more than one of each kind of FU. The IPv6 client, however, has only one of each required FU type and only one internal data bus. Some of the FUs used in the router project were not needed for the client application, because the IPv6 router had considerably higher performance requirements. Thus, the client operation can be carried out with less processing power, implying a smaller amount of FUs and data transport buses. Obviously, this also leads to a smaller area and lower power consumption. A detailed description of the IPv6 router processor and the IPv6 functional units is given in [20].

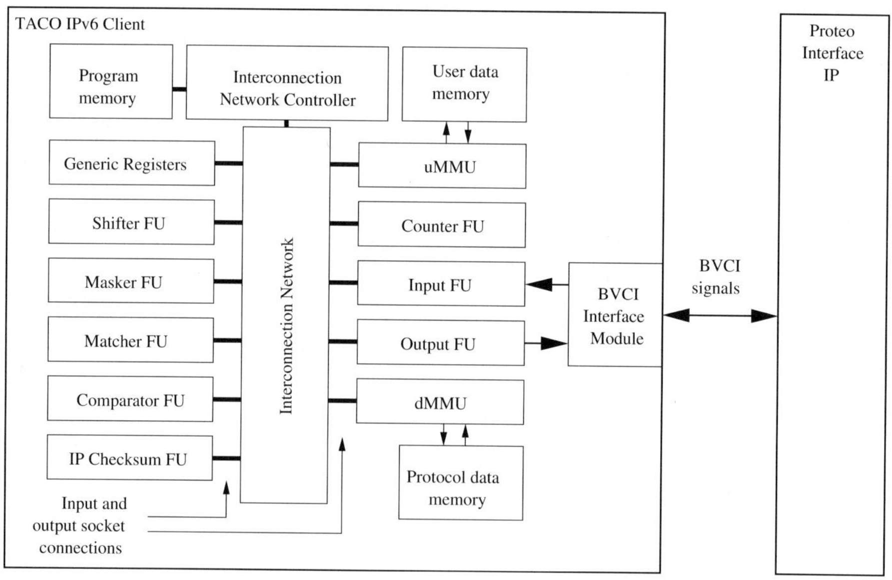

Figure 16.6. The TACO IPv6 Client and its connection to the Proteo network.

To connect the TACO IPv6 client to the PROTEO on-chip network, a Basic Virtual Component Interface (BVCI) [21] module was designed and implemented. This module acts as a bridge between the TACO input and output FUs and the PROTEO network node (see Figure 16.6). The interface module design process was quite straight forward thanks to the well defined protocol processor Input/Output (I/O) interface and the solid BVCI specification.

SystemC simulations of the TACO IPv6 client and its application code showed that the IPv6 client operation requires 1920 clock cycles per datagram. In a 100 Mbps network environment (with a peak transmission rate of 8300 datagrams per second) this means that the minimum clock speed for the TACO IPv6 client is 17 MHz. We proceeded to synthesize the TACO IPv6 client using 0.18 μm CMOS technology and a standard cell library with the target clock speed of 200 MHz. Targeting the synthesis to this clock speed should make it possible for the IPv6 client to operate at network speeds of up to 1 Gbps.

Physical estimations of the IPv6 client were carried out by applying the well known Rent's rule [7] and by assigning a separate Rent's exponent to each block. Additionally, the relative amounts of combinatorial and sequential logic are evaluated based on the functionality of each block. This is important when estimating the power consumption of the block, since different values of the switching activity factor have to be used for different types of logic. The values 0.25 and 0.5 have been assumed for combinatorial and sequential logic, respectively. Further discussion on this method can be found in chapter 3: Global Interconnect Analysis and [1].

The estimates suggested a logic area of 0.30 mm^2 using 200 MHz clock speed. This result reflects the core size without memories. The TACO IPv6 client was synthesized using a standard cell library of 0.18 μm CMOS technology with a target clock speed of 200 MHz. The synthesis results and estimates for different module areas are shown in Table 16.3. Synthesis resulted in a logic area of 0.23 mm^2. When comparing this to the estimated value, it can be stated that the accuracy of the area estimates was $A_{synth} = 0.76\ A_{estim}$. This accuracy is sufficient at the system level, where slight pessimism in estimates is always preferable: if the estimations would suggest a smaller area than what is obtainable through synthesis, original design constraints might not be met. The power consumption estimate reflects the situation of operating at 200 MHz (i.e. at about 1 Gbps network speed). The network speed for the target application is considerably lower, which reduces the power consumption respectively. Also, the estimate expects all functional units to be active at all times, which is a worst-case scenario.

	Estimated average power [mW]	Estimated area [μm^2]	Synthesized area [μm^2]
Matcher	7.1	13 785	12 851
Shifter	11.0	21 313	19 302
Comparator	13.1	25 205	17 197
Masker	6.4	12 346	11 335
Checksum	13.2	25 413	15 595
Counter	12.1	23 354	11 782
Input FU	13.9	26 958	26 064
Output FU	12.6	24 362	17 002
dMMU	7.5	14 633	11 721
uMMU	4.1	7908	7342
Sockets (35)	19.8	37 231	31 720
Network Controller	14.6	28 346	12 660
Total (IPv6 client part)	135.4	270 854	194 571
BVCI Interface wrapper	15.6	30 342	33 991
Total (IPv6 client with BVCI wrapper)	151.0	301 196	228 562

Table 16.3. Estimated power consumption and area, and synthesized area for all modules in the TACO IPv6 Client protocol processor.

4.5 Implementation of the RSA Decryption Block

The RSA decryption block occupies approximately nine square millimeters of silicon area on the 0.18 micron technology used and needs about 650,000 clock cycles for the decryption of a 1024-bit message. Using Wallace tree multipliers allows the non-pipelined circuit to operate at 200MHz giving a throughput of approximately 300kbits/s. With a proper 4-stage pipelining scheme for the multipliers, the data rates needed for smooth videoconferencing can be reached. If more bandwidth is required, the circuitry can be duplicated as many times as needed. The duplicates are easy to run in parallel because the decryption of each 1024-bit fraction of a sequence is independent from the decryption of other fractions of the same sequence.

The selection of the amount of decryption blocks not only depends on the required bandwidth but also on the power consumption goals. The performance provided by a single unit is sufficient for this case study, since the required operating clock frequency of 650MHz can be achieved by means of pipelining. However, the selection of a single unit keeps manufacturing costs down at the expense of power consumption. This happens because pipelining implies more registers to clock and internal gates need to provide high current driving capability to achieve the operating speed. It would thus reduce the power consumption of the device if there were multiple non-pipelined decryption blocks running in parallel with lower clock frequency. Since the implementation of RSA decryption is one of the key design points concerning battery life in mobile applications, the algorithm should not be implemented in software because of the enormous power consumption due to the required memory accesses.

4.6 The COFFEE Processor Implementation for MPEG-2

Performance analysis is often difficult to carry out based solely on an application specification. In many cases, selecting the right CPU is critical for the cost of a NoC, and therefore it is not ideal to select a CPU based on rough estimates about needed performance. Unfortunately, to get reliable figures for comparison, both the CPU (or an accurate model of it) and the final software are needed. These are usually not available when NoC hardware is designed. Performance should not be the only criteria for selecting a CPU. The number and the type of integrated peripheral devices have a major impact on cost, and therefore only the necessary peripherals should be selected to reduce silicon area and power consumption.

In the COFFEE processor core, the selections of the right type of multiplier, the size of the register bank, and the width of the internal buses affect performance and power consumption. Two multipliers are provided in the current version of COFFEE. Table 16.4 summarises the effect of different component choices. As can be seen from the table, the 16-bit multiplier occupies less than 20% of the area needed for the 32-bit multiplier. Reducing the width of the internal buses means in part giving up software compatibility. Software written for the 16-bit version can run on the 32-bit version but not the other way around. The core interface does not change between 16-bit and 32-bit versions with the exception of the data address bus width. Both versions support 32-bit and 16-bit instruction encoding, but the 16-bit version is only able to execute a subset of the instructions available in the 32-bit version. Selecting a register file with a single 32-register bank instead of two such banks reduces area and power consumption considerably. In a single register bank core, the difference compared to a two register bank core is that user mode and privileged mode share the same register bank. Finally, the required peripherals are attached to the core.

Block/Module	Area/mm^2	Power/mW	Delay/ns
Arithmetic Logic Unit	0.108	15	3.86
Core Control Unit	0.024	5	2.13
Interrupt Controller	0.122	5	3.54
32-bit Multiplier	0.234	60	4.82
16-bit Multiplier	0.044	10	1.32
Peripheral Control Block	0.138	30	1.62
Register File (2 Banks of 32 Registers)	0.713	100	1.80
Core with All of the Above Blocks	1.38	225	4.82
Core with 16-bit Multiplier	1.19	175	3.86
Core with Single Bank Register File	1.02	170	4.82
Core w/16-bit Mult. and Single Bank RF	0.83	120	3.86

Table 16.4. Physical characteristics obtained from synthesis for the COFFEE core at 200 MHz.

Since the only task assigned to the COFFEE processor was the decoding of an MPEG-2 bitsream, no operating system was used, and therefore an implementation with a single register bank was chosen. During evaluation, both types of multipliers were shown to provide adequate performance, even though the 16-bit multiplier requires a slightly higher minimum clock frequency than the 32-bit multiplier. For the used application software the required operating frequency was approximately 50MHz with the 32-bit multiplier. The set of selected peripherals was minimal. One of the internal timers of the COFFEE processor core was used to synchronize the frame transmission to the display controller, and a memory mapped register bank was connected to the processor data bus in order to interface the VCI wrapper. The VCI wrapper reserved one of the interrupt lines to signal that requested data had been received by the network node.

4.7 System-Level Performance

For most of the preceding discussion, a specific application was assumed to be run on the studied hardware architecture. However, the platform architecture was designed with a wider scope in mind, aiming at satisfying the needs of a large set of applications. This network cluster could thus serve designers of future systems as reusable platform IP. Design of such very-high abstraction level IP requires a more sophisticated engineering approach when compared to designing on lower abstraction levels.

At first glance, the scope of this study seems quite limited due to the rather high-end nature of the presented application. The flexibility of this platform IP becomes evident by considering the possibilities of modifying and parameterizing the design. For example, in quite many applications with less stringent security requirements, the RSA block could be left out of the system to allow data rates of several gigabits per second with an implementation technology similar to the one used in this study. The direction of communication could also be easily changed if the network cluster was to operate as a transmitter. Good documentation of such reconfigurability is vital for reuse purposes and has to be provided with the design.

The actual simplicity of this case study is due to the steady data flow present in the application domain the platform is designed for. This implies that communication traffic in the system is heavier in one direction than the other since there is only system control information going in the direction opposite that of the data flow. Because the amount of data moving through the system stays virtually the same over time, there are very few idle cycles expected for the processors. Hence, it is not beneficial to perform multi-tasking or multi-threading, and there is little to gain by usage of an operating system.

The results of the synthesis-based performance estimations are summarized in table 16.5 for the various platform components and the platform as a whole. The power consumption figures are rather speculative worst-case estimates and the actual value depends on the data being processed. The platform implementation was not refined to the floor plan level, and without layout or accurate floor plan, the estimation of power consumption in global wires is highly inaccurate. In table 16.5, the power consumption figures for individual blocks include rough estimates about their respective portions of global communication. Delays are given as processing time measured from input to output. The overall delay is dominated by the RSA block, which requires approximately 1 ms at 650 MHz for processing a fraction of an encrypted message. Since human threshold for perception of inconvenient delay in terms of synchronization is about 10 ms, the RSA operating frequency could be safely lowered down to approximately 100 MHz while multiple RSA blocks are required to run in parallel to provide the required bandwidth. This would lower the overall power consumption as discussed earlier in section 4.5.

The PROTEO network implementation was verified to operate with at least a 200 MHz host clock frequency. This frequency gives a network bandwidth of 1.6 Gbps in one direction. The respective peak bandwidth for payload is 1.28 Gbps. Higher speeds could be achievable, but were not verified for this study. Lowering the network node operating speed does not help much in power saving, because consumption is dominated

Block/Module type	Area [mm^2]	Power [mW]	Delay [cycles]	Delay [μs]
TACO	0.20	135	160	0.8
COFFEE	1.02	170	50,000	250
Data Memory (64 kB)	0.82	500	32	0.16
Instruction Memory (128 kB)	1.62	300	32	0.16
RSA Unit	9.05	1,500	656,897	3,284
I/O Buffer (128 kB)	1.64	20	128	0.64
Stream Buffers (4 kB/4 kB)	0.10	30	128/128	0.64/0.64
PROTEO Network	0.25	70	160-256	0.8-1.28
I/F Wrappers	0.10	25	0	0
Total	14.8	2,750	700,000	3,500

Table 16.5. IP block characteristics at 200 MHz: silicon areas obtained through synthesis, power consumption estimates, and delays for processing 1 kilobit of data.

by changes in the data content flowing through the network. Although Cyclic Redundancy Check (CRC) code was used for error detection in the on-chip network, communication faults were not modeled for the presented case study.

The 200 MHz clock frequency was also verified for the TACO protocol processor. With this frequency, the payload bandwidth is at 1.18 Gbps. Like with the PROTEO network, considerably higher speeds might be possible, but have not been verified for the case study. The realized operating frequency allows the platform to be connected to a majority of existing IPv6 networks. Thanks to the buffer memories utilized in the network, no local memory modules were needed to accompany TACO in addition to its own internal caches.

As expected, the RSA block proved to be the dominating bottleneck. In most applications, the system data flow should mainly bypass the RSA unit, and only the critically confidential part of the complete information stream should require the RSA decryption service. A decryption speed of 1 Mbps was reached using a single decryption block. Achieving this required heavy modification of the original design to force operation at the required 650MHz clock frequency. Lower decryption speeds are more desirable for many applications due to the high power consumption of the RSA module. If the whole video stream of the case study application is RSA encrypted, the battery life of the receiving device will be affected. For comparison purposes, if a standard cell phone battery was used for powering the platform, it would last only about an hour based on the given power consumption estimates.

The used MPEG-2 application software was profiled for estimations about the required operating speed of the COFFEE processor and minimum sizes of local memories connected to it. It turned out that a clock frequency of 50 MHz would suffice for the 1 Mbps video stream. The core was verified for operation at 200 MHz, as were the other main modules of the platform, excluding the RSA. This clock frequency would thus allow the MPEG-4 AVC/H.264 standard to be used up to the targeted resolution with a notable headroom. It can also be concluded that with a single COFFEE processor running MPEG-2, a video bandwidth of 4 Mbps could be achieved. For the bit rate of the DVD standard, three processors and an efficient software for multiprocessor decoding would be required. With multiple processors working on the same task, the benefits of an RTOS become evident. RTOS for the COFFEE processor is presently under development. The minimum standard size instruction and data memory modules to be used with the profiled application software turned out to be 128 kB and 64 kB respectively.

Multiple applications from the domain must be considered when data stream buffers of a platform are sized. The I/O buffer was designed to provide one second of video stream buffering for the application. The implemented buffer might be insufficient under unfavourable network conditions as well as for some higher bit rate streaming applications that could be run on the platform. The sizes of the network stream buffers were selected more from the safer side. Data granularity for these buffers is determined by the message size of RSA (1 kilobit). Measured by this the selected 4 kB buffers have a depth of 32 data units allowing relatively loose synchrony between the modules.

Stream	Bandwidth
Raw IPv6 Data Stream (restricted by TACO)	1.18 Gbps
MPEG-2 Video Stream (restricted by COFFEE)	4 Mbps
RSA Encrypted Data Stream (restricted by the RSA unit)	1 Mbps

Table 16.6. Maximum bandwidths for different data streams in the case study platform.

In this study, the performance requirement of the target application was not met without modifying the RSA unit, whereas the other network modules provided throughputs in excess of the target specification. The maximum bandwidths of different data streams in the platform after modification are summarized in table 16.6. As the described platform application is somewhat unusual, it demonstrates the importance of considering a great variety of applications in a general case of platform design. A good NoC platform should not need redesigning for the majority of applications belonging to the same domain.

5. Conclusion

The described case study in NoC platform design provides a simple yet useful example for developers of such systems. The utilization of previously made IP blocks can speed up system development and reduce the size of the design team. The growing complexity of communication design can be tackled by adopting reuse of a parameterizable general purpose interconnection IP with standardized interfaces. System level design efforts are not as specific purpose as they first appear. If the system level hardware description is made as easily modifiable and parameterizable as the lower level designs, it has great reuse potential.

The design of very high abstraction level IP calls for a more general purpose approach to the problem. This means that designers need to think at least twice before making any application specific optimizations. Instead, application domain specific optimizations should be considered from several points of view. The basic engineering skill, that is making tradeoffs, is thus far more complex to utilize efficiently at very high abstraction levels. It follows that design teams will become more and more dependable on designers' personal experience and intelligence, unless efficient EDA tools are developed to aid and speed up the process.

For awareness of the possibilities offered by an IP design, the entire design space should be thoroughly explored and documented. A large design space is characteristic for a good reusable IP. If the design space of feasible implementations is small, there is probably something too specific in the chosen architecture or abstraction level of the hardware description. A restricted design space could also be a result of poor configurability. The ability to control implementation through parameters is very important for EDA tool development. One ambitious research topic could be the development of an automatic design space explorer for platform IPs.

References

[1] T. Ahonen, T. Nurmi, J. Nurmi, and J. Isoaho. Block-wise extraction of rent's exponents for an extensible processor. In *IEEE Computer Society Annual Symposium on VLSI*, pages 193–199, Tampa, Florida, February 2003.

[2] D. I. August, K. Keutzer, S. Malik, and A. R. Newton. A disciplined approach to the development of platform architectures. *Microelectronics Journal*, 33(11), November 2002.

[3] H. Corporaal. *Microprocessor Architectures - from VLIW to TTA*. John Wiley and Sons Ltd., Chichester, West Sussex, England, 1998.

[4] S. Deering and R. Hinden. Internet protocol, version 6 (IPv6) specification. *RFC 2460*, December 1998.

[5] M. Keating and P. Bricaud. *Reuse Methodology Manual for System-on-a-Chip Designs*. Kluwer Academic Publishers, Boston, second edition, June 1999. 312 pp.

[6] J. Kylliäinen, J. Nurmi, and M. Kuulusa. COFFEE-A Core for Free. In *Proceedings of the International Symposium on System-on-Chip*, Tampere, Finland, November 2003.

[7] B. S. Landman and R. L. Russo. On a pin versus block relationship for partitions of logic graphs. *IEEE Transactions on Computers*, C20(12):1469–1479, December 1971.

[8] J. Lilius and D. Truscan. UML-driven TTA-based protocol processor design. In *Proceedings of the 2002 Forum for Design and Specification Languages (FDL'02)*, Marseille, France, September 2002.

[9] J. Lilius, D. Truscan, and S. Virtanen. Fast Evaluation of Protocol Processing Architectures for IPv6 Routing. In *Proceedings of the 2003 Design, Automation and Test in Europe conference (DATE'03)*, Munich, Gemany, March 2003.

[10] J. Muttersbach, T. Villiger, and W. Fichtner. Practical design of globally-asynchronous locally-synchronous systems. In *Proceedings of the 6th International Symposium on Advanced Research in Asynchronous Circuits and Systems*, Eilat,Israel, April 2000.

[11] J. Muttersbach, T. Villiger, H. Kaeslin, N. Felber, and W. Fichtner. Globally-asynchronous locally-synchronous architectures to simplify the design of on-chip systems. In *Proceedings of the 12th Annual IEEE International ASIC/SOC Conference*, Washington DC, USA, September 1999.

[12] OCP-IP Association. *Open Core Protocol Specification Release 1.0.* OCP-IP, 2001. www.ocpip.org.

[13] T. Ristimäki and J. Nurmi. Implementation of a fast 1024-bit RSA encryption on platform FPGA. In *Proceedings of the 6th IEEE International Workshop on Design and Diagnostics of Electronics Circuits and Systems*, Poznan, Poland, April 2003.

[14] R. L. Rivest, A. Shamir, and L. Adleman. A method of obtaining digital signatures and public-key cryptosystems. *Communications of the ACM*, 21(2):120–126, 1978.

[15] I. Saastamoinen, D. Sigüenza-Tortosa, and J. Nurmi. An IP-based on-chip packet-switched network. In A. Jantsch and H. Tenhunen, editors, *Networks on chip*, chapter 10, pages 193–213. Kluwer Academic, 2003.

[16] D. Tabak and G. J. Lipovski. MOVE architecture in digital controllers. *IEEE Transactions on Computers*, 29(2):180–190, February 1980.

[17] S. Virtanen and J. Lilius. The TACO protocol processor simulation environment. In *Proceedings of the 9th International Symposium on Hardware/Software Codesign (CODES'01)*, pages 201–206, Copenhagen, Denmark, April 2001.

[18] S. Virtanen, J. Lilius, T. Nurmi, and T. Westerlund. TACO: Rapid design space exploration for protocol processors. In *the Ninth IEEE/DATC Electronic Design Processes Workshop Notes*, Monterey, CA, USA, April 2002.

[19] S. Virtanen, J. Lilius, and T. Westerlund. A processor architecture for the TACO protocol processor development framework. In *Proceedings of the 18th IEEE NORCHIP Conference*, pages 204–211, Turku, Finland, November 2000.

[20] S. Virtanen, D. Truscan, and J. Lilius. TACO IPv6 router - a case study in protocol processor design. Technical Report 528, Turku Centre for Computer Science, Turku, Finland, April 2003.

[21] VSIA On-Chip Bus Development Working Group. *Virtual Component Interface Standard Version 2 (OCB 2 2.0)*. VSI Alliance^{TM}, April 2001. www.vsi.org.